Geometric Theory of Incompressible Flows with Applications to Fluid Dynamics

Mathematical
Surveys
and
Monographs

Volume 119

Geometric Theory of Incompressible Flows with Applications to Fluid Dynamics

Tian Ma
Shouhong Wang

American Mathematical Society

2000 *Mathematics Subject Classification*. Primary 35Q30, 35Q35, 76D05, 76D10, 37C10, 37C75, 86A10, 86A05; Secondary 46E25, 20C20.

For additional information and updates on this book, visit
www.ams.org/bookpages/surv-119

Library of Congress Cataloging-in-Publication Data

Ma, Tian, 1956–
Geometric theory of incompressible flows with applications to fluid dynamics / Tian Ma, Shouhong Wang.
p. cm. — (Mathematical surveys and monographs ; v. 119)
Includes bibliographical references and index.
ISBN 0-8218-3693-5 (alk. paper)
1. Global analysis (Mathematics) 2. Vector fields. 3. Differential equations, Partial. 4. Manifolds. 5. Fluid dynamics. 6. Geophysics. I. Wang, Shouhong, 1962– II. Title. III. Mathematical surveys and monographs ; no. 119.

QA614.M34 2005
532′.0535—dc22 2005048120

♾ The paper used in this book is acid-free and falls within the guidelines
established to ensure permanence and durability.
Visit the AMS home page at http://www.ams.org/

10 9 8 7 6 5 4 3 2 1 10 09 08 07 06 05

Dedicated to Professor WENYUAN CHEN

Contents

Preface ix

Introduction 1
- 0.1. Representation of Fluid Flows 1
- 0.2. Motivation and Main Objectives 2
- 0.3. The User's Guide 4
- Notes for Introduction 14

Chapter 1. Structure Classification of Divergence-Free Vector Fields 17
- 1.1. Limit Set Theorem 17
- 1.2. Poincaré-Hopf Index Theorem on Manifolds with Boundaries 24
- 1.3. Structural Classification 31
- 1.4. Topological Classification 40
- Notes for Chapter 1 49

Chapter 2. Structural Stability of Divergence-Free Vector Fields 51
- 2.1. Structural Stability of Divergence-Free Vector Fields with Free Boundary Conditions 51
- 2.2. Structural Stability for Divergence-Free Vector Fields with Dirichlet Boundary Conditions 60
- 2.3. Two Dimensional Hamiltonian Structural Stability 71
- 2.4. Block Structure of Hamiltonian Vector Fields 75
- 2.5. Local Structural Stability 77
- Notes for Chapter 2 80

Chapter 3. Block Stability of Divergence-Free Vector Fields on Manifolds with Nonzero Genus 81
- 3.1. Instability on Manifolds with Nonzero Genus 81
- 3.2. Block Structure and Block Stability 87
- 3.3. Structural Evolution of the Taylor Vortices 98
- Notes for Chapter 3 108

Chapter 4. Structural Stability of Solutions of Navier-Stokes Equations 109
- 4.1. Genericity of Stable Steady States 109
- 4.2. Properties for Structurally Stable Solutions on the Reynolds Numbers 114
- 4.3. Asymptotic Hamiltonian Structural Stability 117
- 4.4. Asymptotic Block Stability 123
- 4.5. Periodic Structure of Solutions of the Navier-Stokes Equations 127
- 4.6. Structure of Solutions of the Rayleigh-Bénard Convection 142
- Notes for Chapter 4 155

Chapter 5. Structural Bifurcation for One-Parameter Families of Divergence-Free Vector Fields 157
5.1. Necessary Conditions for Structural Bifurcation 157
5.2. Structural Bifurcation for Flows with No-Normal Flow Boundary Conditions 160
5.3. Structural Bifurcation for Flows with Dirichlet Boundary Conditions 167
5.4. Boundary Layer Separations of Incompressible Flows I 177
5.5. Boundary Layer Separations of Incompressible Flows II 181
5.6. Structural Bifurcation near Interior Singular Points 187
5.7. Genericity of Structural Bifurcations 198
Notes for Chapter 5 201

Chapter 6. Two Examples 203
6.1. Fluid Flow Maps and Double-Gyre Ocean Circulation 203
6.2. Boundary Layer Separation on Driven Cavity Flow 210
Notes for Chapter 6 221

Bibliography 229

Index 233

Preface

We present in this book a geometric theory for incompressible flows and its applications to fluid dynamics. This study was initiated by the authors in the mid-1990s with the original motivation to contribute to the understanding of oceanic dynamics in physical space. The development of the theory and its applications have gone well beyond the original motivation since then.

The main objective of the work presented in this book is to study the stability and transitions of the structure of incompressible flows and their applications to fluid dynamics and geophysical fluid dynamics. This book addresses both kinematic and dynamic theories for incompressible flows and their applications.

Part of this book was used for a one-semester graduate topic course in the Mathematics Department at Indiana University. This book contains six chapters. The first three chapters are devoted to the classification and stability of topological structures of divergence-free vector fields on two-dimensional compact manifolds. The last three chapters deal with the classification, stability and evolution of topological structures of the solutions of hydrodynamical equations such as the Euler equations and the Navier-Stokes equations. Examples with numerical studies are given in the last chapter.

We would like to acknowledge explicitly the singular influence of Professors Ciprian Foias and Roger Temam, in our studies of mathematical fluid mechanics. We have greatly benefitted from discussions with Jerry Bona, Ciprian Foias, Susan Friedlander, Michael Ghil, Robert Glassey, David Hoff, Darryl Holm, James McWilliams, Paul Newton, George Papanicoulaou, Jie Shen, Roger Temam, Cheng Wang, Mohammed Ziane, and Kevin Zumbrun. Our warm thanks to all of them. Also, we are grateful to Wen Masters and Reza Malek-Madani of the Office of Naval Research for their constant support and encouragement. We express our sincerest thanks to John Ewing and Edward Dunne of the American Mathematical Society for their great effort, support, encouragement and confidence in this book project.

Finally, nothing would have been possible without the understanding and patience of Li and Ping, and special thanks go to Jiao, Wayne and Melinda, for all the fun they have brought us.

The research presented in this book was supported in part by grants from the Office of Naval Research and the National Science Foundation.

Tian Ma and Shouhong Wang
Bloomington, April 5, 2005

Introduction

We present in this chapter a brief introduction to both the Eulerian and the Lagrangian representations of incompressible fluid flows. We also present the main motivations and objectives of the studies, and provide a User's Guide on selected topics presented in this book.

0.1. Representation of Fluid Flows

In the Euler representation, the motion of a fluid flow is governed by a set of partial differential equations, such as the Euler or the Navier-Stokes equations. Given an initial velocity, for instance, the velocity field at any future time is the solution of the Euler (without the underlined term) and/or the Navier-Stokes equations (with the underlined term):

$$
\begin{cases} \dfrac{\partial u}{\partial t} + u \cdot \nabla u + \nabla p - \underline{\mu \triangle u} = f, \\ \text{div } u = 0, \\ u|_{t=0} = \varphi. \end{cases} \tag{0.1.1}
$$

Here the spatial domain is $M \subset \mathbb{R}^n$ ($n = 2$ or 3), $x = (x_1, \cdots, x_n)$ is the spatial coordinate system, t is the time variable, $u = u(x,t) = (u_1(x,t), \cdots, u_n(x,t))$ is the velocity field, $p = p(x,t)$ is the pressure function, f is the external forcing, $\varphi = \varphi(x)$ is the initial velocity field, and μ is the viscosity coefficient. The differential operators used in the above equations are the gradient

$$
\nabla = \left(\frac{\partial}{\partial x_1}, \cdots, \frac{\partial}{\partial x_n} \right),
$$

and the divergence operator

$$
\text{div } u = \sum_{i=1}^{n} \frac{\partial u_i}{\partial x_i}.
$$

Both the Euler equations ($\mu = 0$) and the Navier-Stokes equations ($\mu > 0$) are equipped with proper boundary conditions. The typical boundary condition for the Euler equations is the no-normal flow condition

$$
u \cdot n = 0, \quad \forall x \in \partial M, \tag{0.1.2}
$$

where ∂M is the boundary of M, and n is the normal vector of ∂M. The typical boundary conditions for the Navier-Stokes equations are the Dirichlet boundary conditions (no-slip)

$$
u = 0, \quad \forall x \in \partial M, \tag{0.1.3}
$$

or the free slip boundary conditions

$$u \cdot n = 0, \quad \frac{\partial u \cdot \tau}{\partial n} = 0, \quad \forall x \in \partial M, \tag{0.1.4}$$

where τ is the tangential vector of ∂M.

In the case where M is the n-dimensional flat torus $M = \mathbb{R}^n/(2\pi\mathbb{Z})^n$, the following periodic boundary conditions are also used for both the Navier-Stokes equations and the Euler equations

$$u(x + 2k\pi, t) = u(x, t), \quad p(x + 2k\pi, t) = p(x, t), \tag{0.1.5}$$

for any n-tuple of integers $k = (k_1, \cdots, k_n)$.

The Lagrange representation of a fluid flow, on the other hand, amounts to studying the dynamics and trajectories in the 2- or 3-dimensional *physical space the fluid occupies*. Namely, one studies the dynamics of the ordinary differential equations in the physical space:

$$\frac{dx}{dt} = u(x, t), \qquad x|_{t=0} = x_0. \tag{0.1.6}$$

Here u is the velocity field of the fluid, satisfying the Navier-Stokes equations or the Euler equations.

The above dynamical system (0.1.6) defines a flow on M,

$$\Phi : M \times \mathbb{R} \to M, \tag{0.1.7}$$

such that for each t, $\Phi(\cdot, t) : M \to M$ is an isomorphism, and $\Phi(\cdot, 0) = Id : M \to M$ is the identity map on M. A domain $D \subset M$ in a fluid evolves with time to $D_t = \Phi(D, t) = \{ \Phi(x, t) \,|\, x \in D\}$.

For any vector field u on M satisfying one of the boundary conditions given by (0.1.2–0.1.5), the following two statements are equivalent:

(1) the flow is incompressible; i.e., for any subdomain $D \subset M$ with sufficiently smooth boundary,

$$|D_t| = |D|, \quad \forall t \in \mathbb{R},$$

where $|D|$ stands for the volume of D;

(2) u is divergence-free; i.e.,

$$\text{div } u = 0.$$

This is the well-known Liouville theorem, which is the direct consequence of the following computation:

$$\frac{d|D_t|}{dt} = \frac{d}{dt}\left(\int_{D_t} dx\right) = \int_D \frac{\partial J}{\partial t} dx = \int_D (\text{div } u) J dx = \int_{D_t} (\text{div } u) dx,$$

where $J = \det(\partial \Phi_i / \partial x_j)$ is the determinant of the Jacobian matrix of the mapping $\Phi(\cdot, t) : M \to M$.

0.2. Motivation and Main Objectives

The main objective of the work presented in this book is to study the stability and transitions of the structure of incompressible flows and their applications to fluid dynamics and geophysical fluid dynamics.

This program of research contains studies in two interconnected areas:

(A) the development of a global geometric theory of divergence-free vector fields on general two-dimensional compact manifolds, and

(B) the study of the structure (and its stability and transitions) of velocity fields for two-dimensional incompressible fluid flows governed by the Navier-Stokes equations or the Euler equations.

The study in the first area is kinematic in nature. The results and methods developed can be applied naturally to other problems of mathematical physics involving divergence-free vector fields. The main topics in this area include structural classification, structural stability, and structural bifurcation. The study in the second area links the kinematic structure to the dynamics.

This program of study was initiated by the authors in the mid-1990s with the original motivation to understand the dynamics of the ocean currents in the physical space. In particular, one original goal was to classify the topological structure and its transitions of the *instantaneous* velocity field (i.e., streamlines in the Eulerian coordinates), treating the time variable as a parameter.

The development of the theory and its applications have gone well beyond the original motivation since then. One such development is to provide a rigorous theory for boundary layer separation of incompressible fluid flows, which shall be described in the User's Guide in the next section and in Chapter 5. This is a long-standing problem in fluid dynamics and a rigorous theory is much needed.

Furthermore, divergence-free vector fields appear naturally in a large class of evolution partial differential equations describing various physical phenomena displaying complex turbulent behavior. For instance, many such partial differential equations can be written as the convection of a passive scalar density by an incompressible velocity field. More precisely, they are of the form:

$$\begin{cases} \dfrac{\partial q}{\partial t} + \operatorname{div}(qv) = 0, \\ v = L(q), \quad \operatorname{div} v = 0, \end{cases} \tag{0.2.1}$$

where $q(x,t)$ is some scalar density function defined on $M \times \mathbb{R}$ and $v(x,t)$ is an incompressible vector field, which can be recovered from q by solving a partial differential equation system. Thus, L denotes a (not necessarily linear) integro-differential operator. Among typical examples of such systems are 1) the linear transport equation with v being a given incompressible vector field, 2) the quasi-geostrophic model in ocean/atmosphere dynamics, and 3) collisionless kinetic equations such as the Boltzman-Poisson equation of stellar dynamics and the Vlasov-Maxwell equations of plasmas. It is hoped that the ideas, methods, and results presented in this book shall yield new insight into the underlying phenomena as well as answers to some fundamental issues in mathematical physics.

As we know, a healthy mode of research in applied mathematics includes two aspects. The study of the underlying physical problems involves on the one hand applications of the existing mathematical theory to the understanding of the underlying physical problems, and on the other hand the development of new mathematical theories. The program of research addressed in this book can be considered as an attempt at this latter aspect. Namely, motivated by the study of geophysical fluid dynamics problems, the new mathematical theory is developed under close links to the physics, and in return the theory is applied to the physical problems, although more applications are yet to be explored.

0.3. The User's Guide

This section is intended for the more fluid mechanics-oriented reader who does not want to follow all the mathematical details, but rather wants to rapidly reach the main results and methods presented in this book.

As mentioned in the previous section, the theory presented in this book includes both kinematic and dynamic studies. In this User's Guide, we shall selectively present some of the results and methods to be addressed in this book.

Notations.

1. Unless otherwise stated, we always assume throughout this book that M is a two-dimensional (2-D) compact Riemannian orientable manifold with or without boundary. From the fluid mechanics applications, the reader can assume M is either a planar region, or a submanifold of a sphere, or the 2-D torus $\mathbb{T}^2$.
2. On the boundary ∂M of M, τ denotes the tangential vector, and n the normal vector.
3. In this book, unless otherwise stated, we always assume that $r \geq 1$ is an integer. Let $C^r(TM)$ be the space of all r-th differentiable vector fields v on M, and let $C^r_n(TM)$ be the space of all r-th differentiable vector fields v on M such that $v|_{\partial M} \in C^r(T\partial M)$; namely, the restriction of any r-th differentiable vector field $v \in C^r(TM)$ on the boundary ∂M is an r-th differentiable vector field of the tangent bundle of ∂M.
4. In addition, we set
$$\begin{aligned} C^r_n(TM) &= \{v \in C^r(TM) \mid v \cdot n = 0 \text{ on } \partial M\}, \\ D^r(TM) &= \{v \in C^r_n(TM) \mid \text{div } v = 0\}, \\ D^r_0(TM) &= \{v \in D^r(TM) \mid v \text{ is regular }\}, \\ B^r_0(TM) &= \{v \in D^r(TM) \mid v = 0 \text{ on } \partial M\}. \end{aligned}$$
5. A point $p \in M$ is called a singular point of $v \in C^r_n(TM)$ if $v(p) = 0$.
6. A singular point p of $v \in C^r_n(TM)$ is called non-degenerate if the Jacobian matrix $Dv(p)$ is invertible.
7. A vector field $v \in C^r_n(TM)$ is called regular if all singular points of v are non-degenerate.

0.3.1. Structural classification. The first step in this theory is to give a kinematic structural classification of divergence-free vector fields. This is done in the following steps.

1. For regular divergence-free vector fields, the classification is given by Theorems 1.3.3 and 1.3.2. These two theorems amount to saying that the structure of a regular divergence-free vector field consists of a) saddle connections, b) circle cells and circle bands, and c) ergodic sets. When the manifold M is a submanifold of S^2, the structure consists of only the first two categories. Loosely speaking, saddle connections form the frame of the structure, separating the other building blocks (circle cells, circle bands and ergodic sets) from each other. The structure of both circle cells and circle bands are trivially known, while an ergodic set is a pseudo-manifold whose Euler characteristic can be calculated explicitly in terms of saddles points on it; see Theorem 1.3.13 for the general situation and Theorem 1.3.14 for an ergodic set on a torus.

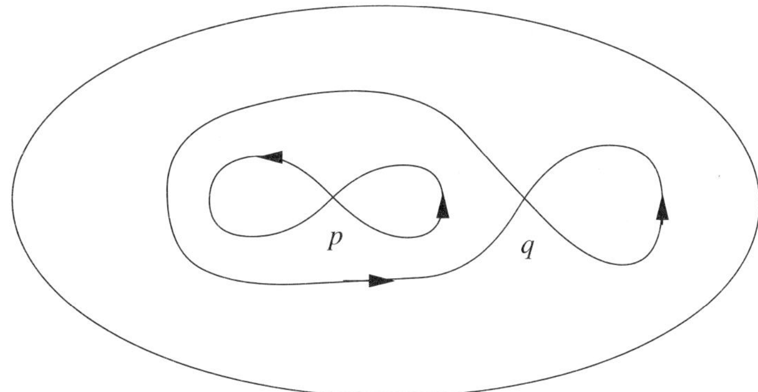

FIGURE 0.3.1. A stable flow pattern with interior saddles being self-connected.

2. The structural classification leads also to a topological classification given by Theorem 1.4.6, which determines the topological equivalency of two divergence-free vector fields with isomorphic "skeletons."

3. The classification of non-regular divergence-free vector fields can be understood through the approximation by regular fields or structurally stable fields.

4. Technically, one of the key ingredients for the classification is the Poincaré-Bendixson theorem for divergence-free vector fields on general two-dimensional manifolds.

0.3.2. Structural stability. The next important issue is the study of structural stability, as it has been the main driving force behind much of the development of dynamical systems theory following the program initiated by S. Smale and others. We are interested in the structural stability of an incompressible vector field with divergence-free vector field perturbations. We call this notion of structural stability the *incompressible structural stability*, or structural stability for brevity.

MAIN RESULTS–THEOREMS 2.1.2 AND 2.2.9. Necessary and sufficient conditions for an incompressible vector field v on a compact manifold $M \subset S^2$ with or without boundary to be structurally stable are:

a) v is regular;
b) all interior saddle points of v are self-connected; and
c) each boundary saddle point is connected to boundary saddles on the same connected component of the boundary.

We note, in addition, that the set of all structurally stable divergence-free vector fields is an open and dense set of $D^r(TM)$.

This theorem is stated for $v \in D^r(TM)$; similar results for $v \in B^r(TM)$ or $B_0^r(TM)$ are also true.

EXAMPLES. We now examine a few flow patterns as shown in Figures 0.3.1–0.3.4. From the stability criteria just mentioned, it is easy to see that both flow patterns given in Figures 0.3.1 and 0.3.2 are structurally stable.

On the other hand, the two flow patterns given in Figures 0.3.3 and 0.3.4 are structurally unstable. The flow pattern given by Figure 0.3.3 does not have saddle points on the boundary. The instability is caused by the saddle connection connecting two interior saddles p and q. With arbitrarily small perturbations with

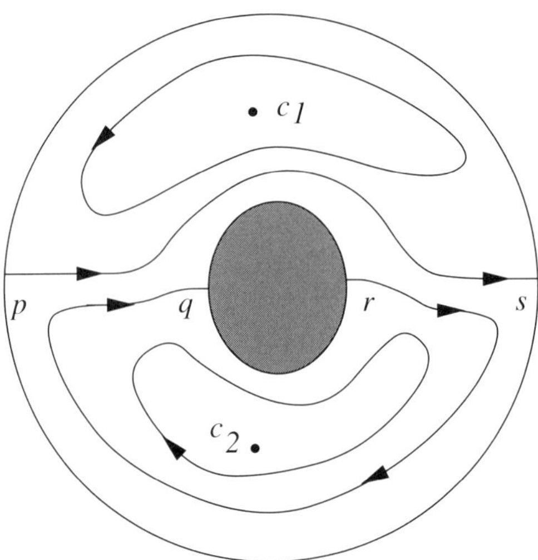

FIGURE 0.3.2. A stable flow pattern with each boundary saddle being connected to saddles on the same connected component of the boundary.

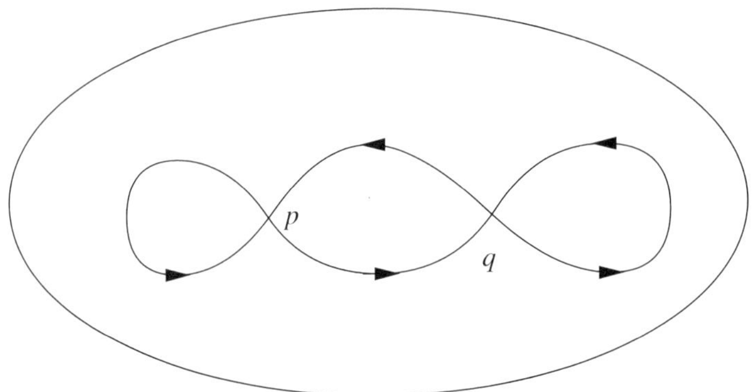

FIGURE 0.3.3. An unstable flow pattern with two interior saddles being connected.

tubular divergence-free vector fields near the saddle point p, the saddle connection will break, and lead to a new stable pattern such as the pattern given by Figure 0.3.1.

In Figure 0.3.4, the flow pattern has four saddles on the boundary. Here the instability comes from the boundary saddle connections between different connected components of the boundary; i.e., p is connected to q, and r is connected to s. As in the previous situation, with arbitrarily small perturbations with tubular divergence-free vector fields surrounding the inner island, both saddle connections (p to q and r to s) break and lead to a new stable pattern as given by Figure 0.3.2.

MAIN TECHNICAL INGREDIENTS. The proof of the structural stability theorem involves two main technical ingredients.

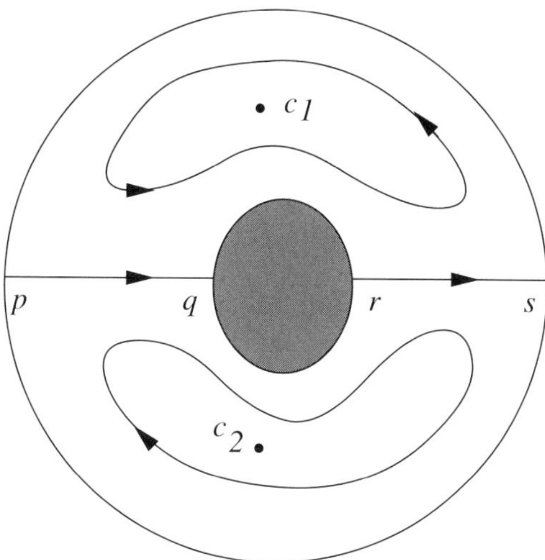

FIGURE 0.3.4. An unstable flow pattern with two boundary saddles on different connected components of the boundary being connected.

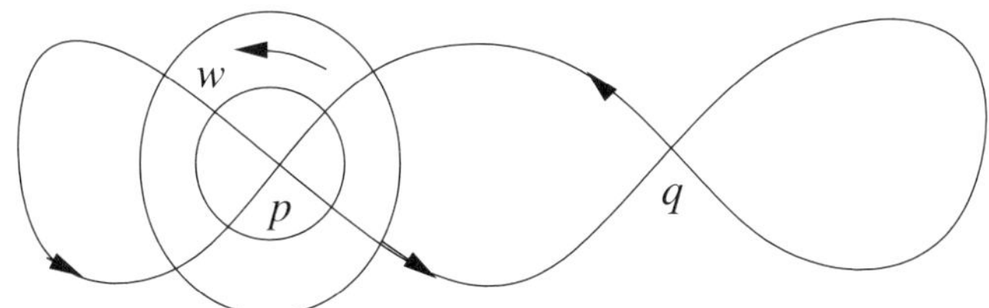

FIGURE 0.3.5. A schematic for breaking interior saddle connections by a hollow vortex.

First, saddle connections are broken by hollow vortices as shown in Figure 0.3.5, where the superposition $v+w$ of the hollow vortex w and the flow v given by Figure 0.3.3 leads to the flow pattern as shown in Figure 0.3.1.

Second, two fields with topologically equivalent skeletons including saddle connections and centers are topologically equivalent.

0.3.3. Block structure and block stability. For incompressible flows defined on a two-dimensional torus or on a two-dimensional manifold M of nonzero genus, we show that no divergence-free vector field is structurally stable with divergence-free vector field perturbations. This suggests quite a different scenario from the case when the manifold is a submanifold of the two-dimensional sphere.

Our next natural step toward developing a kinematic theory as part of our goal of a global geometric theory on two-dimensional manifolds is to go beyond this instability result by introducing two new concepts: block structure and block stability. A novel theory can then be developed using the block structure. For this purpose, we collect a class of divergence-free vector fields, having special block structure, which we call basic vector fields. More precisely, a divergence-free vector field is called a basic vector field if its phase diagram has a block structure that consists of a finite number of flow-invariant retractable blocks and ergodic sets such that the restrictions of the vector field on the retractable blocks are self-connected. Using this idea, the main results include the following assertions:

(i) all basic vector fields form an open and dense set of all divergence-free vector fields,
(ii) the block structure is stable,
(iii) the flow is either periodic or non-trivially recurrent on the ergodic sets, and
(iv) the structural instability is due completely to the ergodicity and/or periodicity on the ergodic sets.

In a nutshell, these results provide a complete classification on the structure and stability of divergence-free vector fields on general two-dimensional orientable compact manifolds, including the torus corresponding to the periodic boundary conditions in the Euler representation of fluid flows. The study outlined here is carried out in detail in Chapter 3, and in Section 4.5.

From the fluid mechanics point of view, it is interesting to consider the case where the manifold M is a two-dimensional torus $\mathbb{T}^2 = \mathbb{R}^2/(2\pi\mathbb{Z})^2$. This case corresponds to the fluid flows in a two-dimensional domain with double periodic boundary conditions; see also discussions in the next subsection.

0.3.4. Boundary layer separation and applications to oceanic boundary currents. The nature of a flow's boundary layer separation from the boundary plays a fundamental role in many physical problems, and often determines the nature of the flow in the interior as well. The main objective of this section is to present a rigorous characterization of the boundary layer separations of two-dimensional incompressible fluid flows. This is a long-standing problem in fluid mechanics, going back to the pioneering work of Prandtl [**84**] in 1904. Basically, in the boundary layer, the shear flow can detach/separate from the boundary, generating a slow backflow and leading to more complicated turbulent behavior; see Figure 0.3.6 (following Goldstein [**31**]). Here, graphs of the velocity against distance from the wall are shown for various sections (the wall being drawn straight, for convenience). The upper chain-dotted line represents the limit of the boundary layer; the lower chain-dotted line, the limit of the small back flow. The stream-lines are shown as broken lines across the figure.

From the theoretical point of view, it is important to characterize the separation. Based on L. Prandtl [**84**], it is observed experimentally that the separation can only occur at the point $\bar{x}$ where the normal derivative of the velocity field vanishes,

$$\frac{\partial u(\bar{x})}{\partial n} = 0,$$

which is the classical Prandtl condition. However, there are *no known* theorems to determine *when*, *where* and *how* the separation occurs; see A. Chorin and J. Marsden [**11**] and W. Jäger, P. Lax and C. Morawetz [**39**].

The main objective of the study presented in this book is to develop a rigorous theory to characterize the separation, and to apply the theory to some fluid problems.

We address here briefly some main aspects of this study, as presented in Chapter 5.

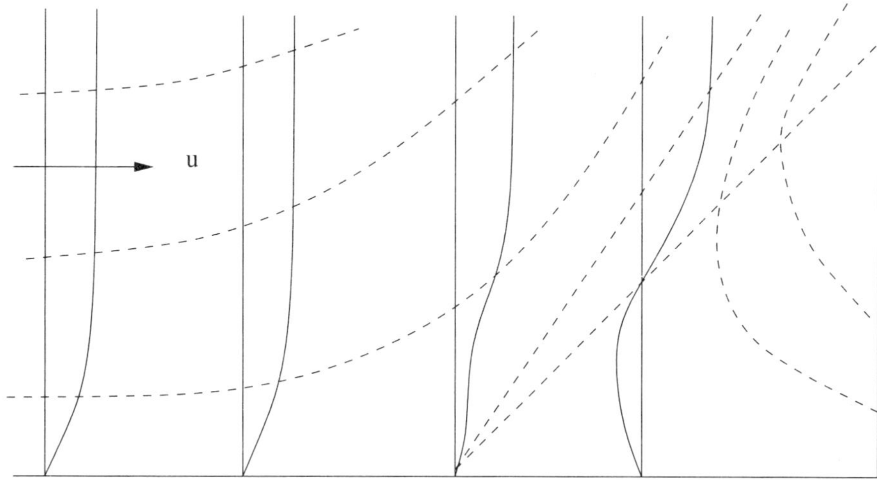

FIGURE 0.3.6. Boundary layer separation of shear flow

For convenience, we assume the boundary ∂M contains a flat part $\Gamma \subset \partial M$, and consider structural bifurcation near a ∂–singular point $\bar{x} \in \Gamma$. For simplicity, we take a coordinate system (x_1, x_2) with $\bar{x}$ at the origin and with Γ given by

$$\Gamma = \{(x_1, 0) \,\big|\, |x_1| \leq \delta_0\}, \qquad \bar{x} = 0,$$

for some $\delta_0 > 0$. Obviously, the tangent and normal vectors on Γ are the unit vectors in the x_1 and x_2 directions, respectively. Let $u \in C^1([0,T], B_0^r(TM))$ $(r \geq 2)$ be a one-parameter family of divergence-free vector fields with the homogeneous Dirichlet boundary condition. In a neighborhood $U \subset M$ of $\bar{x} \in \Gamma$, $u(x,t)$ and its normal derivative

$$v(x,t) = \frac{\partial u(x,t)}{\partial n}$$

can be expressed by

$$\begin{cases} u(x,t) = u^0(x) + (t - t_0)u^1(x) + o(|t - t_0|), \\ v(x,t) = v^0(x) + (t - t_0)v^1(x) + o(|t - t_0|), \\ u^0(x) = u(x, t_0), \qquad u^1(x) = \dfrac{\partial u(x,t)}{\partial t}|_{t=t_0}, \\ v^0(x) = \dfrac{\partial u^0}{\partial n}, \qquad v^1(x) = \dfrac{\partial u^1}{\partial n}. \end{cases} \tag{0.3.1}$$

The first step of the study is to find certain kinematic conditions for structural bifurcation. They are given by Assumption 5.2.1. For convenience, we recount them here.

Assumption 5.2.1. *Let $\bar{x} = 0 \in \Gamma$ be an isolated degenerate ∂-singular point of $u^0(x)$, $u^0 \in C^{k+1}$ near $\bar{x} \in \Gamma$ for some $k \geq 2$. Assume that*

$$\frac{\partial u^0(0)}{\partial n} = 0, \tag{0.3.2}$$

$$\text{ind}(v^0, 0) \neq -\frac{1}{2}, \tag{0.3.3}$$

$$\frac{\partial^{k+1} u_1^0(0)}{\partial^k \tau \partial n} \neq 0, \tag{0.3.4}$$

$$\frac{\partial u^1(0)}{\partial n} \neq 0. \tag{0.3.5}$$

Condition (0.3.2) says that $\bar{x} = 0 \in \Gamma$ is a ∂-singular point of $u^0(x)$, or equivalently, the leading order vorticity vanishes at $\bar{x}$. This is the so-called Prandtl condition, which was suggested by Prandtl to identify possible boundary layer separation points of incompressible flows.

Condition (0.3.3) amounts to saying that the index of v^0 at $\bar{x} = 0$ is different from $-1/2$. Hence, let

$$\text{ind } (v^0, 0) = -\frac{n}{2} \qquad (n \neq 1).$$

Then there are exactly $n \neq 1$ interior orbits of u^0 connected to $\bar{x} \in \Gamma$. This shows that $\bar{x} \in \Gamma$ is a degenerate ∂-singular point of $u^0(x)$, which is necessary for structural bifurcation due to the structural stability theorem.

Condition (0.3.5) amounts to saying that the first order term u^1 of the Taylor expansion for the normal derivative of u is different from zero. Also, this is necessary; otherwise, we need to work on a higher order Taylor expansion, and the corresponding results presented in this book will be true as well. In view of fluid mechanics applications, Condition (0.3.5) is equivalent to nonzero vorticity for u^1, which is a necessary condition for the bifurcation. In addition, it is easy to see that

$$\frac{\partial u_1^1(0)}{\partial x_2} = \frac{\partial u_1^1(0)}{\partial n} \neq 0,$$

which shows that the acceleration of fluid in the tangential direction at p near the boundary layer is nonzero.

Condition (0.3.4) is a technical condition, and amounts to saying that the tangential component u_1^0 of the leading-order term is Taylor expandable. For fluid flows satisfying the Navier-Stokes equations, local analyticity of the solutions implies that this condition is always satisfied.

Structural Bifurcation. Under the above assumption, the following kinematic criteria hold true:

(1) The flow structure of u near $\bar{x}$ and t_0 is fully classified, and depends on $\text{ind}(v^0, \bar{x})$.
(2) In particular, structural bifurcation occurs at t_0.
(3) The case where $\text{ind}(v^0, \bar{x}) = 0$ corresponds to the boundary layer separation. In this case, t_0 is the time instant when the separation occurs, and the separation appears exactly as shown in Figure 0.3.7.

Technically speaking, the results are achieved with delicate analysis of the flow structure near the boundary for both the free boundary and the Dirichlet boundary

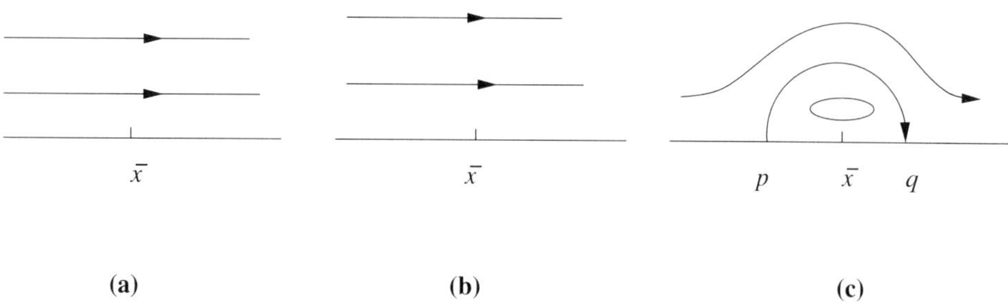

FIGURE 0.3.7. Boundary layer separation of shear flow.

conditions. The first step is to classify the flow structure and its transitions near the boundary for flows with only no normal flow boundary conditions; see Section 5.2. Second, we analyze the detailed flow structure in the boundary layer for flows with the Dirichlet boundary condition; see Section 2.2. Third, we make connections between the structure of the original velocity fields and the structure of the normal derivative of the velocity field; see Sections 5.3–5.5 .

REATTACHMENT OF THE SEPARATION. In Figure 0.3.7, we see reattachment of the stream lines to the boundary, although the fluid particle trajectories may not be reattached to the boundary. One could think the fluid particle moves into the interior of the domain as the bubbles (of the stream lines) amplify as the time evolves. This fact is proved in [**29, 30**], and supported by numerical simulations for the driven cavity flow in Section 6.1. Also, the reattachment is consistent with experiments as shown in van Dyke's book [**16**].

The numerical simulation for cavity driven flow presented in Section 6.1 demonstrates the boundary separation, and clearly shows the reattachment.

VORTICITY "CRISIS." Consider a shear flow as shown in Figure 0.3.7(a) and (b). This amounts to saying that there exists a neighborhood $\mathcal{O}$ of $\bar{x}$ such that the vorticity $\omega = -\frac{\partial u_1}{\partial x_2} + \frac{\partial u_2}{\partial x_1}$ is

$$\omega \leq 0, \quad \forall x \in \mathcal{O}, \quad t \leq t_0. \tag{0.3.6}$$

Then Assumption 5.3.1 with $k = 2$ is equivalent to the following:

$$\omega = 0, \quad \frac{\partial \omega}{\partial \tau} = 0, \quad \frac{\partial^2 \omega}{\partial \tau^2} < 0, \quad \frac{\partial \omega}{\partial t} > 0 \quad \text{at } (\bar{x}, t_0). \tag{0.3.7}$$

In other words, the separation takes place where the vorticity reaches the "crisis" as shown in Figure 0.3.8. This is consistent with physical studies; see [**31**], among others.

ADVERSE PRESSURE GRADIENT. In addition, we can rigorously prove that under the vorticity crisis condition in (0.3.7), the adverse pressure gradient holds true at the critical separation point, i.e.,

$$\frac{\partial p(\bar{x}, t_0)}{\partial \tau} > 0. \tag{0.3.8}$$

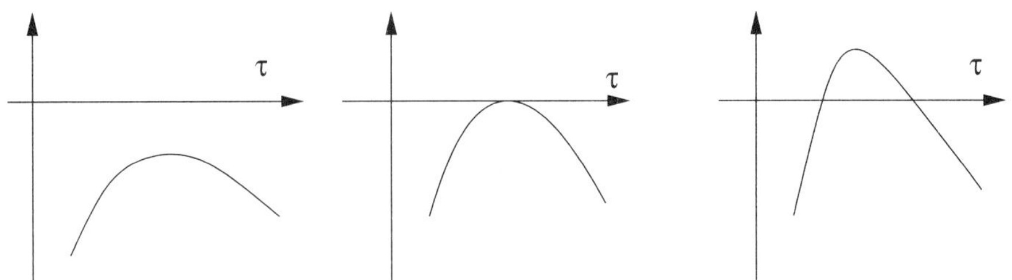

FIGURE 0.3.8. Vorticity "Crisis".

SEPARATION TIME AND LOCATION. The critical location and time of the separation satisfy the following identity:

$$\frac{\partial \varphi_\tau(\bar{x})}{\partial n} = \int_0^{t_0} \left\{ R^{-1}[\nabla \times \triangle u + k \triangle u \cdot \tau] + \lambda(\nabla \times f + k f_\tau) \right\} (\bar{x}, t) dt. \tag{0.3.9}$$

CURVED BOUNDARY. Finally, we remark that the same results hold true for structural bifurcation near a curved boundary. In that situation, we only need to replace Condition (0.3.3) by the geometrical condition: There are n ($n \neq 1$) interior orbits of u^0 connected to $\bar{x}$.

0.3.5. Stability and transitions of the solutions of the Navier-Stokes equations. As we mentioned earlier, one ultimate goal of the study is to apply the kinematic theory to study the dynamics. The study along this direction belongs to Area (B) mentioned in Section 0.2. In Section 0.3.3, we have illustrated some progress made in this area of study for boundary layer separation. In this guide, we address two examples, which provide links between the kinematic theory and the dynamics, and, as always, we refer the interested readers to the rest of the book for further details.

GENERICITY OF STRUCTURALLY STABLE FIELDS. Mathematically speaking, it is important to study the genericity of structurally stable solutions of the Navier-Stokes equations. This is done in the following three steps in this book.

First, at the kinematic level, we show that all structurally stable divergence-free vector fields form an open and dense subset of the space of all divergence-free vector fields; see Theorems 2.1.3 and 2.2.9.

Second, for steady state solutions of the Navier-Stokes equations, there is an open and dense set $\mathcal{F}$ of forcing in a proper function space such that for any forcing $f \in \mathcal{F}$, all corresponding steady states are structurally stable.

Third, for the time-dependent solutions of the Navier-Stokes equations, it is shown that given initial data, there is an open and dense set $\mathcal{F}$ of forcing in a proper function space such that for any forcing $f \in \mathcal{F}$, the time-dependent solution $u(x,t)$ of the Navier-Stokes equations are structurally stable for a residual (hence, dense) set of the time interval $[0,\infty)$. This result justifies the structural bifurcation of the solutions of the Navier-Stokes equations.

PERIODIC STRUCTURE OF SOLUTIONS OF THE NAVIER-STOKES EQUATIONS. Another example demonstrating the link between the kinematic theory and dynamics is the classification of the periodic structure and its transitions for the solutions of the Navier-Stokes equations with double periodic conditions. In fact, this link is potentially useful for classifying the vortex structure related to the defect of the Ginzburg-Landau equations of superconductivity.

Technically, we proceed as follows. First, the structure of all eigenvectors of the corresponding Stokes problem is classified using block structure, and is linked to the typical structure of the Taylor vortices. Then the structure of the solutions of the Navier-Stokes equations forced either by eigenmodes or by potential forcing is classified. In particular, the structure of the solutions of the two-dimensional Navier-Stokes equation in the spectral manifolds as studied by C. Foias and J. Saut [**21, 22**] is classified as well.

STRUCTURE OF BIFURCATED SOLUTIONS OF THE RAYLEIGH-BÉNARD CONVECTION. Last, but not least, an important aspect of the link between the kinematic theory and dynamics of fluid flows is the connection between *dynamic bifurcation* of the partial differential equations models of fluid flows and the *structural bifurcation* of the flow structure in the physical spaces. This program of study was initiated recently, using a dynamic bifurcation theory newly developed by the authors. In Section 4.6, we briefly describe the theory and its applications to the Rayleigh-Bénard convection problem.

Convection is the well-known phenomena of fluid motion induced by buoyancy when a fluid is heated from below. It is, of course, familiar as the driving force in atmospheric and oceanic phenomena, and in the kitchen! The Rayleigh-Bénard convection problem was originated in the famous experiments conducted by H. Bénard in 1900. Bénard investigated a fluid, with a free surface, heated from below in a dish, and noticed a rather regular cellular pattern of hexagonal convection cells. In 1916, Lord Rayleigh [**87**] developed a theory to interpret the phenomena of Bénard experiments. He chose the Boussinesq equations with some boundary conditions to model Bénard's experiments, and linearized these equations using normal modes. He then showed that the convection would occur only when the non-dimensional parameter, called the Rayleigh number,

$$R = \frac{g\alpha\beta}{\kappa\nu} h^4, \tag{0.3.10}$$

exceeds a certain critical value, where g is the acceleration due to gravity, α the coefficient of thermal expansion of the fluid, $\beta = |dT/dz| = (\bar{T}_0 - \bar{T}_1)/h$ the vertical temperature gradient with $\bar{T}_0$ the temperature on the lower surface and $\bar{T}_1$ the temperature on the upper surface, h the depth of the layer of the fluid, κ the thermal diffusivity and ν the kinematic viscosity.

Mathematically speaking, this is a bifurcation problem for the governing partial differential equations. Since Rayleigh's pioneering work, there have been intensive studies for this problem; see among others Chandrasekhar [**10**] and Drazin and Reid [**15**] for linear theories, and Kirchgässner [**43**], Rabinowitz [**86**], and Yudovich [**106, 107**], and the references therein for nonlinear theories. Most, if not all, known results on bifurcation and stability analysis of the Rayleigh-Benard problem are restricted to the bifurcation and stability analysis when the Rayleigh number crosses a simple eigenvalue in certain subspaces of the entire phase space obtained by imposing certain symmetry.

It is clear that a complete nonlinear bifurcation and stability theory for this problem should at least include:

(1) a bifurcation theorem when the Rayleigh number crosses the first critical number for all physically sound boundary conditions,
(2) asymptotic stability of bifurcated solutions, and
(3) the structure/patterns and their stability and transitions in the physical space.

The main difficulties for such a complete theory are twofold. The first is due to the high nonlinearity of the problem, as in other fluid problems, and the second is due to the lack of a theory to handle bifurcation and stability when the eigenvalue of the linear problem has even multiplicity.

To achieve the first two parts of this objective, a new notion of bifurcation, attractor bifurcation, was introduced in Ma and Wang [**65, 60**]; see also the new book [**66**] by the authors. The method and results presented in this book, including in particular a structural stability theorem, provide the needed tools for studying the third part of the objective.

The main results are stated as follows.

First, we show that as the Rayleigh number R crosses the first critical value R_c, the Boussinesq equations bifurcate from the trivial solution to an attractor $\mathcal{A}_R$, with dimension between $m-1$ and m. Here, the first critical Rayleigh number R_c is defined to be the first eigenvalue of the linear eigenvalue problem, and m is the multiplicity of this eigenvalue R_c. In comparison with known results, the bifurcation theorem, obtained in the present book, holds true for all cases with any multiplicity m of the critical eigenvalue R_c for the Bénard problem under any set of physically sound boundary conditions. As the trivial solution becomes unstable as the Rayleigh number crosses the critical value R_c, $\mathcal{A}_R$ does not contain this trivial solution.

Second, as an attractor, the bifurcated attractor $\mathcal{A}_R$ has asymptotic stability in the sense that it attracts all solutions with initial data in the phase space outside of the stable manifold, with co-dimension m, of the trivial solution.

As Kirchgässner indicated in [**43**], an ideal stability theorem would include all physically meaningful perturbations and establish the local stability of a selected class of stationary solutions, and today we are still far from this goal. On the other hand, fluid flows are normally time dependent. Therefore, bifurcation analysis for steady state problems provides in general only partial answers to the problem, and is not enough for solving the stability problem. Hence, it appears that the right notion of asymptotic stability after the first bifurcation should be best described by the attractor near, but excluding, the trivial state. It is one of our main motivations for introducing attractor bifurcations, and it is hoped that the stability of the bifurcated attractor obtained in this book provides a method toward an ideal stability theorem.

Third, we show that in the two-dimensional case, for any initial data outside of the stable manifold of the trivial solution, the solution of the Boussinesq equations will have the roll structure as t is sufficiently large.

Notes for Introduction

1. This chapter presents some brief introduction to fluid dynamics. Although the material presented is self-contained, interested readers are referred to the following sources. There are, of course, many others.

- For topics on classical fluid mechanics: G. I. Batchelor [**6**], S. Goldstein [**31**], L. D. Landau and E. M. Lifshitz [**45**], and C. Truesdell [**103**];

- For topics on mathematical fluid mechanics: A. J. Chorin and J. E. Marsden [**11**], P. Constantin and C. Foias [**13**], C. Doering and J. D. Gibbon [**14**], S. Friedlander [**24**], J. L. Lions [**46**], P. L. Lions [**49, 50**], A. Majda and A. Bertozzi [**69**], R. Temam [**102**];
- For topics on geophysical fluid flows: J. Pedlosky [**80**], J. P. Peixoto and A. H. Oort [**81**], M. Ghil and S. Childress [**27**], R. Salmon [**91**], J. L. Lions, R. Temam and S. Wang [**47, 48**]; and
- For topics on N-vortex studies: H. Aref [**4**], P. Newton [**75**].

2. The User's Guide provides a general outline of the work presented in this book. Readers are referred to the later chapters for details. It is impossible to explain all the notations and concepts in this User's Guide, and readers are referred to chapters, as well as to the Index at the end of the book, for details.

CHAPTER 1

Structure Classification of Divergence-Free Vector Fields

As we mentioned earlier, this book addresses studies of incompressible flows in two areas: A) the development of a global geometric theory of divergence-free vector fields on general two-dimensional compact manifolds with or without boundaries, and B) the study of the structure and its transitions of velocity fields for two-dimensional incompressible fluid flows governed by the Navier-Stokes equations or the Euler equations.

This chapter is the first step toward a complete kinematic theory for incompressible flows as described in Area A), which is a crucial step for linking the kinematics to dynamics as described in Area B).

The main topics of this chapter are structural and topological classifications in terms of simpler objects, which play the role of building blocks in the classification. These building blocks include circle cells, circle bands, saddle connections for structural classification and the "skeleton" needed for the topological classification.

1.1. Limit Set Theorem

1.1.1. Some basic concepts and lemmas. As mentioned earlier, unless otherwise explicitly stated, let M always be a two-dimensional, orientable, differentiable Riemannian manifold with boundary ∂M.

Let $C^r(TM)$ be the space of all r-th differentiable vector fields v on M, and let $C^r_n(TM)$ be the space of all r-th differentiable vector fields v on M with no-normal flow condition

$$v \cdot n = 0 \qquad \text{on } \partial M.$$

Unless otherwise stated, we always assume that $r \geq 1$.

Consider $v \in C^r_n(TM)$ $(r \geq 1)$. Let $p \in M$ and (φ, U) be a local coordinate system near p. Then $D\varphi \circ v \circ \varphi^{-1}$ is a vector field in $V \equiv \varphi(U) \subset \mathbb{R}^2$.

A point $p \in M$ is a singular point of v if $v(p) = 0$; otherwise, p is called a regular point. An interior singular point $p \in \overset{\circ}{M}$ is called non-degenerate if $v(p) = 0$ and the Jacobian matrix

$$J(D\varphi \circ v \circ \varphi^{-1})|_{q=\varphi(p)}$$

is nonsingular; otherwise, $p \in \overset{\circ}{M}$ is called degenerate.

For a boundary point $p \in \partial M$, we take a local coordinate (φ, U) near p such that

$$\begin{cases} \varphi : U \to \mathbb{R}^2_+ = \{x \in \mathbb{R}^2 \mid x_n \geq 0\}, \\ \varphi(U \cap \partial M) = \partial \mathbb{R}^2_+, \\ \varphi(p) = 0. \end{cases} \tag{1.1.1}$$

Under this local coordinate system, $v \in C_n^r(TM)$ can be expressed as follows:

$$\begin{cases} u = D\varphi \circ v \circ \varphi^{-1} : \mathbb{R}^2_+ \to \mathbb{R}^2, \\ u(x) = (u_1(x), u_2(x)), \\ u_2(x) = 0, \ \forall\, x \in \partial\mathbb{R}^2_+. \end{cases} \tag{1.1.2}$$

DEFINITION 1.1.1. *Let $p \in \partial M$ be a singular point of a vector field $v \in C_n^r(TM)$, i.e., $v(p) = 0$. p is called non-degenerate if the Jacobian matrix $J(u(0))$, given by*

$$J(u(0)) = \begin{pmatrix} \dfrac{\partial u_1(0)}{\partial x_1} & \dfrac{\partial u_1}{\partial x_2} \\ 0 & \dfrac{\partial u_2(0)}{\partial x_2} \end{pmatrix},$$

is non-singular, where $u(x)$ is defined by (1.1.2).

DEFINITION 1.1.2. *A vector field $v \in C_n^r(TM)$ $(r \geq 1)$ is called regular if all singular points of v are non-degenerate.*

Let $v \in C_n^r(TM)$, $p \in M$ be a non-degenerate singular point of v, and λ_1, λ_2 the two (nonzero) eigenvalues of the Jacobian $J(v)(p)$. Then p is called

(1) a saddle point if λ_1 and λ_2 are real and $\lambda_1 \cdot \lambda_2 < 0$,
(2) a center if λ_1 and λ_2 are pure imaginary numbers,
(3) a focus if both $\mathrm{Re}\lambda_1$ and $\mathrm{Re}\lambda_2$ are negative, or
(4) a node if both $\mathrm{Re}\lambda_1$ and $\mathrm{Re}\lambda_2$ are positive.

For $v \in C_n^r(TM)$, let $\gamma = \{\Phi(x,t)\}_{t\in\mathbb{R}}$ be the orbit passing through $x \in M$ at $t = 0$ of the flow generated by v. The following terminologies are standard:

(1) the orbit γ is called a closed orbit if there is a time $T_0 > 0$ such that for any $t \in \mathbb{R}$, $\Phi(x,t) = \Phi(x, t+T_0)$;
(2) the ω-limit set $\omega(x)$ and the α-limit set $\alpha(x)$ of the orbit $\{\Phi(x,t)\}_{t\in\mathbb{R}}$ are defined by

$$\begin{aligned} \omega(x) &= \{y \in M \mid \text{ there exist } t_n \to \infty \quad \text{such that } \Phi(x,t_n) \to y\}, \\ \alpha(x) &= \{y \in M \mid \text{ there exist } t_n \to -\infty \quad \text{such that } \Phi(x,t_n) \to y\}; \end{aligned}$$

(3) if $\alpha(x)$ is a single point p, then p is called the starting point of the orbit γ;
(4) if $\omega(x)$ is a single point q, then q is called the ending point of the orbit γ; and
(5) if γ is a closed orbit and $\omega(p) = \gamma$ for any point p in a small neighborhood of γ, then γ is called a limiting cycle of v.

One of the main objectives of this book is to study the structure of divergence-free vector fields, which arise in many problems of mathematical physics, in particular in fluid mechanics. Motivated by the no-normal flow conditions of fluid flows on boundaries, we consider only vector fields with vanishing normal components on the boundary. Hence we set

$$D^r(TM) = \{v \in C_n^r(TM) \mid \ \mathrm{div}\ v = 0\}.$$

The differential operator div is the divergence operator of a vector field, which can be defined in terms of the Levi-Civita connection.

Elements in $D^r(TM)$ are called divergence-free vector fields on M. A divergence-free vector field $v \in D^r(TM)$ can be characterized by the flow $\Phi(\cdot,\cdot)$ generated by v:

$$\Phi : M \times \mathbb{R} \to M$$

such that for each time t fixed, $\Phi(\cdot, t) : M \to M$ is a diffeomorphism, preserving volumes in M.

The following lemma is basic for the studies hereafter.

LEMMA 1.1.3. *Let $v \in D^r(TM) (r \geq 1)$, and let $U \subset M$ be the set consisting of all closed orbits of v. The following assertions hold true.*

(1) *U is open;*

(2) *For each connected component $\mathcal{C}$ of ∂U, if $\mathcal{C} \cap \overset{\circ}{M} \neq \phi$, then $\mathcal{C}$ must contain singularities of v; and*

(3) *If $\mathcal{C} \subset \partial M$, then either $\mathcal{C}$ is a closed orbit, or $\mathcal{C}$ contains at least a singular point of v.*

PROOF. Let $\gamma \subset \overset{\circ}{M}$ be a closed orbit. By the tubular neighborhood theorem, there exists a tubular neighborhood V of γ in M such that V is homeomorphic to $S^1 \times (-1, 1)$, and v has no singularities in V.

Then we only have to show that for any $x_0 \in \gamma$, there is a neighborhood D of x_0 in M such that if $x \in D$, the orbit $\Phi(x, t)$ is closed. To this end, let $\Sigma \subset V$ be an open arc transversal to v containing x_0. By the existence of the Poincaré mapping (see Proposition 1.2, p. 94 of Palis and Melo [**77**]), the Poincaré mapping P_Σ is a C^r-diffeomorphism from a neighborhood Σ_1 of x_0 in Σ onto an open set in Σ, and x_0 is a fixed point of the Poincaré mapping.

For any $x \in \Sigma_1 \subset V$, if the orbit $\Phi(x, t)$ is not closed, then the orbit returns to intersect with Σ_1 at x_1. Without loss of generality, we assume that $x_1 \in (x_0, x) \subset \Sigma_1$; see Figure 1.1.1.

Let C be the closed curve defined by

$$C = \{\Phi(x,t) \mid 0 \leq t \leq t_1\} \cup [x, x_1],$$

where $[x, x_1] \subset \Sigma$, $\Phi(x, 0) = x$, $\Phi(x, t_1) = x_1$. Let A be the domain enclosed by C and γ. Since Σ is transversal to v which points inward into A, the positive orbit of x_1 is contained in A. Hence for any $t > 0$,

$$\Phi(C, t) \subset A,$$

which implies that for any $t > 0$,

$$|\Phi(A, t)| < |A|, \tag{1.1.3}$$

which contradicts $v \in D^r(TM)$ being divergence-free. Here, $|A|$ stands for the area of A. Hence, $\{\Phi(x,t)\}_{t\in\mathbb{R}}$ is a closed orbit.

The above argument shows in fact that for any $x \in \Sigma_1 \subset V$, $\{\Phi(x,t)\}_{t\in\mathbb{R}}$ is a closed orbit. Hence, it is easy to see that there is a neighborhood of γ containing only closed orbits.

If γ intersects with ∂M, then $\gamma \subset \partial M$, due to v being tangential to ∂M. Then we can prove in the same fashion that there is an open neighborhood of γ containing only periodic orbits. It follows that U is open, i.e., Assertion (1) holds true. The proofs of Assertions (2) and (3) are trivial; we omit the details.

The proof is complete. □

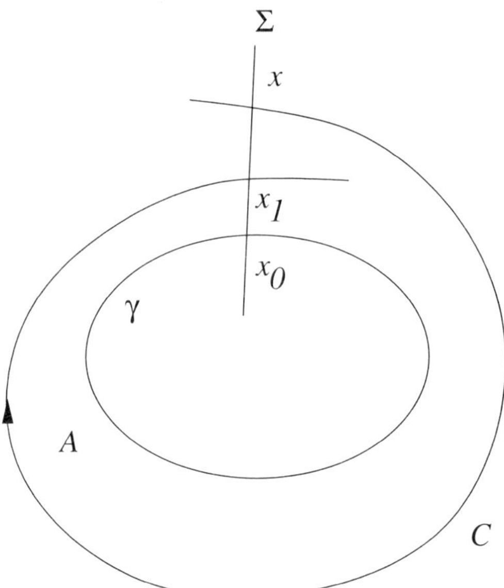

FIGURE 1.1.1. Schematic illustrating the proof of Lemma 1.1.3.

LEMMA 1.1.4. *Let $v \in D^r(TM)$ $(r \geq 1)$. Then each interior non-degenerate singular point of v is either a center or a saddle point. A non-degenerate singularity on the boundary ∂M must be a saddle point.*

PROOF. As we know, an interior non-degenerate singular point of a vector field is either a center, a saddle, a node, or a focus. As in the proof of Lemma 1.1.3, for a divergence-free vector field, neither a node nor a focus is allowed. □

1.1.2. Poincaré-Bendixson theorem for divergence-free vector fields. We start with the following classical Poincaré-Bendixson theorem for general vector fields.

THEOREM 1.1.5 (Poincaré-Bendixson Theorem). *Let $M \subset S^2$ be a two-dimensional compact submanifold, let $v \in C^r_n(TM)$ $(r \geq 1)$ be regular, and let $x_0 \in M$ be a regular point of v. Then the ω-limit set $\omega(x_0)$ (resp. the α-limit set $\alpha(x_0)$) must be one of the following types:*

1. *a singular point,*
2. *a closed orbit, or*
3. *a union of singular points $\{p_1, \cdots, p_n\}$ and orbits such that if an orbit $\gamma \subset \omega(x_0)$ (resp. $\gamma \subset \alpha(x_0)$), then $\alpha(\gamma) = p_i$ and $\omega(\gamma) = p_j$ for some i and j.*

As a corollary of the Poincaré-Bendixson Theorem and Lemmas 1.1.3–1.1.4, we immediately obtain the following refined version of the Poincaré-Bendixson Theorem for divergence-free vector fields.

THEOREM 1.1.6. *Let $M \subset S^2$ be a two-dimensional compact submanifold, let $v \in D^r(TM)$ $(r \geq 1)$ be regular, and let $x_0 \in M$ be a regular point of v. Then the ω-limit set $\omega(x_0)$ (resp. the α-limit set $\alpha(x_0)$) is either a saddle point, or a closed orbit, which is a non-limiting cycle.*

1.1.3. Limit set theorem. In this subsection, we generalize the Poincaré-Bendixson Theorem to divergence-free vector fields on general two-dimensional compact manifolds.

DEFINITION 1.1.7. *Let $v \in D^r(TM)$.*

(1) *An orbit with its end points is called a saddle connection if its α- and ω-limit sets are saddle points.*

(2) *A set Ω is called a closed domain if $\Omega = closure(\overset{\circ}{\Omega})$.*

Let $V \subset M$ be the set of all closed orbits and centers of v. By Lemma 1.1.3, V is an open set in M. We set

$$K = \overline{M - \bar{V}}.$$

Then the following lemma is obvious.

LEMMA 1.1.8. *Let $v \in D^r(TM)$ $(r \geq 1)$ be regular.*

(1) *If $K = \emptyset$, then $\overline{V} = M$ and $M - V$ consists of saddle connections;*

(2) *If $K \neq \emptyset$, then K is a closed domain, i.e., closure$(\overset{\circ}{K}) = K$.*

The main theorem in this section is the following limit set theorem on general two-dimensional compact manifolds with or without boundaries.

THEOREM 1.1.9 (Limit Set Theorem). *Let $v \in D^r(TM)$ $(r \geq 1)$ be regular, and let $x_0 \in M$ be a regular point of v. Then the α-limit set $\alpha(x_0)$ (resp. ω-limit set $\omega(x_0)$) is either one of the following:*

(1) *a non-limiting cycle, closed orbit,*
(2) *a saddle point, or*
(3) *a connected closed domain $\Omega \subset K = \overline{M - \bar{V}}$, with $\partial\Omega$ consisting of saddle connections.*

We now introduce a basic lemma, which will be used often in the remainder of the book as well as in the proof of the Limit Set Theorem here. This lemma can be proved in the same fashion as the proof of Lemma 1.1.3.

LEMMA 1.1.10. *Let $v \in D^r(TM)$ $(r \geq 1)$ be regular, and let $\Gamma \subset M$ be a closed curve consisting of saddle connections of v. Let there exist an open set $D \subset M$ such that Γ is a connected component of ∂D, and for any saddle point $p \in \Gamma$, there are no orbits connected to p in $\overset{\circ}{D}$. Then for any $x \in \overset{\circ}{D}$ sufficiently close to Γ, the orbit $\{\Phi(x,t)\}_{t\in\mathbb{R}}$ is closed.*

PROOF OF THEOREM 1.1.9. *Step 1.* By Lemma 1.1.4, a non-degenerate interior singular point of a regular vector field $v \in D^r(TM)$ must be either a center or a saddle, and a non-degenerate boundary singular point must be a saddle. Hence, if $\alpha(x_0)$ (resp. $\omega(x_0)$) is a singular point, then it must be a saddle.

By the Poincaré–Bendixson theorem for divergence-free vector fields and Lemma 1.1.10, it suffices then to prove Assertion (3). Without loss of generality, we assume that the closed domain $K = \overline{M - \bar{V}}$ is connected.

Step 2. Since the number of saddle connections is finite, we only need to consider the case where $x_0 \in K$ such that $\omega(x_0)$ is not a saddle point. It suffices then to prove that $\omega(x_0) = \text{cl}\,\overset{\circ}{\omega}(x_0)$, the closure of the interior of $\omega(x_0)$.

If we assume otherwise, i.e., $\omega(x_0) \neq \text{cl}\,\overset{\circ}{\omega}(x_0)$, then $\text{cl}\,\overset{\circ}{\omega}(x_0) \subset \omega(x_0)$. We need to derive a contradiction.

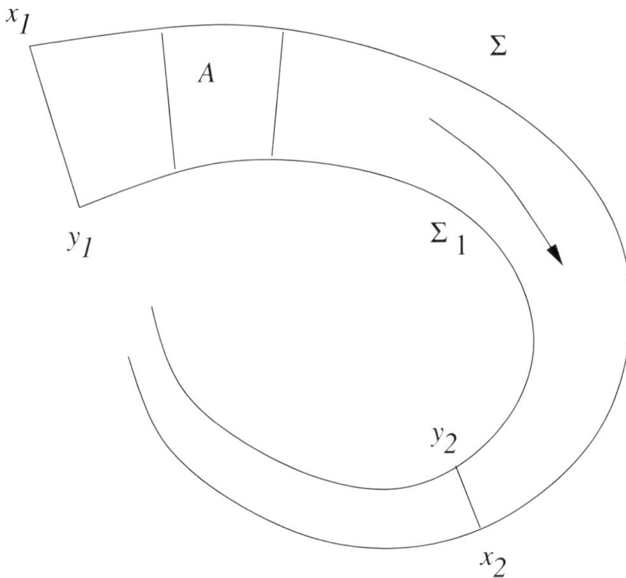

FIGURE 1.1.2. Schematic illustrating the proof of Theorem 1.1.9.

To this end, we define a set $N = \overset{\circ}{K} - \omega(x_0)$. Obviously, N is open and nonempty; otherwise $K = \omega(x_0)$, a contradiction to the assumption that $\omega(x_0)$ is not a closed domain.

Moreover, N is an invariant set, since both $\overset{\circ}{K} = M - \bar{V}$ and $\omega(x_0)$ are invariant. Hence ∂N is also an invariant set consisting of orbits and saddle points of v.

Step 3. Claim: each connected component of ∂N is of finite length. To see this, let $\Sigma \subset \partial N$ be a connected component of ∂N having infinite length.

Since there are only a finite number of saddle points of v, we choose $\widetilde{x} \in \Sigma \subset \partial N$ such that the orbit $\{\Phi(\widetilde{x}, t)\}_{0 \leq t < \infty} \subset \Sigma$ is of infinite length. Let $x_1 = \Phi(\widetilde{x}, t_1)$ $(0 \leq t_1 < \infty)$ and let $\Gamma \subset M$ be an arc transversal to v starting from x_1 and entering N. Since M is compact and the orbit $\Phi(\widetilde{x}, \cdot)$ is of infinite length, there is a time t_2 $(t_1 < t_2)$ such that $x_2 = \Phi(\widetilde{x}, t_2) \in \Gamma$ and $\Phi(\widetilde{x}, t) \notin \Gamma$ for any $t_1 < t < t_2$. For simplicity, let $(x_1, x_2) \subset \Gamma$ be the open segment from x_1 to x_2. Then for any $x \in (x_1, x_2) \cap N$, there is a time $\tau > 0$ such that either $\Phi(x, \tau) \in \Gamma$, $\Phi(x, \tau) \notin (x_1, x_2)$ and $\Phi(x, t) \notin \Gamma$ for any $0 < t < \tau$, or $\Phi(x, \tau) \in (x_1, x_2)$ and $\Phi(x, t) \notin \Gamma$ for any $0 < t < \tau$. It implies that there is a point $y \in (x_1, x_2)$ such that $y \in \partial N$, $(x_1, y) \subset N$ and $(x_1, y) \cap \partial N = \emptyset$. Among all those arcs Γ, there is an arc Γ_1 such that the length of the open segment $(x_1, y_1) \subset \Gamma_1$ is the smallest. We call the length of the line segment (x_1, y_1) the width of N at x_1, and y_1 the dual point of x_1.

Since the area of M is finite, the width $h(t)$ of N at the point $\Phi(\widetilde{x}, t) \subset \Sigma \subset \partial N$ converges to zero as $t \to \infty$; see Figure 1.1.2.

Let

$$\Sigma_1 = \{y_t \in \partial N \mid y_t \text{ is the dual point of } \Phi(\widetilde{x}, t), 0 \leq t < \infty\}.$$

Obviously, Σ_1 is of infinite length. Let $\widetilde{N} \subset N$ be the subdomain enclosed by the line segment $[x_1, y_1]$, $\widetilde{\Sigma} = \{x = \Phi(\widetilde{x}, t) | t_1 \leq t < \infty\}$ and the dual points of $\widetilde{\Sigma}$. Without loss of generality, we assume that there is no singular point of v in $\widetilde{N}$.

It is easy to see that for any $\varepsilon > 0$ sufficiently small, there is a time $T > 0$ such that for any $t > T$, the area $|\widetilde{N}_t| < \varepsilon$. Here $\widetilde{N}_t \subset N$ is the subdomain enclosed by the line segment $[x_t, y_t]$, $\widetilde{\Sigma}_t = \{x = \Phi(\widetilde{x}, \tau) | t \leq \tau < \infty\}$ and the dual points of $\widetilde{\Sigma}_t$, where $x_t = \Phi(\widetilde{x}, t)$ and y_t is the dual point of x_t.

Consider a domain $A \subset \widetilde{N}$ with $|A| > \epsilon_0 > 0$. Then there is a time $T_0 > 0$ such that for any $t > T_0$, $|\widetilde{N}_t| < \varepsilon_0$. Since there is no singular point of v in $\widetilde{N}$, there exists $T_1 > 0$ sufficiently large such that when $t > T_1$, $\Phi(A, t) \subset \widetilde{N}_{T_0}$. Hence

$$|\Phi(A, t)| \leq |\widetilde{N}_{T_0}| < \varepsilon_0,$$

a contradiction to the flow Φ being area-preserving.

Step 4. By Step 3, it is easy to see that each connected component of ∂N consist of saddle connections. Therefore, both N and ∂N contain only a finite number of connected components. Hence,

$$\partial N = \{x \in M \quad | \quad \exists x_n \in \overset{\circ}{N}, x_n \to x\},$$

thanks to N being an open set with a finite number of connected components.

Step 5. Let

$$L_1 = \omega(x_0) \cap \partial K,$$

$$L_2 = \omega(x_0) - \overset{\circ}{\omega}(x_0) - L_1.$$

We claim that $L_2 \subset \partial N$.

For any $x \in L_2$, we have $x \in \overset{\circ}{K}$. For any sufficiently small neighborhood $\mathcal{O}(x)$ of x, we have $\mathcal{O}(x) \subset \overset{\circ}{K}$.

We now prove that

$$\mathcal{O}(x) \cap N \neq \emptyset.$$

Otherwise, noticing that $N = \overset{\circ}{K} - L_2 - \overset{\circ}{\omega}(x_0)$, we have

$$\mathcal{O}(x) \subset L_2 \cup \overset{\circ}{\omega}(x_0), \qquad \mathcal{O}(x) \cap \overset{\circ}{\omega}(x_0) \neq \emptyset.$$

There are two possibilities. First, $\mathcal{O}(x) \subset \overset{\circ}{\omega}(x_0)$, which contradicts $x \notin \overset{\circ}{\omega}(x_0)$. Second,

$$\mathcal{O}(x) - \ \mathrm{cl}[\mathcal{O}(x) \cap \overset{\circ}{\omega}(x_0)]$$

is a nonempty open subset of L_2, a contradiction to $\overset{\circ}{L}_2 = \emptyset$.

Therefore by Step 4 we showed that $x \in \bar{N}$ and $x \notin N$. Namely, $L_2 \subset \partial N$.

Step 6. By Steps 2–5, we conclude that

$$\omega(x_0) - \overset{\circ}{\omega}(x_0) \neq \partial \overset{\circ}{\omega}(x_0),$$

$$\omega(x_0) = \overset{\circ}{\omega}(x_0) \cup L,$$

with L consisting of saddle connections. This is a contradiction to $\Phi(x_0, \cdot)$ being nontrivially recurrent.

The proof is complete. □

REMARK 1.1.11. Results similar to the Limit Set Theorem are not true for general vector fields on general two-dimensional manifolds. The Cherry flow, for instance, provides a nice counterexample; see **[77]**.

REMARK 1.1.12. A more detailed topological structure of the closed domain in the Limit Set Theorem will be studied in next section.

1.2. Poincaré-Hopf Index Theorem on Manifolds with Boundaries

The classical Poincaré–Hopf index theorem was proved by Hopf [**38**] in 1926 after earlier partial results by Brouwer and Hadamard with a two-dimensional version given by Poincaré in 1885. The classical index theorem and its known extensions (see among others, [**32, 72**]) are usually presented in the cases where either the manifold is compact without boundary or the vector field has no zeroes on the boundary with some additional conditions.

Motivated by its applications in later sections (e.g., Sections 1.3, 3.1, 3.2, and 6.1), we present in this section an extension of the classical Poincaré–Hopf index theorem to vector fields on a compact manifold with boundary, which are tangential to the boundary, and may vanish on the boundary.

1.2.1. Indices on manifolds with boundaries. In this section, we assume that M is an m-dimensional differentiable manifold with boundary ∂M, and as before, let $C^r(TM)$ be the space of all r-th differentiable vector fields v on M, and let $C^r_n(TM)$ be the space of all r-th differentiable vector fields v on M satisfying the no-normal flow condition:

$$v \cdot n = 0 \qquad \text{on } \partial M,$$

where $v_n = v \cdot n$, and n is the unit normal vector on ∂M.

Let $p \in \overset{\circ}{M}$, and let (η, U) be a local coordinate system near p such that

$$\eta(p) = 0 \ \in V = \eta(U) \subset \mathbb{R}^m.$$

If $v(p) = 0$, then p is called a singular point of v. If p is an isolated singular point of $v \in C^r(TM)(r \geq 1)$, then $x = 0$ is an isolated zero point of $u : V \to \mathbb{R}^m$, where

$$u = D\eta \cdot v \cdot \eta^{-1} \ : \ V \to \mathbb{R}^m. \tag{1.2.1}$$

DEFINITION 1.2.1. *Let $p \in \overset{\circ}{M}$ be an isolated singular point of $v \in C^0(TM)$. We define the index of p as follows:*

$$\text{ind}(v, p) = \deg(u, 0), \tag{1.2.2}$$

where u is given by (1.2.1) and deg$(u, 0)$ is the Brouwer degree of u at $x = 0$.

When $v \in C^r(TM)$ $(r \geq 1)$, we denote by $J(u(0))$ the Jacobi matrix of u at $x = 0$. If $J(u(0))$ is non-singular, then p is called a non-degenerate singular point of v. If p is a non-degenerate singular point of v, then (1.2.2) is equivalent to the following:

$$\text{ind } (v, p) = \text{sign det } J(u(0)). \tag{1.2.3}$$

We now define the index of a boundary singular point of v. We shall divide the definition procedure in several steps. The basic ideas are as follows:

a) to extend the manifold near the boundary such that there is a tubular neighborhood of the boundary in the extended manifold,
b) to extend the vector field across the boundary, and
c) to define the index of the vector field at the singular point on the boundary by considering the singular point as an interior singular point of the extended vector field on the extended manifold.

Step 1. Extension of M. We start with a manifold M that can be embedded into an Euclidean space $\mathbb{R}^N$. We extend the manifold across the boundary such that the extended manifold contains a tubular neighborhood of the boundary ∂M.

Since $M \subset \mathbb{R}^N$, for every point $x \in \partial M$, M has a unit inward normal vector $n_x \in T_x M$. Notice that $M \subset \mathbb{R}^N$ is a Riemannian manifold with induced Riemannian metric. For any point $x \in M$ and any vector $v \in T_x M$, there exists a unique geodesics starting from x, such that v is its tangent vector at x. For any $x \in \partial M$, let $z(\alpha; x)$ $(\alpha \geq 0)$ be the point $z \in M$ which lies on the geodesics starting from x, whose tangent vector at x is n_x, and the arc-length from z to x is α. Obviously, $z(0; x) = x$.

Since ∂M is a compact $C^r(r \geq 1)$ manifold, there exists a real number $\lambda > 0$ such that for any $x, y \in \partial M$, $x \neq y$, the vector $-\lambda n_x$ does not intersect with the vector $-\lambda n_y$, and the geodesic interval

$$L_x(0, \lambda) \overset{def}{=} \{z(\alpha; x) \quad | \quad 0 \leq \alpha \leq \lambda\}$$

does not intersect with the geodesic interval $L_y(0, \lambda)$. Without loss of generality, we assume that $\lambda = 1$. Then it is easy to see that

$$\widetilde{M} = M \bigcup \{z(\alpha; x) \quad | \quad x \in \partial M, \ -1 < \alpha \leq 0\}$$

is a C^1 manifold. Here

$$\{z(\alpha; x) = x + \alpha n_x \mid x \in \partial M, \ -1 < \alpha \leq 0\}$$

is the segment between x and $x - n_x$ in $\mathbb{R}^N$.

Let S be an open set in $\widetilde{M}$ defined by

$$S = \{z(\alpha; x) \quad | \quad x \in \partial M, \ -1 < \alpha < 1\} \subset \widetilde{M},$$

and we define a mapping

$$\left\{ \begin{array}{l} f : \partial M \times (-1, 1) \to S \subset \widetilde{M}, \\ f(x, \alpha) = z(\alpha; x) \quad \forall (x, \alpha) \in \partial M \times (-1, 1). \end{array} \right. \tag{1.2.4}$$

Obviously, f is a homeomorphism and satisfies

$$f(x, 0) = x, \qquad \lim_{\alpha \to -1} f(x, \alpha) = z(-1; x) = x - n_x \in \partial \widetilde{M}.$$

In particular, we have the following lemma.

LEMMA 1.2.2. *Let* $M \subset \mathbb{R}^N$ *be an* m*-dimensional compact* $C^r(r \geq 1)$ *manifold with boundary. Then* M *can be extended across its boundary* ∂M *to a* C^1 *manifold* $\widetilde{M}$ *in* $\mathbb{R}^N$ *such that*

(1) $M \subset \widetilde{M}$;
(2) *the closure of* $\widetilde{M}$ *is a compact* C^1 *manifold with boundary, and* $\partial \widetilde{M}$ *is homeomorphic to* ∂M;
(3) ∂M *has a tubular neighborhood in* $\widetilde{M}$. *Furthermore, there exists a* C^1 *embedding*

$$f : \partial M \times (-1, 1) \to \widetilde{M}$$

such that

$$f(x, 0) = x, \qquad \lim_{\lambda \to -1} f(\partial M, \lambda) = \partial \widetilde{M}.$$

Step 2. Extension of vector fields. Let $\{\eta_i, U_i\}_{i\in\Lambda}$ be an atlas of the manifold M. A vector field $v \in C^0(TM)$ is represented by a set $\{u_i, V_i\}_{i\in\Lambda}$ of local vector fields in $\mathbb{R}^m$, where

$$u_i = D\eta_i \circ v \circ \eta_i^{-1} : V_i \to \mathbb{R}^m, \ V_i \stackrel{def}{=} \eta_i(U_i) \subset \mathbb{R}^m. \tag{1.2.5}$$

Since ∂M is an $m-1$ dimensional C^r manifold, we choose a C^r atlas $\{h_i, \Omega_i\}$ of ∂M, such that

$$h_i : \Omega_i \to I^{m-1} \tag{1.2.6}$$

are homeomorphisms, where

$$I^k = \{(x_1, \dots, x_k) \in \mathbb{R}^k \mid |x_i| < 1, \ 1 \le i \le k\}.$$

Then $\{\widetilde{h}_i, \Omega_i \times (-1,1)\}_{i\in\Lambda}$ is a C^1 atlas of $\partial M \times (-1,1)$, the tubular neighborhood of ∂M in $\widetilde{M}$. Here the mapping $\widetilde{h}_i : \Omega_i \times (-1,1) \to I^m$ is defined by

$$\widetilde{h}_i(x,\alpha) = \{h_{i,1}(x), \dots, h_{i,m-1}(x), \alpha\} \quad \forall (x,\alpha) \in \Omega_i \times (-1,1), \tag{1.2.7}$$

where $h_i(x) = \{h_{i,1}(x), \dots, h_{i,m-1}(x)\}$ are as in (1.2.6).

Then

$$\Phi = \{\eta_j, U_j\}_{j\in\Gamma} \bigcup \{\widetilde{h}_i, \Omega_i \times (-1,1)\}_{i\in\Lambda}$$

is a C^1 atlas of $\widetilde{M}$, and

$$\Psi = \{\eta_j, U_j\}_{j\in\Gamma} \bigcup \{\widetilde{h}_i \mid_{\Omega_i\times[0,1)}, \Omega_i \times [0,1)\}_{i\in\Lambda}$$

is a C^r atlas of M. The atlas Φ is the extension of Ψ.

Let $\{u_i, V_i\}$ be the set of local vector fields on $\mathbb{R}^m$ of $v \in C^0(TM)$ under atlas Ψ. In the following, we will construct a set $\{\widetilde{u}_i, \widetilde{V}_i\}$ of local vector fields on $\mathbb{R}^m$, such that $\{\widetilde{u}_i, \widetilde{V}_i\}$ is the reflective extension of $\{u_i, V_i\}$.

For $\Phi = \{\eta_j, U_j\}_{j\in\Gamma} \cup \{\widetilde{h}_i, \Omega_i \times (-1,1)\}_{i\in\Lambda}$, we set

$$\widetilde{u}_i(x) = \begin{cases} u_i(x) & \text{for } x \in \widetilde{V}_i = \eta_i(U_i), \\ u_i(x_1, \dots, x_m) & \text{for } x \in I^m_+ = \widetilde{h}_i(\Omega_i \times [0,1)), \\ u_i^-(x_1, \dots, -x_m) & \text{for } x \in I^m_- = \widetilde{h}_i(\Omega_i \times (-1,0]), \end{cases} \tag{1.2.8}$$

where $x = (x_1, \cdots, x_m)$, $u_i^-(x) = \{u_{i,1}(x), \dots, u_{i,m-1}(x), -u_{i,m}(x)\}$ are the reflective extensions of $u_i(x)$ from the plane $\partial\mathbb{R}^m_+ = \{(x_1, \dots, x_{m-1}, 0) \in \mathbb{R}^{m-1}\}$. Since $v|_{\partial M} \in C^0(T\partial M)$, $u_{i,m}(x_1, \dots, x_{m-1}, 0) = 0$, where $u_{i,m}$ are the m-th components of local vector fields $u_i(x)$. By the atlas $\{\widetilde{h}_i, \Omega_i \times (-1,1)\}$ being C^1 and u_i being C^0, we see that the reflective extension (1.2.8) are C^0 local vector fields on $\mathbb{R}^m$. Hence, we get a set $\{\widetilde{u}_i, \widetilde{V}_i\}$ of local vector fields on $\mathbb{R}^m$.

Then it is easy to conclude that $\{\widetilde{u}_i, \widetilde{V}_i\}$ provides a C^0 vector field $\widetilde{v}$ on $\widetilde{M}$ with $\widetilde{v}|_M = v$. Namely, $\widetilde{v}$ is a continuous extension of v.

Step 3. Definition of the index of a boundary singular point. When $M \subset \mathbb{R}^N$, we define the index for a boundary singular point v as follows:

DEFINITION 1.2.3. *Let $M \subset \mathbb{R}^N$ be a $C^r (r \ge 1)$ compact manifold with boundary ∂M and $v \in C^0_n(TM)$. If $p \in \partial M$ is an isolated singular point of v, then the index of v at p is defined by*

$$\text{ind}(v,p) = \frac{1}{2}\text{ind}(\widetilde{v},p), \tag{1.2.9}$$

where $\widetilde{v}$ is the reflective extension of v.

For a general m-dimensional C^r manifold M, by the Whitney embedding theorem, there exists an $N \geq m$ such that M is embedded into $\mathbb{R}^N$. Let $\varphi : M \to N = \varphi(M) \subset \mathbb{R}^N$ be a C^r $(r \geq 1)$ embedding and $v \in C_n^0(TM)$. The C^r homeomorphism φ induces a vector field $v_\varphi \in C_n^0(TN)$ from $v \in C_n^0(TM)$:

$$v_\varphi = D\varphi \circ v \circ \varphi^{-1} : N \to TN. \tag{1.2.10}$$

We call v_φ the induced vector field of v by φ.

DEFINITION 1.2.4. *Let M be a $C^r(r \geq 1)$ compact manifold with boundary ∂M, $v \in C_n^0(TM)$. If $p \in \partial M$ is an isolated singular point of v, then the index of v at p is*

$$\text{ind}(v, p) = \text{ind}(v_\varphi, q), \;\; q = \varphi(p), \tag{1.2.11}$$

where v_φ is the induced vector field of v by the embedding $\varphi : M \to \mathbb{R}^N$.

To justify the above definition, we have to show that (1.2.11) is independent of the embedding φ. To this end, we introduce the following well-known lemma.

LEMMA 1.2.5. *Let M and N be two n-dimensional $C^r(r \geq 1)$ manifolds, and let $\varphi : M \to N$ be a C^r homeomorphism. Let $v \in C_n^0(TM)$, and let $v_\varphi \in C_n^0(TN)$ be the induced vector field of v by φ. If $p \in M$ is an interior isolated singular point of v, then $q = \varphi(p)$ is an isolated singular point of v_φ, and*

$$\text{ind}(v, p) = \text{ind}(v_\varphi, q).$$

Then the following lemma justifies the definition (1.2.11):

LEMMA 1.2.6. *The definition (1.2.11) is independent of the embedding $\varphi : M \to \mathbb{R}^N$.*

PROOF. Let $\varphi_i : M \to \mathbb{R}^N$ $(i = 1, 2)$ be two C^r embeddings, $M_i = \varphi_i(M)$. Then M_1 is C^r homeomorphic to M_2 and the homeomorphism $\varphi : M_1 \to M_2$ is $\varphi = \varphi_2 \circ \varphi_1^{-1}$. Hence, by (1.2.10), if $v \in C_n^0(TM)$, then φ takes v_{φ_1} to v_{φ_2}. Since $\varphi(\partial M_1) = \partial M_2$, the homeomorphism φ can be extended to a homeomorphism $\widetilde{\varphi} : \widetilde{M_1} \to \widetilde{M_2}$, where $\widetilde{M_i}$ are the tubular neighborhood extensions of $M_i (i = 1, 2)$, namely $\widetilde{\varphi}|_{M_1} = \varphi$. By Lemma 1.2.2, we see that

$$\text{ind}(\widetilde{v}_{\varphi_1}, p_1) = \text{ind}(\widetilde{v}_{\varphi_2}, p_2), \;\; \varphi(p_1) = p_1.$$

Hence (1.2.11) is independent of the embedding.

The proof is complete. □

REMARK 1.2.7. The conclusion of Lemma 1.2.5 also holds true for the index of a boundary singular point.

1.2.2. Poincaré–Hopf Index Theorem for Manifolds with Boundaries. The main theorem in this section is the following Poincaré–Hopf index theorem on a compact differentiable manifold with boundary.

THEOREM 1.2.8 (Index Theorem). *Let M be a C^r $(r \geq 1)$ compact manifold with boundary and $v \in C_n^0(TM)$. If v has a finite number of singular points $p_i \in M (1 \leq i \leq I)$, then the following formula holds true:*

$$\sum_{i=1}^{I} \text{ind}\,(v, p_i) = \begin{cases} \chi(M) & \text{if dim } M = \text{even}, \\ 0 & \text{if dim } M = \text{odd}, \end{cases} \tag{1.2.12}$$

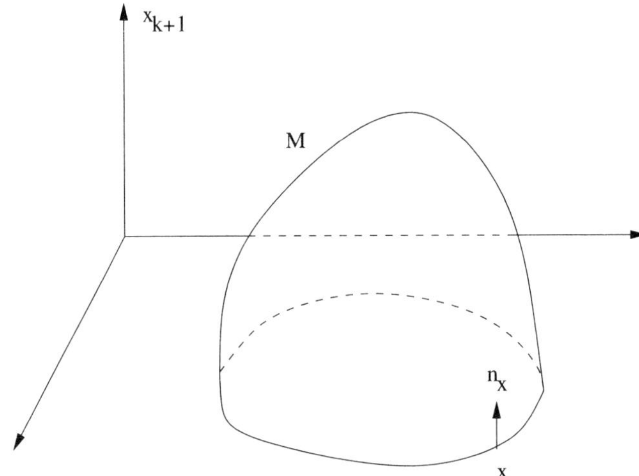

FIGURE 1.2.1. Embedding of M into $\mathbb{R}^{k+1}$.

where $\chi(M)$ is the Euler characteristic of M.

PROOF. We divide the proof into a few steps.

Step 1. First, we will prove a lemma which is necessary for us to prove the index theorem.

LEMMA 1.2.9. *Let M be an m-dimensional $C^r(r \geq 1)$ compact manifold with boundary. Then there exists a number $k \geq n$ such that M can be C^r embedded in $\mathbb{R}^{k+1}$, $\partial M \subset \partial\mathbb{R}_+^{k+1} = \mathbb{R}^k$ and $\overset{\circ}{M} \subset \overset{\circ}{\mathbb{R}}{}_+^{k+1}$. Moreover, for any $x \in \partial M$, the vector n_x starting from x and parallel to the x_{k+1}-axis is the normal vector of ∂M at $x \in \partial M$; see Figure 1.2.1. Here $\mathbb{R}_+^{k+1} = \{x \in \mathbb{R}^{k+1} | x_{k+1} \geq 0\}$.*

PROOF. By the Whitney embedding theorem, there exists a number $k \geq m$ such that M can be C^r embedded into $\mathbb{R}^k$. Without loss of generality, we assume that $M \subset \mathbb{R}^k$.

Since M is a C^r compact manifold with boundary, there exists a finite open covering $\{U_i\}_{i=1,\dots,N}$ of M. Let $\{U_i, \eta_i\}_{i=1,\dots,N}$ be the local coordinates, such that domains U_i with $U_i \cap \partial M \neq \emptyset$ are rectangular. Namely, when $U_i \cap \partial M \neq \emptyset$, the mappings $\eta_i : U_i \to I^m$ are C^r homeomorphisms, and

$$\eta_i|_{U_i \cap M} : U_i \cap M \to I_+^m,$$
$$\eta_i|_{U_i \cap \partial M} : U_i \cap \partial M \to I^m \cap \partial\mathbb{R}_+^m,$$

such that η_i take the outward normal vectors on ∂M of M to the outward normal vectors at $\partial\mathbb{R}_+^m$ of I_+^m.

For the covering $\{U_i\}$, we define functions $f_i : U_i \to \mathbb{R}^1$ as follows:

$$f_i(x) = \begin{cases} 1 & \text{when } x \in U_i \subset \overset{\circ}{M}, \\ g \circ \eta_i(x) & \text{when } x \in U_i,\ U_i \cap \partial M \neq \emptyset, \end{cases}$$

where $g : I^n \to R$ defined by

$$g(x) = \begin{cases} 0 & \text{for } x \in I^m_-, \\ \sqrt{1-(x_n-1)^2} & \text{for } x \in I^m_+. \end{cases}$$

By the partition of unity theorem, there exists functions $\{\zeta_i\}$ subordinate to the covering $\{U_i\}$ of M such that

(1) $\zeta_i \in C_0^\infty(U_i)$, $1 \leq i \leq N$,

(2) $0 \leq \eta_i \leq 1$, and $\sum_{i=1}^N \zeta_i(x) = 1$, $\forall x \in M$.

Thus we define a function $f : M \to \mathbb{R}^1$ by

$$f(x) = \sum_{i=1}^N \zeta_i(x) f_i(x), \ x \in M.$$

It is easy to see that f has the properties:

$$\begin{aligned} & f(x) > 0 && \forall x \in \overset{\circ}{M}, \\ & f(x) = 0 && \forall x \in \partial M, \\ & \frac{\partial f(x)}{\partial n} = \infty && \forall x \in \partial M, \end{aligned}$$

where n are the outward normal vectors at ∂M of M, and f is C^∞ in $\overset{\circ}{M}$.

Let $x_{k+1}(x) = f(x)$ for $x \in M \subset \mathbb{R}^k$. We get a mapping $\varphi : M \to \mathbb{R}^{k+1}$,

$$\varphi(x) = \{x_1, \ldots, x_k, f(x)\}, \text{ for } x = (x_1, \ldots, x_k) \in M \subset \mathbb{R}^k.$$

Obviously, φ is an embedding. Since $M \subset \mathbb{R}^k$ is C^r, and $\frac{\partial f}{\partial n} |_{\partial M} = \infty$ only illustrates that the normal vectors at the boundary of the graph of f are orthogonal with the hyperplane $\mathbb{R}^k \subset \mathbb{R}^{k+1}$, it has no influence on the differentiability of the graph of f. Hence the embedding $\varphi : M \to \mathbb{R}^{k+1}$ is C^r, and satisfies the properties required by the lemma.

The proof is complete. □

Step 2. Let $\varphi : M \to \mathbb{R}^{k+1}$ be the C^r embedding as in Lemma 1.2.8, and let φ be expressed by

$$\varphi(z) = \{x_1(z), \ldots, x_k(z), x_{k+1}(z)\}, \ z \in M,$$

where $x_{k+1}(z) > 0$ for $z \in \overset{\circ}{M}$, and $x_{k+1}(z) = 0$ for $z \in \partial M$.

From the embedding φ, we obtain another C^r embedding, referred to as the reflective embedding of φ, $\varphi_- : M \to \mathbb{R}^{k+1}$, defined by

$$\varphi_-(z) = \{x_1(z), \ldots, x_k(z), -x_{k+1}(z)\}, \ \forall\, z \in M.$$

Obviously, $\varphi(M)$ is C^r homeomorphic to $\varphi_-(M)$, and

$$\varphi(x) = \varphi_-(x) \text{ for } x \in \partial M. \tag{1.2.13}$$

Hence it is legitimate to define a quotient space $M_+ \# M_-$ by

$$M_+ \# M_- = \Big(\varphi(M) \bigcup \varphi_-(M)\Big) \Big/ \{\varphi(x) = \varphi_-(x) | x \in \partial M\},$$

where $M_+ = \varphi(M)$ and $M_- = \varphi_-(M)$. Since M_+ and M_- have the same normal vectors on the common boundary, $M_+ \# M_-$ is a C^1 closed compact manifold.

For $v \in C_n^r(TM)(r \geq 1)$, under the embeddings φ and φ_-, one can get respectively the induced vector fields $v_\varphi \in C_n^r(TM_+)$ and $v_{\varphi_-} \in C_n^r(TM_-)$. By (1.2.13), we have

$$v_\varphi|_{\partial M_+} = v_{\varphi_-}|_{\partial M_-}. \tag{1.2.14}$$

By Lemma 1.2.5 and Remark 1.2.7, we have

$$\sum_{i=1}^{I} \mathrm{ind}\{v, p_i\} = \sum_{i=1}^{I} \mathrm{ind}\{v_\varphi, \varphi(p_i)\} = \sum_{i=1}^{I} \mathrm{ind}\{v_{\varphi_-}, \varphi_-(p_i)\}. \tag{1.2.15}$$

On the manifold $M_+ \# M_-$, by (1.2.14), we can induce a C^0 vector field $v^* \in C^0(T(M_+\#M_-))$ as follows.

$$v^*(x) = \begin{cases} v_\varphi(x), & \text{for } x \in M_+, \\ v_{\varphi_-}(x), & \text{for } x \in M_-. \end{cases}$$

Since φ and φ_- are mutual reflective embeddings, $v_\varphi(x)$ and $v_{\varphi_-}(x)$ are mutual reflective vector fields. Hence, by the definition of indices of singular points on the boundary, for an isolated singular point $p \in \partial M$ of v, we have

$$\mathrm{ind}(v_\varphi, \varphi(p)) = \mathrm{ind}(v_{\varphi_-}, \varphi_-(p)) = \frac{1}{2}\mathrm{ind}(v^*, \varphi_-(p)). \tag{1.2.16}$$

When $p \in \overset{\circ}{M}$ are the isolated singular points of v, by Lemma 1.2.2, we have

$$\mathrm{ind}(v_\varphi, \varphi(p)) = \mathrm{ind}(v_{\varphi_-}, \varphi_-(p)) = \mathrm{ind}(v^*, \varphi(p)) = \mathrm{ind}(v^*, \varphi_-(p)). \tag{1.2.17}$$

Hence, from (1.2.15–1.2.17), we can get

$$\begin{aligned} \sum_j \mathrm{ind}(v^*, q_j) &= \sum_{i=1}^{I} \mathrm{ind}(v_\varphi, \varphi(p_i)) + \sum_{i=1}^{I} \mathrm{ind}(v_{\varphi_-}, \varphi_-(p_i)) \\ &= 2\sum_{i=1}^{I} \mathrm{ind}(v, p_i). \end{aligned}$$

By the Poincaré–Hopf index theorem on a closed compact manifold for continuous vector fields (see [**72, 44**]), we have

$$\sum_j \mathrm{ind}(v^*, q_j) = \chi(M_+\#M_-).$$

Hence, we obtain

$$\sum_{i=1}^{I} \mathrm{ind}(v, p_i) = \frac{1}{2}\chi(M_+\#M_-). \tag{1.2.18}$$

Step 3. It suffices then to prove that

$$\chi(M\#M) = \begin{cases} 2\chi(M) & \text{for } \dim M = \text{even}, \\ 0 & \text{for } \dim M = \text{odd}, \end{cases} \tag{1.2.19}$$

which follows from the following general formula

$$\chi(N_1 \cup N_2) = \chi(N_1) + \chi(N_2) - \chi(N_1 \cap N_2)$$

for two manifolds N_1 and N_2.

The proof is complete. □

1.3. Structural Classification

We study in this section the structural classification of divergence-free vector fields on a two-dimensional compact orientable manifold M. The classification theorem will be crucial for the study of the structural and block stabilities in the next chapter.

1.3.1. Structural classification.

We start with the following definition.

DEFINITION 1.3.1. *Let $v \in D^r(TM)$.*

(1) *Let $p \in M$ be a center; then, there is an open neighborhood C of p, such that for any $x \in C$ ($x \neq p$), the orbit $\{\Phi(x,t)\}_{t\in\mathbb{R}}$ is closed. The largest such neighborhood C of p is called a circle cell of v.*
(2) *Let $B \subset M$ be an open set, such that for any $x \in B$, the orbit $\{\Phi(x,t)\}_{t\in\mathbb{R}}$ is closed, and each connected component Σ of ∂B is not a single point. Then B is called a circle band of v.*
(3) *A closed domain $F \subset M$ is called an ergodic set of $v \in D^r(TM)$ if for any $x \in F$ with $\omega(x)$ not a singular point of v, then $\omega(x) = F$.*

By Lemma 1.1.3, Theorem 1.1.6 and the Limit Set Theorem, Theorem 1.1.9, we immediately obtain the following two structural classification theorems.

THEOREM 1.3.2 (Structural Classification Theorem I). *Let $v \in D^r(TM)(r \geq 1)$ be regular. Then the topological structure of v consists of a finite number of connected components, which are of the following types:*

(1) *circle cells, which are homeomorphic to open disks,*
(2) *circle bands, which are homeomorphic to open annuli,*
(3) *ergodic sets, and*
(4) *saddle connections.*

THEOREM 1.3.3 (Structural Classification Theorem II). *Let $M \subset S^2$ be a two-dimensional compact manifold, and let $v \in D^r(TM)(r \geq 1)$ be regular. Then the topological structure of v consists of a finite number of connected components of the following types:*

(1) *circle cells,*
(2) *circle bands, and*
(3) *saddle connections.*

1.3.2. Structure of ergodic sets.

In this subsection, we study topological properties of ergodic sets.

Let M be a two-dimensional compact manifold with boundary, and let ∂M have r connected components ($r \geq 1$). The genus $g = g(M)$ of M is defined as follows:

$$g = g(M) = 1 - \frac{1}{2}(\chi(M) + r),$$

where $\chi(M)$ is the Euler characteristics of M.

LEMMA 1.3.4. *Let M_1 and M_2 be two two-dimensional compact orientable manifolds with the same genus and the same Euler characteristic. Then M_1 is homeomorphic to M_2.*

To characterize a two-dimensional compact orientable manifold with boundary, we introduce a concept of standard manifolds.

DEFINITION 1.3.5. *Let N be a compact manifold without boundary and with genus $k \geq 0$. A sub-manifold $M \subset N$ is called a standard manifold of genus k with boundary if each connected component of ∂M is retractable to a point in N.*

By Lemma 1.3.4, it is easy to obtain the following lemma.

LEMMA 1.3.6. *Any two-dimensional compact orientable manifold of genus k with boundary must be homeomorphic to a standard manifold of genus k with boundary. Hence if $M_1 \subset M$ is a submanifold, then $g(M_1) \leq g(M)$.*

We need a more general concept of a topological set with genus g.

DEFINITION 1.3.7. *Let N be a compact orientable manifold without boundary and with genus $k \geq 0$. A closed domain $\Omega \subset N$ is called an (orientable) pseudo-manifold with genus g if*

(1) *Ω is connected and $\partial\Omega$ is homeomorphic to a union of a finite number of circles S^1, each of which has a finite number of common points with the others, and*
(2) *there exists a submanifold $M \subset N$ such that $\Omega \subset M$ and M is retractable to Ω in N.*

The genus g of Ω is defined to be the genus of M, and M is called an extended manifold of Ω. If $g = 1$, then Ω is called a pseudo-torus.

REMARK 1.3.8. The genus g of Ω is independent of the extended manifold. In fact, if $M_1, M_2 \subset N$ are two extended manifolds of Ω, then as M_1, M_2 are retractable to Ω in N, we have

$$H_1(M_1, \partial M_1) = H_1(M_2, \partial M_2). \tag{1.3.1}$$

Let g_i be the genus of M_i, and let b_i be the Betti numbers of $H_1(M_i, \partial M_i)$ $(i = 1, 2)$. Then (1.3.1) implies $b_1 = b_2$. Hence $g_1 = g_2$.

REMARK 1.3.9. By definition, the genus g of a pseudo-manifold Ω is given by

$$g = 1 - \frac{1}{2}(\chi(\Omega) + r),$$

where $r \geq 0$ is the number of connected components of the boundary of the extended manifold M of Ω.

REMARK 1.3.10. The topological structure of a pseudo–manifold Ω is essentially the same as that of its extended manifold. Therefore, by Lemma 1.3.4, the topological properties of Ω can be characterized by a standard manifold with the same genus and the same Euler characteristic as Ω.

REMARK 1.3.11. For a pseudo–manifold Ω, $\partial\Omega$ is homeomorphic to a union of r_1 circles S^1. In general, r_1 is not equal to the number r of connected components of the boundary of an extended manifold M of Ω. For example, for the pseudo–torus Ω and its extended manifold M as shown in Figure 1.3.1, $r_1 = 2$ and $r = 1$. □

If $p \in \overset{\circ}{M}$ is an interior saddle point of v, then p is connected by four orbits, two of which are stable, and the other two are unstable. If $p \in \partial M$ is a boundary saddle point, then p is connected by three orbits with two on the boundary and one in the interior of M.

DEFINITION 1.3.12. *Let Ω be an ergodic set of v. Saddle points on Ω can be divided into the following three categories; see Figure 1.3.2:*

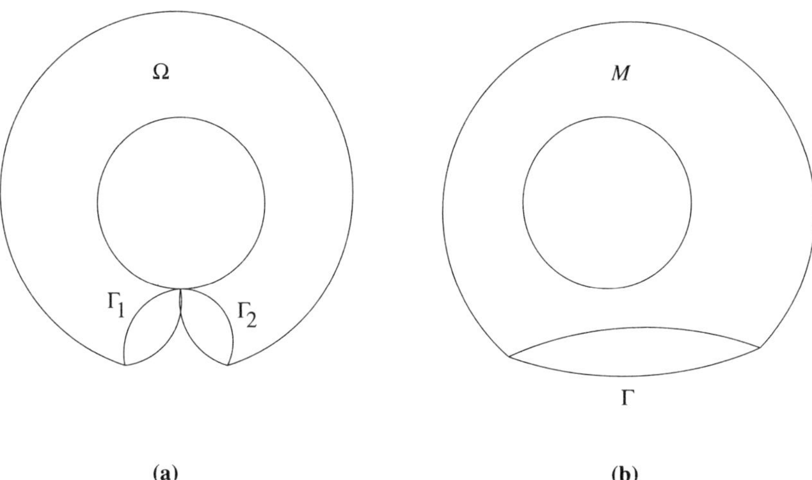

FIGURE 1.3.1. a). A pseudo-torus Ω with two circles Γ_1 and Γ_2; b). the extended manifold M of Ω with boundary component Γ.

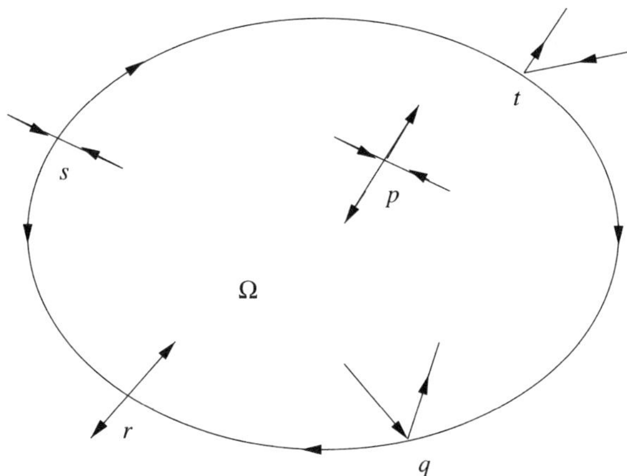

FIGURE 1.3.2. Two Ω–interior saddles p and p, two Ω–boundary saddles r and s, and one Ω–exterior saddle t.

(1) *A saddle point $p \in \Omega$ of v is an Ω-interior saddle if there are four orbits connecting p in Ω;*
(2) *A saddle point $p \in \Omega$ of v is an Ω-boundary saddle if there are only three orbits connecting p in Ω; and*
(3) *A saddle point $p \in \Omega$ of v is an Ω-exterior saddle point if there are only two orbits connecting p in Ω.*

The following theorem characterizes the topological structure of ergodic sets.

THEOREM 1.3.13. *Let the genus of M be $k \geq 1$, let $v \in D^r(TM)$ $(r \geq 1)$ be regular and let $\Omega \subset M$ be an ergodic set of v. Then*

(1) *Ω is homeomorphic to a pseudo-manifold with the genus g of Ω such that*

$$1 \leq g \leq k; \tag{1.3.2}$$

(2) *The Euler characteristic of* Ω *is given by*

$$\chi(\Omega) = -s - \frac{b}{2}, \tag{1.3.3}$$

where s *is the number of* Ω*-interior saddles and* b *is the number of* Ω*-boundary saddles of* v *in* Ω*;*

(3) *The genus of* Ω *satisfies*

$$g = 1 + \frac{1}{2}(s + \frac{1}{2}b - r), \tag{1.3.4}$$

where r *is the number of connected components of the boundary of the extended manifold* M_1.

PROOF. By the Limit Set Theorem in the previous subsection, Ω is a connected closed domain in $K = \overline{M - \overline{V}}$, where $V \subset M$ is the set of all closed orbits and centers of v.

We divide the proof into a few steps.

Step 1. We claim that Ω *is a pseudo-manifold.* It is clear that $\overline{V} \subset M$ consists of circle cells, circle bands and saddle connections. It is easy to see that each connected component of $\partial\overline{V}$ is homeomorphic to $\bigvee_n S^1$, which is defined to be the quotient space

$$\bigvee_n S^1 = S^1/\{x_i = y_i \quad | \quad 1 \le i \le n\}.$$

Here x_i, $y_i \in S^1$ are $2n$ $(n \ge 0)$ distinct points on the circle S^1.

On the other hand, by the Limit Set Theorem, $\partial\Omega$ consists of saddle connections, and $\partial\Omega = \Gamma_1 \cup \Gamma_2$, where $\Gamma_1 \subset \partial\bar{V} \cup \partial M$ and Γ_2 are the common boundaries of two ergodic sets of v if $\Gamma_2 \cap \overset{\circ}{K} \neq \emptyset$. Hence, $\partial\Omega$ has the boundary properties of a pseudo–manifold.

It suffices then to prove that there exists a manifold M_1 and a topological set $\Omega_1 \subset M_1$ such that M_1 is retractable to Ω_1 and Ω is homeomorphic to Ω_1.

Let Σ_i $(i = 1, \cdots, m)$ be the connected components of $\partial\Omega$. For each Σ_i, there is a diffeomorphism between Σ_i and $S[x^i_j, y^i_j; j = 1, \cdots, n_i]$. It is easy to see that the self-intersection points corresponding to $x^i_j = y^i_j$ $(j = 1, \cdots, n_i)$ are all Ω-interior saddles on Σ_i. The local topological structure of each Ω-interior saddle on $\partial\Omega$ is one of the two types illustrated in Figure 1.3.3(a)–(b).

Let A_i $(i = 1, \cdots, m)$ be a closed domain with boundary $\partial A_i = \cup_{j=1}^{k_i} \sigma_i^j \cup \sigma_i^0$ such that

1) each σ_i^j $(j = 1, \cdots, k_i)$ is homeomorphic to S^1,
2) σ_i^0 is homeomorphic to Σ_i with the homeomorphism $h_i : \sigma_i^0 \to \Sigma_i$, and
3) A_i is retractable to σ_i^0 .

An example of A_i is depicted in Figure 1.3.4, where p_1, p_2 and p_5 are of the type given in Figure 1.3.3(a), while p_3, p_4 and p_6 are of the type in Figure 1.3.3(b).

Then we define

$$M_1 = [\sum_{i=1}^{m} A_i + \Omega]/[h_1, ..., h_m],$$

and it is easy to see that M_1 is a two-dimensional compact manifold retractable to Ω.

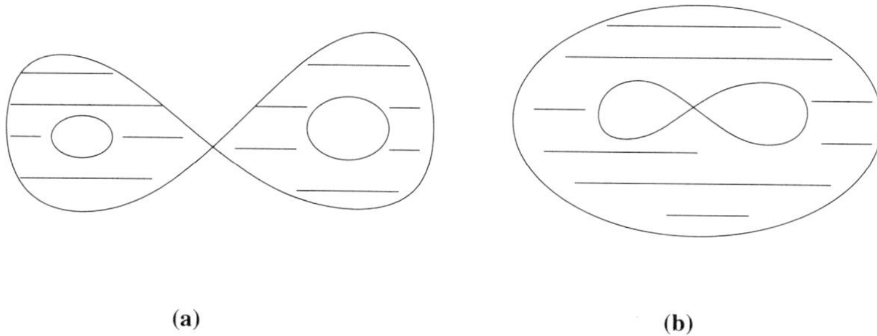

FIGURE 1.3.3. Local topological structure of an Ω-interior saddle on $\partial\Omega$.

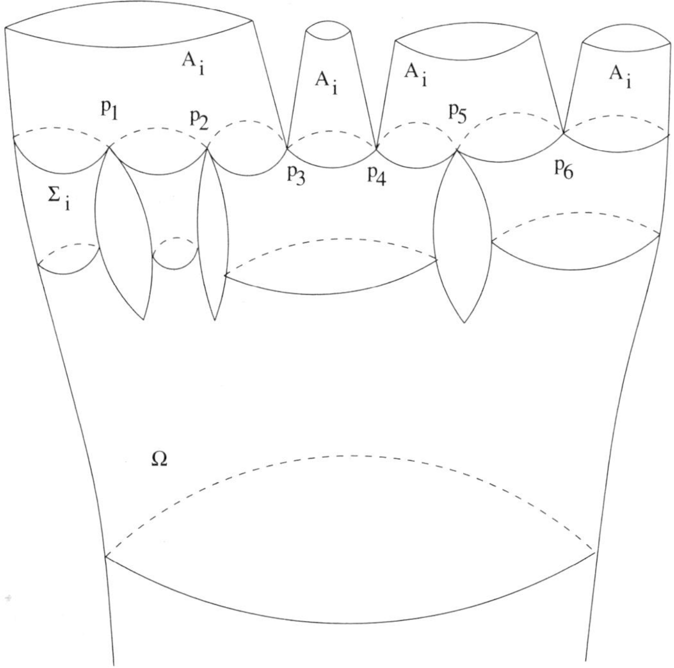

FIGURE 1.3.4. Schematic illustrating an extended manifold of Ω.

Step 2. It is easy to see that

$$g = g(\Omega) = g(M_1) \leq g(M) = k. \tag{1.3.5}$$

By the Poincaré–Bendixson Theorem, on an orientable manifold with genus $k = 0$, there is no ergodic set. Therefore

$$g \geq 1. \tag{1.3.6}$$

Then (1.3.2) follows from (1.3.5) and (1.3.6).

Step 3. To prove (1.3.3), we extend the vector field v on Ω to its extended manifold. To this end, let $\omega = v|_\Omega$ be the restriction of v on Ω. Then ω is a vector field on Ω which is tangent to $\partial\Omega$. Let M_1 be an extended manifold of Ω. Then

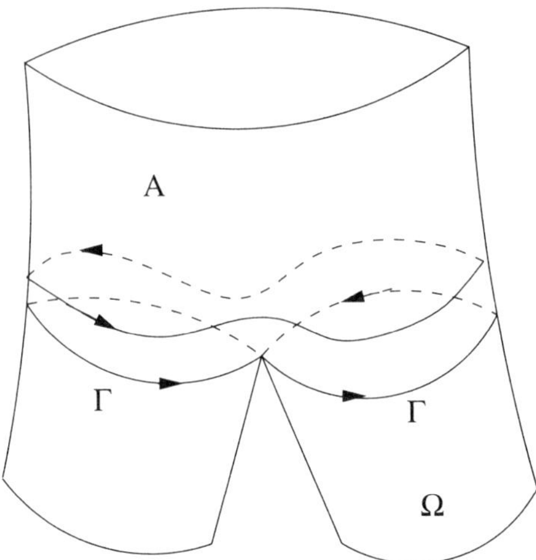

FIGURE 1.3.5. Schematic illustrating the extended vector field.

we need to extend the vector field ω on Ω to a vector field u on M_1 such that u is tangent to ∂M_1 (it is not necessary for u to be divergence-free).

A connected component $\Gamma \subset \partial\Omega$ is called a boundary component of $\partial\Omega$ if Γ has an open neighborhood $A \subset M_1$ (see Figure 1.3.4) such that $\overline{A}$ is retractable to Γ in M_1. For each boundary component Γ of $\partial\Omega$, we extend the vector field w in two different cases:

Case a). There are no Ω-boundary saddles on Γ. Since w is tangent to Γ, Γ consists of singular points and orbits of w. We call Γ an orbit curve if the singular points $p_1, \cdots, p_n \in \Gamma$ divide Γ into m orbits $\gamma_1, \cdots, \gamma_m$ in order such that the limit sets of the orbits satisfy

$$\omega(\gamma_i) = \alpha(\gamma_{i+1}), \quad 1 \le i \le m-1,$$

or $i \in Z_m$ for Γ is a closed curve. Here $Z_m = Z/m+1$ is the mod m group.

Obviously, in this case, Γ is a closed orbit curve. It implies that w can be extended to the closed domain $A \subset M_1$ corresponding to Γ given in Step 1 by setting up tubular flows in A such that A becomes circle bands of the extended vector field; see Figure 1.3.5.

Case b). Since Ω-boundary saddles appear in pairs, let there be $2n$ $(n \ge 1)$ Ω-boundary saddles $\{p_i \mid 1 \le i \le 2n\}$ on Γ. Let A be the corresponding closed domain constructed in Step 1 with boundary $\partial A = \cup_{j=1}^k \sigma^j \cup \sigma^0$ such that

1) each σ^j $(j = 1, \cdots, k)$ is homeomorphic to S^1,
2) σ^0 is homeomorphic to Γ with homeomorphism $h : \sigma^0 \to \Gamma$, and
3) A is retractable to σ^0 .

Without loss of generality, we assume that $\sigma^0 = \Gamma$ and $h = id$. Then we choose $2n$ points $\{q_i \mid 1 \le i \le 2n\}$ on $\cup_{j=1}^k \sigma^j$ and $2n$ simple curves $\{\gamma_i \mid 1 \le i \le 2n\}$, corresponding to $\{p_i \mid 1 \le i \le 2n\}$ as in Figure 1.3.6. Meanwhile we obtain cells D_i with ∂D_i consisting of four arc segments, γ_i, γ_{i+1}, $\widetilde{p_ip_{i+1}} \in \Gamma$ and $\widetilde{q_iq_{i+1}} \in \Sigma$. We prescribe flow orbits on γ_i and on $\widetilde{q_iq_{i+1}}$ as in Figure 1.3.6 such that $\alpha(\gamma_i) = \omega(\widetilde{p_ip_{i+1}}) = p_i$, $\omega(\gamma_i) = \alpha(\widetilde{q_iq_{i+1}}) = q_i$, $\alpha(\gamma_{i+1}) = \omega(\widetilde{q_iq_{i+1}}) = q_{i+1}$, and $\omega(\gamma_{i+1}) = \alpha(\widetilde{p_ip_{i+1}}) = p_{i+1}$. Hence, ∂D_i is an extended orbit. Then we can

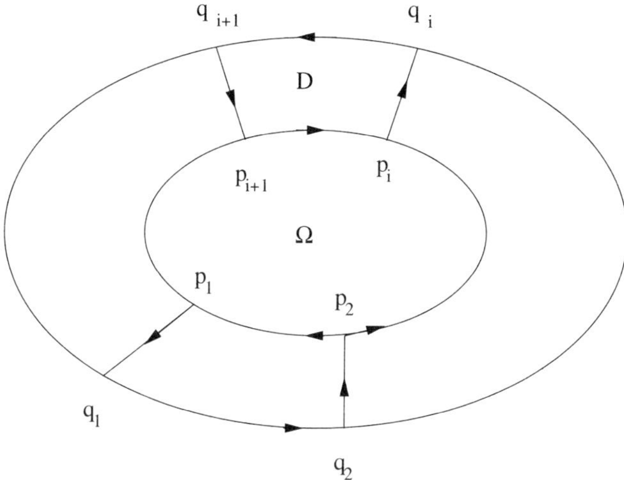

FIGURE 1.3.6. Extension of the vector field to D.

define a vector field u_i on $\overline{D}_i$ such that the orbits of u_i on ∂D_i are as prescribed, and D_i is a circle cell of u_i; see Figure 1.3.6. Thus we obtain the extension u of w on $\overline{A}$ such that $u = u_i$ for $x \in \overline{D}_i$ $(1 \leq i \leq 2m)$.

Step 4. Proof of (1.3.3). By the Poincaré–Hopf index theorem on manifolds with boundary in Section 1.2.2, we have

$$\sum_{j=1}^{N} \operatorname{ind}(u, x_j) = \chi(M_1) = \chi(\Omega), \tag{1.3.7}$$

where $x_j \in M_1$ $(1 \leq j \leq N)$ are all singular points of u.

Now we examine each term on the left hand side of (1.3.7). First, if $x_j \in \partial\Omega$ is an Ω-exterior saddle, then x_j is a degenerate singular point of u, which is connected only by two orbits, and we have

$$\operatorname{ind}(u, x_j) = 0 \quad \text{for any } \Omega\text{-exterior saddle } x_j \in \partial\Omega. \tag{1.3.8}$$

An Ω-interior saddle $x_j \in \partial\Omega$ is also a saddle point of u and

$$\operatorname{ind}(u, x_j) = -1 \quad \text{for any } \Omega\text{-interior saddle } x_j \in \partial\Omega. \tag{1.3.9}$$

If $x_j \in \partial\Omega$ is an Ω-boundary saddle, then by the extension in case b) above, x_j is a saddle point of u on $\overset{\circ}{M}_1$. Hence

$$\operatorname{ind}(u, x_j) = -1 \quad \text{for any } \Omega\text{-boundary saddle } x_j \in \partial\Omega. \tag{1.3.10}$$

Since Ω is an ergodic set of v, there are no centers on Ω. Therefore, we have

$$\sum_{\substack{j=1,\cdots,N \\ x_j \in \overset{\circ}{\Omega}}} \operatorname{ind}(u, x_j) = -\text{ number of saddles of } v \text{ in } \overset{\circ}{\Omega}\,. \tag{1.3.11}$$

It follows from (1.3.7–1.3.11) that

$$\sum_{j=1}^{N}\operatorname{ind}(u,x_j) = -s-2B+\sum_{\substack{j=1,\cdots,N\\ x_j\in\partial M_1}}\operatorname{ind}(u,x_j)+\sum_{\substack{j=1,\cdots,N\\ x_j\in\mathring{A}_i,i=1,\cdots,m}}\operatorname{ind}(u,x_j) \tag{1.3.12}$$

where s is the number of Ω-interior saddles of v, and $b = 2B$ is the number of Ω-boundary saddles of v.

By the extension of w, when the boundary component Γ of $\partial\Omega$ has no Ω-boundary saddles, there are no singular points in $\mathring{A}$ corresponding to Γ and on the boundary $\partial M_1 \cap \partial A$. When Γ has $2m$ ($m \geq 1$) boundary saddles, there are $2m$ centers in A, and $2m$ singular points on $\partial M_1 \cap \partial A$. Obviously, the $2m$ singular points of u on $\partial M_1 \cap \partial A$ are boundary saddle points of u on ∂M_1. By the definition of an index of a boundary singular point, we have

$$\operatorname{ind}(u,y) = -\frac{1}{2}, \quad \text{for } y \in \partial M_1 \text{ being a boundary saddle point of } u.$$

Therefore, we have

$$\sum_{\substack{j=1,\cdots,N\\ x_j\in\mathring{A}_i,i=1,\cdots,m}}\operatorname{ind}(u,x_j) = 2B, \tag{1.3.13}$$

$$\sum_{\substack{j=1,\cdots,N\\ x_j\in\partial M_1}}\operatorname{ind}(u,x_j) = -B. \tag{1.3.14}$$

Then (1.3.3) follows from (1.3.7) and (1.3.12–1.3.14).

The proof is complete. □

1.3.3. Structure of ergodic sets on tori. When M is a torus with or without boundary, the topological structure of an ergodic set of M is simpler. We recall that $\bigvee_m S^1 \subset \mathbb{T}^2$ is the connected topological set that consists of m circles S^1 with exactly $m-1$ common points between them, and $\bigvee_m S^1$ encloses m closed disks in $\mathbb{T}^2$.

THEOREM 1.3.14. *Let $M \subset \mathbb{T}^2$ be a torus with or without boundary, and let $v \in D^r(TM)$ ($r \geq 1$) be regular. Let $\Omega \subset M$ be an ergodic set of v. Then*

(1) *Ω is a pseudo-torus;*
(2) *each connected component of $\partial\Omega$ is homeomorphic to $\bigvee_m S^1$ ($m \geq 1$), and is connected by exactly two saddle orbits in $\mathring{\Omega}$; and*
(3) *$\mathring{\Omega}$ does not contain saddle points of v.*

PROOF. Assertion (1) is obvious; we only have to prove Assertions (2) and (3). We divide the proof into a few steps.

Step 1. For any simple closed curve $\gamma \in \mathbb{T}^2$, if γ is not homological to zero in the homology $H_1(\mathbb{T}^2)$, then $\mathbb{T}^2 - \gamma$ is homeomorphic to an open submanifold of S^2. By the Poincaré–Bendixson theorem, $\partial\Omega$ must not contain a closed curve that is not homological to zero in $H_1(M, \partial M)$. It means that each simple closed curve in $\partial\Omega$ must enclose a closed disk in $\mathbb{T}^2$. Again, by the Poincaré–Bendixson theorem, each connected component Σ of $\partial\Omega$, which consists of m circles S^1, cannot have more than $m-1$ intersections between them. Hence $\Sigma = \bigvee_m S^1$.

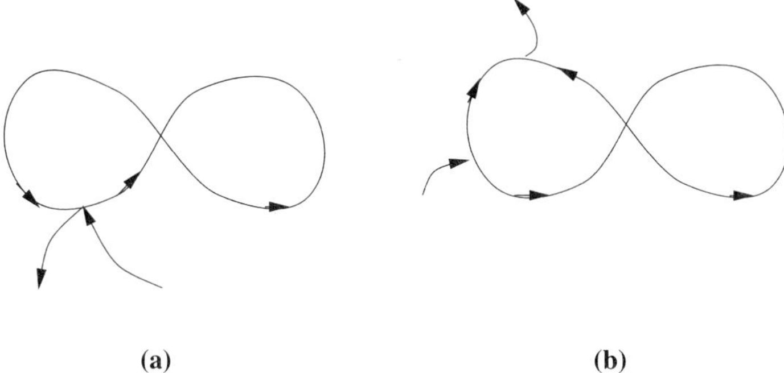

FIGURE 1.3.7. Orbits in $\overset{\circ}{\Omega}$ connected to saddle points on Γ_i appear in pairs.

Step 2. Assume that $\partial\Omega$ has n connected components, i.e.,

$$\partial\Omega = \sum_{i=1}^{n} \Gamma_i.$$

In Step 1, we showed that Γ_i is homeomorphic to $\bigvee_{m_i} S^1$. Then it is easy to see that

$$\chi(\Omega) = -\sum_{i=1}^{n} m_i.$$

Step 3. The number of intersection points in Γ_i is $m_i - 1$. Hence, the total number of intersection points on $\partial\Omega$ is

$$\begin{aligned} s_1 &= \sum_{i=1}^{n}(m_i - 1) \\ &= -\chi(\Omega) - n \\ &= (\text{by } (1.3.3)) \\ &= s + \frac{b}{2} - n. \end{aligned}$$

Hence

$$s - s_1 + \frac{b}{2} = n, \tag{1.3.15}$$

where n is the number of connected components of $\partial\Omega$.

Step 4. Claim: For each $i = 1, \cdots, n$, there are at least two orbits in $\overset{\circ}{\Omega}$ connected to Γ_i. First, there is at least one orbit connected to Γ_i in $\overset{\circ}{\Omega}$; otherwise, by Lemma 1.1.10, there exist closed orbits in $\overset{\circ}{\Omega}$, a contradiction. Second, it is obvious to see that such orbits in $\overset{\circ}{\Omega}$ appear in pairs and are connected to saddle points on Γ_i; see Figure 1.3.7.

Then by (1.3.15) there are exactly two orbits connected to Γ_i in $\overset{\circ}{\Omega}$, and there are no saddle points in $\overset{\circ}{\Omega}$.

The proof is complete. □

1.4. Topological Classification

1.4.1. Saddle connection diagrams. Let $M \subset S^2$ be a two-dimensional compact manifold and let $v \in D^r(TM)$ be regular. Let $V \subset M$ be the open set consisting of all closed orbits and centers of v. By Theorem 1.3.3, $M = \overline{V}$ and $M - V$ consists of saddle connections of v. V is the union of a finite number of circle cells and circle bands. Namely,

$$V = \sum_{i=1}^{I} C_i + \sum_{j=1}^{J} B_j,$$

where C_i are circle cells and B_j are circle bands.

The set $N = M - V$ is called the saddle connection set of v. Obviously, N has a finite number of connected components, i.e.,

$$N = \sum_{k=1}^{K} N_k.$$

Let Γ be the set consisting of connected components of the boundaries of circle cells and circle bounds, the connected components of the saddle connection set N of v. Namely,

$$\Gamma = \Gamma(v) = \{\partial C_i, \partial B_j, N_k \quad | \quad 1 \leq i \leq I,\ 1 \leq j \leq J,\ 1 \leq k \leq K\}. \tag{1.4.1}$$

We call Γ the Γ-set of v. To determine the phase diagram of v in terms of the saddle connection sets, we need to introduce a partial order in the Γ-set. To this end, we first define a reference object and interiors of each element of Γ.

1. *Definition of a reference object* R: If $\partial M \neq \emptyset$, we choose a connected component $R = \Sigma$ of ∂M as the reference object. If $\partial M = \emptyset$, then $M = S^2$; therefore, by the Poincaré–Hopf index theorem, a regular vector field $v \in D^r(TM)$ must have a center. Hence we choose a center, i.e., $R = p$, as the reference object.

2. Using the reference object R, we define an interior $\overset{\circ}{\sigma}$ of each element $\sigma \in \Gamma$ by

$$\overset{\circ}{\sigma} = \{x \in M |\ \text{each } C^0 \text{ path connecting } x \text{ and } y \in R \text{ intersects with } \sigma\}. \tag{1.4.2}$$

Obviously, the interior $\overset{\circ}{\sigma}$ of $\sigma \in \Gamma$ is a closed domain in M, and $\sigma \subset \overset{\circ}{\sigma}$.

We are now in position to define a partial order in Γ.

DEFINITION 1.4.1. *For $\sigma_1, \sigma_2 \in \Gamma_R$, we define $\sigma_1 < \sigma_2$ if either*

a) *$\sigma_1 \subset \overset{\circ}{\sigma}_2$ and $\overset{\circ}{\sigma}_1 \neq \overset{\circ}{\sigma}_2$, or*
b) *$\sigma_1 \subseteq \sigma_2$ and $\sigma_2 \not\subseteq \sigma_1$.*

The Γ-set is said to be a saddle connection diagram relative to the reference object R, denote by Γ_R, if the Γ-set is equipped with the above partial order $\leq$.

To justify the above definition, we hereafter prove two lemmas that illustrate the topological structure of the interior of $\sigma \in \Gamma$ and the compatibility of Conditions a) and b) in Definition 1.4.1.

LEMMA 1.4.2. 1) *For any $\sigma \subset \Gamma$, $\overset{\circ}{\sigma}$ = a union of a finite number of closed disks in M, which intersect each other exactly at a single point,*
2) *if $\sigma_1 \subset \sigma_2$, then $\overset{\circ}{\sigma}_1 \subset \overset{\circ}{\sigma}_2$.*

PROOF. Conclusion 2) is obvious, and we only have to prove Conclusion 1). For any $\sigma \subset \Gamma$, σ is either a topological set of type $\bigvee_n S^1$, or a connected component of the saddle connection set N. Therefore σ is a union of a finite number of simple closed curves. Since $M \subset S^2$, by the Jordan curve theorem, there exists a sub-union of simple closed curves in σ, each of which intersects with another exactly at one point, such that all the other simple closed curves are in the interior of the sub-union. It means that Conclusion 1) holds true. □

LEMMA 1.4.3. *For a saddle connection diagram Γ_R, the relations in a) and b) in Definition 1.4.1 satisfy:*

1) *if $\sigma_1 < \sigma_2$, then $\sigma_2 \not\leq \sigma_1$,*
2) *if $\sigma_1 \leq \sigma_2$ and $\sigma_2 \leq \sigma_3$, then $\sigma_1 \leq \sigma_3$,*
3) *if $\sigma_1 < \sigma_2$ and $\sigma_1 < \sigma_3$, then either $\sigma_2 \leq \sigma_3$ or $\sigma_3 \leq \sigma_2$.*

PROOF. Conclusions 1) and 2) are obvious; we only need to prove 3).

Since $\sigma_1 < \sigma_2$ and $\sigma_1 < \sigma_3$, $\overset{\circ}{\sigma}_2 \cap \overset{\circ}{\sigma}_3 \neq \emptyset$. If $\sigma_2 \cap \sigma_3 = \emptyset$, then it must be either $\sigma_2 \subset \overset{\circ}{\sigma}_3$ or $\sigma_3 \subset \overset{\circ}{\sigma}_2$, and $\overset{\circ}{\sigma}_2 \neq \overset{\circ}{\sigma}_3$. Hence it suffices to prove the case of $\sigma_2 \cap \sigma_3 \neq \emptyset$.

If $\sigma_2 \subset \sigma_3$ (or $\sigma_3 \subset \sigma_2$), then Conclusion 3) has held. If we assume otherwise, then σ_2 and σ_3 must only be the connected components of the boundary of the circle cells and circle bands.

Since $M \subset S^2$ (or $M \subset P^2$), each circle S^1 in $\overset{\circ}{M}$ encloses an open disk $D \subset M$, or an open annulus $A \subset M$. It implies that each connected component of the boundary of a circle cell or a circle band is homeomorphic to $V_n S^1$, which consists of n circles S^1 and has exactly $n-1$ intersections between the circles. Therefore σ_2 and σ_3 are respectively homeomorphic to $\bigvee_n S^1$ and $\bigvee_m S^1$.

On the other hand, if σ is a boundary of a circle cell, and σ is homeomorphic to $\bigvee_n S^1$ with $n > 1$, then there is a circle C in σ such that the other $n-1$ circles are in the interior of C. For a circle band B, ∂B has exactly two connected components σ_1 and σ_2; if $\sigma_1 \subset \overset{\circ}{\sigma}_2$, then σ_1 is called an inner boundary and σ_2 is an outer boundary of B. If σ is an outer boundary of a circle band and homeomorphic to $\bigvee_n S^1$ with $n > 1$, then there is a circle C in σ, called the exterior circle of σ, such that all the other $n-1$ circles of σ are in the interior of C. If σ is an inner boundary of a circle band, then none of the n circles in σ is in the interior of another circle.

We now return to prove conclusion 3).

Let σ_2, σ_3 be the boundaries of circle cells. We have $\sigma_2 \subset \overset{\circ}{\sigma}_3$ (or $\sigma_3 \subset \overset{\circ}{\sigma}_2$); otherwise $\overset{\circ}{C}_2 \cap \overset{\circ}{C}_3 = \Sigma$ where C_2, C_3 are respectively the exterior circle of σ_2 and σ_3, and Σ is a point or an arc segment, which means $\overset{\circ}{\sigma}_1 \not\subset \overset{\circ}{\sigma}_2 \cap \overset{\circ}{\sigma}_3 = \overset{\circ}{C}_2 \cap \overset{\circ}{C}_3$, a contradiction.

Let σ_2 be the boundary of a circle cell and let σ_3 be the boundary of a circle band. When σ_3 is an outer boundary, the exterior circle of σ_2 must be in the interior of the exterior circle of σ_3, hence $\sigma_2 \subset \overset{\circ}{\sigma}_3$. When σ_3 is an inner boundary, the exterior circle C_2 of σ_2 satisfies either $C_2 \subset \sigma_3$, or $C_2 \subset \overset{\circ}{\sigma}_3$ and $\overset{\circ}{C}_2 \neq \overset{\circ}{\sigma}_3$.

Let σ_2, σ_3 be the boundaries of the circle band. If σ_2 and σ_3 are inner boundaries, then by $\sigma_2 \cap \sigma_3 \neq \emptyset$, we have $\sigma_2 = \sigma_3$. If σ_2 is an inner boundary and σ_3 is an outer boundary, then the exterior circle C_3 of σ_3 is in σ_2, i.e., $C_3 \subset \sigma_2$. By Lemma 1.4.1 and the assumption that $\sigma_2 \not\subset \sigma_3$ and $\sigma_3 \not\subset \sigma_2$, we have that $\sigma_3 \subset \overset{\circ}{\sigma}_2$ and

$\overset{\circ}{\sigma}_3 \neq \overset{\circ}{\sigma}_2$. If σ_2 and σ_3 are the exterior boundary, then for the outer circles C_2 and C_3 of σ_2 and σ_3, we have either $C_2 \subset \overset{\circ}{C}_3$ or $C_3 \subset \overset{\circ}{C}_2$; namely, $\sigma_2 \subset \overset{\circ}{\sigma}_3$ or $\sigma_3 \subset \overset{\circ}{\sigma}_2$.

The proof is complete. □

DEFINITION 1.4.4. *Let $v_1, v_2 \in D^r(TM)$ $(r \geq 1)$ be two regular vector fields, and let Γ_{R_1} and Γ_{R_2} be saddle connection diagrams of v_1 and v_2. Γ_{R_1} and Γ_{R_2} are called isomorphic if there exists a one-to-one and onto map $h : \Gamma_{R_1} \longrightarrow \Gamma_{R_2}$ such that*

a) *for $\sigma_1, \sigma_2 \in \Gamma_{R_1}$, $\sigma_1 \leq \sigma_2$ if and only if $h(\sigma_1) \leq h(\sigma_2)$,*
b) *for each $\sigma \in \Gamma_{R_1}$, there exists a homeomorphism $\psi : \sigma \to h(\sigma)$ preserving the orbit orientation,*
c) *for any two homeomorphisms $\psi_i : \sigma_i \to h(\sigma_i)$ $(i = 1, 2)$, if $\sigma_1 \subset \sigma_2$, then $\psi_2|_{\sigma_1} = \psi_1$.*

1.4.2. Topological classification theorem on S^2.

DEFINITION 1.4.5. *Two vector fields $u, v \in D^r(TM)$ are called topologically equivalent if there exists a homeomorphism $\phi : M \to M$ that takes the orbits of u to the orbits of v, preserving orientation.*

The main theorem in this section is the following topological classification theorem of regular divergence-free vector fields on the manifolds $M \subset S^2$ (or $M \subset P^2$) with or without boundary.

THEOREM 1.4.6 (Topological classification). *Let $M \subset S^2$, and let $v_1, v_2 \in D^r(TM)$ $(r \geq 1)$ be two regular vector fields. Then v_1 and v_2 are topologically equivalent if and only if there exist saddle connection diagrams Γ_{R_1} and Γ_{R_2} of v_1 and v_2 respectively such that Γ_{R_1} and Γ_{R_2} are isomorphic.*

REMARK 1.4.7. This theorem can be considered not only as a topological classification theorem, which sets a one-to-one correspondence between these equivalent classes of vector fields and their saddle connection diagrams, but also as a topological structure theorem, which amounts to saying that the topological structure of a regular divergence-free vector field is completely determined by the saddle connections.

REMARK 1.4.8. For topological classifications of general vector fields on two-dimensional manifolds, see [**83**].

1.4.3. A preliminary lemma. In order to prove Theorem 1.4.6, we first introduce a lemma that will also be useful in the proof of the structural stability theorem in next chapter.

DEFINITION 1.4.9. *Let $v_1, v_2 \in D^r(TM)$ be two vector fields, and C_1 and C_2 (resp. B_1 and B_2) are, respectively, the circle cells (resp. the circle bands) of v_1 and v_2. We say that ∂C_1 (resp. ∂B_1) is homeomorphic to ∂C_2 (resp. ∂B_2) if there is a homeomorphism $\Psi : \partial C_1 \to \partial C_2$ (resp. $\Psi : \partial B_1 \to \partial B_2$) such that Ψ takes the singular points on ∂C_1 (resp. ∂B_1) of v_1 to the singular points on ∂C_2 (resp. ∂B_2) of v_2, and Ψ preserves the orientation.*

LEMMA 1.4.10. *Let C_1 and C_2 be two circle cells (resp. let B_1 and B_2 be two circle bands), and let $\varphi_0 : \partial C_1 \to \partial C_2$ (resp. $\varphi_0 : \partial B_1 \to \partial B_2$) be the homeomorphism on boundary. Then there exists a homeomorphism $\varphi : \overline{C_1} \to \overline{C_2}$ (resp.*

$\varphi : \overline{B_1} \to \overline{B_2}$) such that φ takes the closed orbits in C_1 (resp. B_1) to the closed orbits in C_2 (resp. B_2), preserving orientation, and $\varphi|_{\partial C_1} = \varphi_0$ (resp. $\varphi|_{\partial B_1} = \varphi_0$).

PROOF. We divide the proof into three steps.

Step 1. We first consider the case where C_1 *and* C_2 *are two circle cells.*

Let $p \in C_1$, $q \in C_2$ be the centers in C_1 and C_2, and let $a_i \in \partial C_1$, $b_i \in \partial C_2$ ($i = 1, \dots, m$) be the singular points on ∂C_1 and ∂C_2 ($m \geq 0$) such that $\varphi_0(a_i) = b_i$ ($i = 1, \dots, m$).

Note that ∂C_1 and ∂C_2 are homeomorphic to the topological sets of S_k^1 type, possibly with some self-intersections at the singular points a_i and b_i ($i = 1, \dots, m$). If there exist self-intersections on ∂C_1 and ∂C_2, as $\varphi_0 : \partial C_1 \to \partial C_2$ is a homeomorphism, φ_0 takes the self-intersections on ∂C_1 to the self-intersections on ∂C_2.

Assume that a_j and $b_j = \varphi_0(a_j)(1 \leq j \leq k)$ are the self-intersections on ∂C_1 and ∂C_2 , $k \leq m$. If we regard every self-intersection $a_j \in \partial C_1$ (resp. $b_j \in \partial C_2$) as two different singular points a_{j1} and a_{j2} (resp. b_{j1} and b_{j2}) with $\varphi_0(a_{ji}) = b_{ji}$ ($i = 1, 2$), then $\overline{C_1}$ and $\overline{C_2}$ can be regarded as two closed disks with $m + k$ singular points on their boundaries.

We take N different points $p_i \in \partial C_1$ and $q_i \in \partial C_2$ respectively, $1 \leq i \leq N$, $N \geq \max\{m + k, 3\}$, such that the $m + k$ singular points on ∂C_1 and ∂C_2 are contained in $\{p_i | 1 \leq i \leq N\}$ and $\{q_j | 1 \leq j \leq N\}$ respectively, and such that

$$\varphi_0(p_i) = q_i, \quad 1 \leq i \leq N. \tag{1.4.3}$$

With the center $p \in \overline{C_1}$ and the given points $\{p_i | 1 \leq i \leq N\} \subset \partial C_1$, we triangulate the disk $\overline{C_1}$ as follows:

For each given point p_i, we can take a simple curve $\widetilde{pp_i}$, which connects the center p and p_i in $\overline{C_1}$, such that $\widetilde{pp_i}$ intersects with every closed orbit in C_1 at exactly one point. Three curves $\widetilde{pp_i}$, $\widetilde{pp_{i+1}}$ and $\widetilde{p_ip_{i+1}}$ enclose a triangle in C_1, denoted by $\triangle pp_ip_{i+1}$, satisfying the following properties:

a) the curve $\widetilde{p_ip_{i+1}} \subset \partial C_1$ connects two given adjacent points p_i and p_{i+1} on ∂C_1;
b) for any point $a \in \widetilde{pp_i}$ ($a \neq p, p_i$), there exists a simple curve $l \subset \triangle pp_ip_{i+1}$ such that l is the interaction of the triangle $\triangle pp_ip_{i+1}$ and the closed orbit passing through point a. l has the point a as one of its end points, and a point $b \in \widetilde{pp_{i+1}}$ as the other end point. Moreover, l intersects respectively with $\widetilde{pp_i}$ and $\widetilde{pp_{i+1}}$ exactly at one point a and b; the curve l is called an orbit curve starting from point a; and
c) for any $x \in \triangle pp_ip_{i+1}$, there is a unique orbit curve l passing through x.

The above triangulation of C_1 satisfying a)–c) is called a *normal triangulation,* and each triangle $\triangle pp_ip_{i+1}$ is called a *normal triangle.*

In the same fashion for the center $q \in C_2$ and the given points $\{q_i | 1 \leq i \leq N\}$, we have a normal triangulation $\{\triangle qq_iq_{i+1} | 1 \leq i \leq N\}$ of C_2.

By the definition of normal triangulations of circle cells and (1.4.3), it is easy to see that the existence of the homeomorphism $\varphi : \overline{C_1} \to \overline{C_2}$ required by the lemma is equivalent to the existence of the following homeomorphisms:

$$\varphi_i : \triangle pp_ip_{i+1} \to \triangle qq_iq_{i+1}, \quad 1 \leq i \leq N, \tag{1.4.4}$$

where φ_i satisfy that

1) $\varphi_i(p) = q$, $\varphi_i(p_i) = q_i$, $\varphi_i(p_{i+1}) = q_{i+1}$;

2) the restriction of φ_i on the two sides $\widetilde{pp_i}$ and $\widetilde{p_ip_{i+1}}$ of $\triangle pp_ip_{i+1}$ is equal to the following homeomorphism:

$$\begin{cases} \varphi_0^* : \widetilde{p_ip_{i+1}} \to \widetilde{q_iq_{i+1}}, \\ \varphi_i^* : \widetilde{pp_i} \to \widetilde{qq_i}, \end{cases} \tag{1.4.5}$$

where $\varphi_0^* = \varphi_0|_{\widetilde{p_ip_{i+1}}}$, and φ_i^* is a given homeomorphism; and

3) for each point $a \in \widetilde{pp_i}$, φ_i maps the orbit curve starting from a in $\triangle pp_ip_{i+1}$ to the orbit curve starting from $\varphi_i(a)$ in $\triangle qq_iq_{i+1}$.

In order to prove the existence of the homeomorphisms (1.4.4), we consider a planar standard triangle. Let $\triangle ABC$ be a planar triangle with the length of every side one. The triangle $\triangle ABC$ is called a canonical triangle.

For each normal triangle $\triangle pp_ip_{i+1}$, there exists a homeomorphism as follows:

$$\begin{cases} h : \triangle pp_ip_{i+1} \to \triangle ABC, \\ h(\widetilde{pp_i}) = \overline{AB}, \quad h(\widetilde{pp_{i+1}}) = \overline{AC}, \quad h(\widetilde{p_ip_{i+1}}) = \overline{BC}, \end{cases} \tag{1.4.6}$$

where $\triangle ABC$ is a canonical triangle. In the canonical triangle $\triangle ABC$, the homeomorphic image of $\triangle pp_ip_{i+1}$ has the following properties:

d) For any $x \in \widetilde{pp_i}$, there exists a simple curve $L \subset \triangle ABC$, starting from $h(x) \in \overline{AB}$, which is the image of $l \subset \triangle pp_ip_{i+1}$, where l is an orbit curve starting from x, i.e., $L = h(l)$. We still call the curve L an orbit curve starting from $h(x)$;

e) For each point $x \in \triangle ABC$, there is a unique orbit curve L passing through x.

Then it is easy to show that the existence of homeomorphisms (1.4.4) is equivalent to the existence of the following homeomorphisms,

$$\Psi : \triangle ABC \to \triangle ABC, \tag{1.4.7}$$

where Ψ satisfies

4) $\Psi(A) = A,\ \Psi(B) = B,\ \Psi(C) = C$;

5) the restriction of Ψ in the two sides $\overline{AB}$ and $\overline{BC}$ equals to two given homeomorphism,

$$\begin{cases} \Psi_1^* : \overline{AB} \to \overline{AB} \text{ with } \Psi_1^*(A) = A,\ \Psi_1^*(B) = B \\ \Psi_2^* : \overline{BC} \to \overline{BC} \text{ with } \Psi_2^*(B) = B,\ \Psi_2^*(C) = C; \end{cases} \tag{1.4.8}$$

6) for each $x \in \overline{AB}$, Ψ maps the orbit curve L starting from x in $\triangle ABC$ to the straight line passing through $\Psi_1(x)$ and paralleling to $\overline{BC}$.

The homeomorphism (1.4.7) satisfying 4)–6) is called a standard homeomorphism.

In fact, if the standard homeomorphisms (1.4.7) exist, then we have the following homeomorphic diagram:

$$\begin{array}{ccc} \triangle pp_ip_{i+1} & \xrightarrow{\varphi_i} & \triangle qq_iq_{i+1} \\ \downarrow h_1 & & \downarrow h_2 \\ \triangle ABC & & \triangle ABC \\ \downarrow \Psi_1 & & \downarrow \Psi_2 \\ \triangle ABC & \xrightarrow{id} & \triangle ABC \end{array}$$

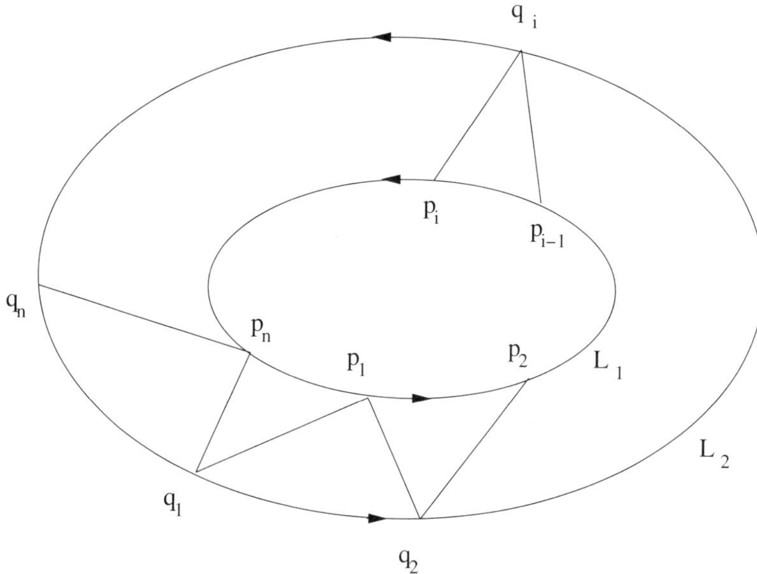

FIGURE 1.4.1. A triangulation of a circle band.

where h_1, h_2 are the homeomorphisms as in (1.4.7), Ψ_1 and Ψ_2 are the standard homeomorphisms, and *id* is the identical map. On the sides $\overline{AB}$, $\overline{BC}$, if we arbitrarily take the homeomorphisms Ψ_{21}^* and Ψ_{22}^* as in (1.4.8) and let

$$\Psi_{11}^* = \Psi_{21}^* \circ h_{21} \circ \varphi_i^* \circ h_{11}^{-1} : AB \to AB,$$

$$\Psi_{12}^* = \Psi_{22}^* \circ h_{22} \circ \varphi_0^* \circ h_{12}^{-1} : BC \to BC,$$

where φ_i^*, φ_0^* are the homeomorphisms as in (1.4.5), and $h_{11} = h_1|_{AB}$, $h_1|_{BC}$, $h_{21} = h_2|_{AB}$, $h_{22} = h_2|_{BC}$, then we infer from the homeomorphic diagram above that the homeomorphisms $\varphi_i = h_2^{-1} \circ \Psi_2^{-1} \circ \Psi_1 \circ h$ satisfy properties 1)–3).

Therefore, it suffices to prove the existence of the standard homeomorphism (1.4.7), and we will proceed with the proof in Step 2 below.

Step 2. Consider now the case where B_1 and B_2 are two circle bands. By Theorem 1.3.3, a circle band is homeomorphic to an open annulus.

Notie that $\varphi_0 : \partial B_1 \to \partial B_2$ is a homeomorphism, and each of ∂B_1 and ∂B_2 is homeomorphic to a topological set of S_k^1 type. In the same fashion as in Step 1, we regard $\overline{B_1}$ and $\overline{B_2}$ as two closed annuli, each of which has N_1 given points on one connected component of the boundary, and N_2 given points on the other component. Let $\partial B_1 = L_1 \cup L_2$, $\partial B_2 = L'_1 \cup L'_2$, and the given points $\{p_i | 1 \le i \le N_1\} \subset L_1, \{q_i | 1 \le i \le N_2\} \subset L_2$, $\{p'_i | 1 \le i \le N_1\} \subset L'_1$, $\{q'_i | 1 \le i \le N_2\} \subset L'_2$. Then

$$\varphi_0(p_i) = p'_i,\ \varphi_0(q_j) = q'_j,\ 1 \le i \le N_1,\ 1 \le j \le N_2.$$

Without loss of generality, we assume $N_1 = N_2 = N$. With the given points $\{p_i | 1 \le i \le N\}$ and $\{p_j | 1 \le j \le N\}$, we triangulate $\overline{B_1}$ as follows (see Figure 1.4.1):

We take a simple curve $\widetilde{q_1 p_1}$, which connects q_1 and p_1 in $\overline{B_1}$, then again take a simple curve $\widetilde{p_1 q_2}$, then $\widetilde{q_2 p_2}, \ldots$ and so on, until finally, we have the curve $\widetilde{p_n q_1}$.

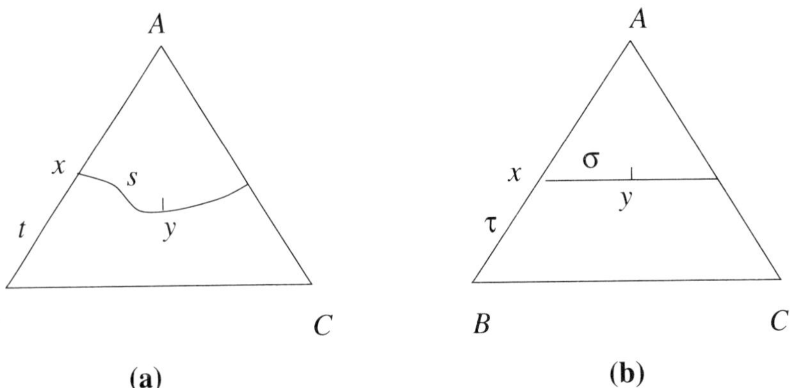

FIGURE 1.4.2. (a) Orbit curve coordinate system; (b) the standard coordinate system.

Each simple curve $\widetilde{p_i q_j} (j = i, i+1)$ intersects with any closed orbit in B_1 exactly at one point. It is easy to see that every triangle $\triangle q_i p_{i-1} p_i$(or $\triangle p_i q_i q_{i+1}$) satisfies the properties a)–c) above. Hence we get a normal triangulation to the circle band B_1. Similarly we also get a normal triangulation to B_2. Thus the existence of the homeomorphism $\varphi : \overline{B_1} \to \overline{B_2}$ required by the lemma is equivalent to the existence of the standard homeomorphism (1.4.7).

Step 3. The existence of the standard homeomorphism. For a normal triangle $\triangle_i$, under the homeomorphism (1.4.6), $h : \triangle_i \to \triangle ABC$, we can set a coordinate system in the standard triangle $\triangle ABC$, called an orbit curve coordinate system. We denote by $(t, s)_c$ the orbit curve coordinate system in $\triangle ABC$, where $t(0 \leq t \leq 1)$ is the distance from a point x in the side $\overline{AB}$ to the point B (see Figure 1.4.2(a)), and s is the arc length of the orbit curve starting from x to a point $y \in \triangle ABC$.

By the properties d)–e), we can see that for any $y \in \triangle ABC$, y can be uniquely expressed by the orbit curve coordinate $y = (t, s)_c$ with $0 \leq t \leq 1,\ 0 \leq s \leq L(t)$, where $L(t)$ is the length of the orbit curve starting from $(t, 0)_c$.

We denote by $(\tau, \sigma)_s$ the coordinate system of $\triangle ABC$, for any $y \in \triangle ABC$, $y = (\tau, \sigma)_s$, where $\tau(1 \leq t \leq 1)$ is the distance from $x \in \overline{AB}$ to the point B, and σ is the distance from y to x, which are in the straight line passing through y and paralleling the side $\overline{BC}$; see Figure 1.4.2(b).

We set the standard homeomorphism (1.4.7) as follows:

$$\Psi = \Psi_2 \circ \Psi_1 : (t, s)_c \to (\tau, \sigma)_s$$

where $\Psi_1 : (t, s)_c \to (\tau, \sigma)_s$ is defined as follows

$$\Psi_1(t, s)_c = \left(\Psi_1^*(t), \frac{1-t}{L(t)} \cdot s \right)_s,$$

where Ψ_1^* is defined as (1.4.8), and $L(t)$ is the length of the orbit curve starting from $(t, 0)_c$.

Obviously, the homeomorphism Ψ_1 takes the orbit curve starting from $(t,0)_c$ in $\triangle ABC$ to the straight line passing through $(\Psi_1^*(t),0)_s$ and paralleling $\overline{BC}$; moreover,

$$\begin{cases} \Psi_1|_{\overline{BC}} = id : \overline{BC} \to \overline{BC} \\ \Psi_1|_{\overline{AB}} = \Psi_1^* : \overline{AB} \to \overline{AB}. \end{cases}$$

And $\Psi_2 : (\tau,\sigma)_s \to (\tau,\sigma)_s$ is defined as follows

$$\Psi_2(\tau,\sigma)_s = \left(\tau, (1-\tau)\Psi_2^*(\frac{\sigma}{1-\tau})\right)_s, \quad 0 \le \sigma \le 1-\tau, \quad 0 \le \tau \le 1, \tag{1.4.9}$$

where Ψ_2^* is defined as in (1.4.8). By (1.4.8), we can see that $\Psi_2^*(0) = 0$, $\Psi_2^*(1) = 1$, and

$$\begin{cases} \Psi_2|_{\overline{AB}} = id : \overline{AB} \to \overline{AB} \\ \Psi_2|_{\overline{BC}} = \Psi_2^* : \overline{BC} \to \overline{BC}. \end{cases}$$

Moreover, we obtain from (1.4.9) that Ψ_2 maps the straight line parallel to $\overline{BC}$ to itself. □

1.4.4. Proof of the classification theorem. We divide the proof into a few steps.

Step 1. Let Γ_{R_1} and Γ_{R_2} be the saddle connection diagrams of v_1 and v_2, and let Γ_{R_1} be isomorphic to Γ_{R_2}. We need to prove that the map $h : \Gamma_{R_1} \to \Gamma_{R_2}$ takes the boundaries of circle cells (resp. the boundaries of circle bands, or the connected components of the saddle connection set) to the boundaries of circle cells (resp. to the boundaries of circle bands, or the connected components of the saddle connection set).

Let $\sigma \in \Gamma_{R_1}$, $\widetilde{\sigma} = h(\sigma)$. If σ is a connected component of the saddle connection set, as σ is homeomorphic to $\widetilde{\sigma}$, then $\widetilde{\sigma}$ is also a connected component of the saddle connection set of v_2. Therefore, it suffices to prove that if σ is the boundary of a circle cell, then $\widetilde{\sigma}$ is also the boundary of a circle cell.

For $\sigma_0 \in \Gamma_R$, if there are n elements $\sigma_1, ..., \sigma_n \in \Gamma_R$ such that

$$\sigma_n < \cdots < \sigma_1 < \sigma_0 \tag{1.4.10}$$

and between any two adjacent elements σ_{i+1} and σ_i $(0 \le i \le n-1)$ there is no element $\alpha \in \Gamma_R$ such that $\sigma_{i+1} < \alpha < \sigma_i$ or $\alpha < \sigma_n$, then (1.4.10) is called a branch of σ_0. Obviously, σ_0 may have many branches. The largest number n in the branches of σ_0 is called the level number of σ_0, and the ith element σ_i in each branch of σ_0 is said to be in the ith level of σ_0.

Obviously, for the two isomorphic diagrams Γ_{R_1} and Γ_{R_2}, $\sigma \in \Gamma_{R_1}$ and $h(\sigma) \in \Gamma_{R_2}$ have the same level number and the same number of elements in any i^{th} level of σ and $h(\sigma)$, $(1 \le i \le n)$.

With the definition of interior in (1.4.2), for each element $\sigma \in \Gamma_R$, we can define an orientation of σ as follows. By Lemma 1.3.1, $\partial \mathring{\sigma} = \bigvee_k S^1$, which is a closed co-direction orbit curve. The orientation of σ (either left rotation or right rotation) is determined by whether the co-direction orbits on $\partial \mathring{\sigma}$ are clockwise or counter-clockwise relative to the interior $\mathring{\sigma}$.

Now we return to the proof of the previous problem. Let $\sigma \in \Gamma_{R_1}$ be a boundary of a circle cell, and let $\widetilde{\sigma} = h(\sigma) \in \Gamma_{R_2}$ be a boundary of a circle band. We shall follow a contradiction. If $\widetilde{\sigma}$ is an outer boundary of a circle band, and homeomorphic to $\bigvee_n S^1$ $(n \ge 1)$, then by Definition 1.4.1, σ is homeomorphic to $\widetilde{\sigma}$ preserving

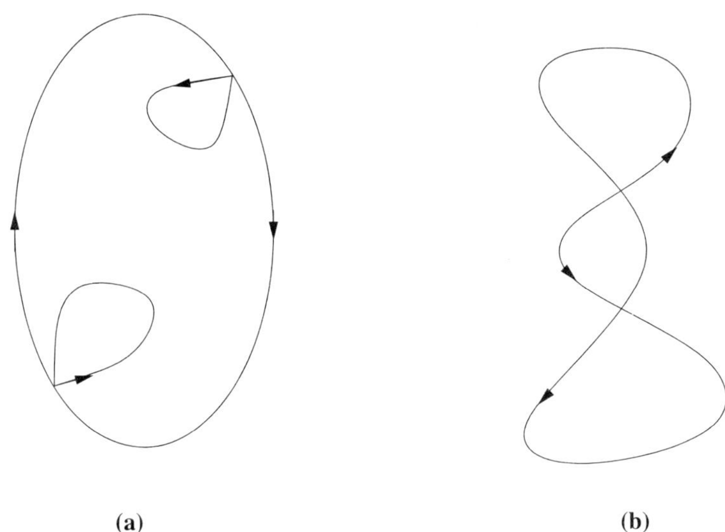

FIGURE 1.4.3. (a) A boundary of a circle cell; (b) an inner boundary of a circle band.

orientation. Hence σ is also homeomorphic to $\bigvee_n S^1$. On the other hand, σ and $\widetilde{\sigma}$ have exterior circles such that the other $n-1$ circles in $\bigvee_n S^1$ are in the circles. This means that there are $n-1$ elements in the first level of σ, and n elements in the first level of $\widetilde{\sigma}$ because there is an inner boundary in the interior of $\widetilde{\sigma}$. It is a contradiction to the earlier statement that σ and $\widetilde{\sigma}$ should have the same number of elements in any same level. If $\widetilde{\sigma}$ is an inner boundary, then none of the n circles in $\bigvee_n S^1$ of $\widetilde{\sigma}$ is in the interior of another circle in $\bigvee_n S^1$. Therefore, the n circles in $\bigvee_n S^1$ of $\widetilde{\sigma}$ have the same orientation (see Figure 1.4.3(b)). But the exterior circle C of σ has a different orientation to the other $n-1$ circles in $\bigvee_n S^1$ of σ (see Figure 1.4.3(a)). It is a contradiction to an earlier statement that σ is homeomorphic to $\widetilde{\sigma}$, preserving orientation.

Step 2. Completion of the proof of the classification theorem. By b) and c) in Definition 1.4.4, it is easy to see that for each element $\sigma \in \Gamma_{R_1}$, the homeomorphism $\psi : \sigma \to \widetilde{\sigma} = h(\sigma)$ takes the singular points on σ to the singular points on $\widetilde{\sigma}$. Let $\sigma_1, ..., \sigma_m \in \Gamma_{R_1}$ be all the boundary of circle cells and circle bands in M, and let A_{σ_i} $(1 \leq i \leq m)$ be the circle cell or circle band with $\partial A_{\sigma_i} = \sigma_i$. By Theorem 1.3.3, $\sum_{i=1}^m \overline{A}_{\sigma_1} = M$. Hence, we can define a global homeomorphism as follows:

$$\begin{aligned} &\rho : M \to M, \\ &\rho(x) = \rho_i(x) && \text{for } x \in \overline{A}_{\sigma_i}, \\ &\rho_i|_{\sigma_i} = \psi_i : \sigma_i \to \widetilde{\sigma}_i && \text{the given homeomorphism,} \\ &\rho_i : \overline{A}_{\sigma_i} \to \overline{A}_{\widetilde{\sigma}_i} && \text{defined as in Lemma 1.4.10.} \end{aligned}$$

It is easy to see that ρ takes the orbits of v_1 to the orbits preserving orientation. Hence if Γ_{R_1} and Γ_{R_2} are isomorphic, then v_1 and v_2 are topologically equivalent.

Obviously, if v_1 and v_2 are topologically equivalent, then there exist saddle connection diagrams Γ_{R_1} and Γ_{R_2} of v_1 and v_2 such that Γ_{R_1} is isomorphic to Γ_{R_2}.

The proof is complete. □

Notes for Chapter 1

1. The classical Poincaré-Bendixson theorem was recalled without the proof in Theorem 1.2.5, whose proof can be found in many classical ordinary differential equations books; see, among others, J. Guckenheimer and P. J. Holmes [**33**], J. K. Hale [**34**], J. Palis and W. de Melo [**77**], and S. Wiggins [**105**].

The classical Poincaré-Bendixson theorem requires that the manifold be a submanifold of the 2-D sphere S^2. The limit set theorem is its generalization to divergence-free vector fields on general two-dimensional manifolds, and is obtained by the authors in [**56**].

2. We refer the reader to J. Milnor [**72**] for the classical Poincaré–Hopf index theorem. Its generalization to vector fields defined on manifolds with boundary given and proved in Section 1.2 is based on Ma and Wang [**55**].

3. The structural classification and the structure of ergodic sets given in Section 1.3 are based on Ma and Wang [**56, 58**].

CHAPTER 2

Structural Stability of Divergence-Free Vector Fields

The next important question toward a kinematic theory for incompressible flows as described in Area A in Section 0.2 is to study stability, and the right notion of stability is the structural stability with divergence-free vector field perturbations. We call this notion of structural stability the incompressibly structural stability, or structural stability for brevity.

The study of structural stability has been the main driving force behind much of the development of dynamical systems theory, following the program initiated by S. Smale and others.

Notice that the divergence-free condition completely changes the general features of structurally stable fields as compared to the situation when this condition is not present. The latter case was studied in a classical paper of M. Peixoto [**82**] in 1962. This chapter is devoted to the (incompressibly) structural stability for incompressible flows in the following cases:

(1) divergence-free vector fields with no-normal flow boundary conditions,
(2) divergence-free vector fields with the Dirichlet (no-slip) boundary conditions, and
(3) Hamiltonian vector fields.

2.1. Structural Stability of Divergence-Free Vector Fields with Free Boundary Conditions

Again, as mentioned before, unless otherwise explicitly stated, we always assume that M is a two-dimensional orientable $C^r (r \geq 1)$ compact Riemannian manifold.

2.1.1. Structural stability theorems. The main motivation is to study the structural stability of the solutions of the Navier-Stokes equations. Therefore, we study the structural stability of divergence-free vector fields with different boundary conditions. To this end, we recall that

$$\begin{aligned} B^r(TM) &= \left\{ v \in D^r(TM) | \quad \frac{\partial v_\tau}{\partial n} = 0 \text{ on } \partial M \right\}, \\ B^r_0(TM) &= \{ v \in D^r(TM) | \quad v = 0 \text{ on } \partial M \}, \\ B^r_\varphi(TM) &= \{ v \in D^r(TM) | \quad v = \varphi \text{ on } \partial M \}, \end{aligned}$$

where $v_n = v \cdot n$, $v_\tau = v \cdot \tau$, n and τ are the unit normal and tangent vectors on ∂M, respectively. If $r = k + \alpha$ with $k \geq 0$ an integer and $0 < \alpha < 1$, then $v \in C^r(TM)$ means that $v \in C^k(TM)$ and all derivatives of v up to order k are α-Hölder continuous.

DEFINITION 2.1.1. *Let X be either $D^r(TM)$, or $B^r(TM)$, or $B^r_0(TM)$, or $B^r_\varphi(TM)$. A divergence-free vector field $v \in X$ is called structurally stable in X if there exists an open neighborhood $\mathcal{O} \subset X$ of v such that for any $u \in \mathcal{O}$, u and v are topologically equivalent.*

Consider a regular divergence-free vector field $v \in D^r(TM)$. We know that each saddle point in the interior $p \in \overset{\circ}{M}$ is connected by four orbits of v, two of which are stable orbits, and the other two are unstable. If both the ω-limits and the α-limits of these four orbits are the saddle point p, then the saddle point p is called self-connected.

The main theorem in this section is the following global structural stability theorem for divergence-free vector fields.

THEOREM 2.1.2. *Let $M \subset S^2$ be a two-dimensional compact manifold, and let X be either $D^r(TM)$ or $B^r(TM)$ $(r \geq 1)$. Then $v \in X$ is structurally stable in X if and only if*

(1) *v is regular,*
(2) *all interior saddle points of v are self-connected, and*
(3) *each boundary saddle point is connected to boundary saddles on the same connected component of the boundary.*

Moreover, all structurally stable vector fields in X form an open and dense set of X.

A few remarks are now in order.

REMARK 2.1.3. When $M \subset K^2$ (or P^2) is a two-dimensional compact non-orientable manifold with or without boundary, results similar to Theorem 2.1.2 hold true as well.

REMARK 2.1.4. The structural stability for general vector fields on a two-dimensional manifold was proved in 1962 by Peixoto.

THEOREM 2.1.5 (Peixoto, [**82**]). *A C^r vector field v on a two-dimensional compact manifold M^2 is structurally stable if and only if*

(i) *the number of singular points and closed orbits is finite, and all are hyperbolic,*
(ii) *there are no orbits connecting saddles, and*
(iii) *the non-wandering set consists of singular points and closed orbits.*

The first condition in Theorem 2.1.2 requires only regularity of the field and so it does not exclude centers which are not hyperbolic and excluded by (i) above. The second condition is of a completely different nature than the corresponding one in the Peixoto theorem. Namely, condition (ii) above excludes the possibility of saddle connections. In contrast, condition (2) amounts to saying that all interior saddles are self-connected! Namely, the interior saddles occur in graphs whose topological form is that of the number 8, the singularities themselves being hyperbolic. The third condition in Theorem 2.1.2 deals with singularities on the boundary.

Moreover, a direct consequence of the Peixoto structural stability theorem and Theorem 2.1.2 is that no divergence-free vector field is structurally stable under general C^r vector field perturbations. Such a drastic change in the stable configurations is explained by the fact that divergence-free fields preserve volume and so attractors and sources can never occur for these fields. In particular, this makes the restriction in the third condition that saddles in the boundary must be connected with saddles in the boundary on the same connected component quite natural.

REMARK 2.1.6. One of important ingredients of S. Smale's program on dynamical systems studies structural stability of the Morse-Smale systems in higher dimensional manifolds. More precisely, a Morse-Smale system is one for which:

(1) the number of singular points and closed orbits is finite and each is hyperbolic;
(2) all stable and unstable manifolds intersect transversally; and
(3) the non-wandering set consists of singular points and closed orbits alone.

THEOREM 2.1.7 (Palis and Smale [**78**]). *A Morse-Smale system is structurally stable.*

Many questions remain open; see [**3, 33, 42, 77, 89**] for details.

REMARK 2.1.8. In [**88**], C. Robinson studied some generic properties of Hamiltonian systems. In particular, he showed nonstructural stability for high dimensional Hamiltonian systems.

The proof of Theorem 2.1.2 occupies the remaining part of this section.

2.1.2. Construction of tubular incompressible flows. We construct a special class of tubular divergence-free vector fields, which are crucial in order to break the saddle connections in the proof hereafter.

Let M be a two-dimensional compact orientable Riemannian manifold, let $\{g_{ij}\}$ be the Riemanian metric under a local coordinate system on M and let $g = \det(g_{ij})$. There is a global two-form on M defined by

$$\omega = \sqrt{g} dx_1 \wedge dx_2.$$

Let $J : T^*M \to TM$ be the symplectic isomorphism induced by the symplectic form ω. A vector field v is called a Hamiltonian vector field if $v = J\nabla H$ for some Hamitonian function H defined on M.

LEMMA 2.1.9. *Let $L \subset M$ be a C^r simple closed curve. Then there exists $\delta > 0$ such that*

(1) *for any $0 < h < \delta$, there exists an open annulus $B \subset M$ with L being one connected component of its boundary, and h being its width;*
(2) *there exists a divergence-free vector field $v \in X$ with*

$$v|_{M-B} = 0, \quad v|_B \neq 0,$$

and B is a circle band of v; and
(3) *the tubular divergence-free vector field v is a Hamiltonian vector field if and only if $L \subset M$, as a closed chain, is homological to zero in $H_1(M, \partial M)$.*

PROOF. *Step 1. Construction of B.* Because L is homological to zero as a closed chain and L is homeomorphic to S^1, there exists a submanifold $N \subset M$ with $L \subset \partial N$.

By the Collar Neighborhood theorem, L has a collar $C \subset N$, i.e., C is homeomorphic to $L \times [0, 1)$.

For any $x \in L$, there is a unit normal vector $n_x \in T_xM$ of L pointing inward to C. Since M is a Riemannian manifold with the Levi-Civita connection on M, there exists a unique geodesics starting from x such that n_x is its tangent vector at x. Let $z(\alpha, x)$ be the point $z \in C \subset M$ that lies on the geodesics such that the arclength from z to x is α. Obviously $z(0, x) = x$.

Since L is a compact $C^r (r \geq 1)$ simple closed curve, there exists a real number $\delta > 0$ such that for any $0 < h < \delta$ and $x, y \in L (x \neq y)$, the geodesic interval

$$L_x(0, h) \stackrel{\text{def}}{=} \{z(\alpha; x) \mid 0 \leq \alpha \leq h\}$$

does not intersect with the geodesic interval $L_y(0, h)$, and $L_x(0, h) \subset C$.

Set

$$B \stackrel{\text{def}}{=} \{z(\alpha; x) \mid 0 < \alpha < h, \ x \in L\}.$$

It is easy to see that $B \subset C$ is homeomorphic to an open annulus, and $L \subset \partial B$.

Step 2. Definition of v. Let $H \in C^r(\bar{B})$ be a function such that

$$\begin{cases} H(z) = f(\alpha) & \text{if } z = z(\alpha, x) \in \bar{B}, \ 0 \leq \alpha \leq h, \ x \in L, \\ f : [0, h] \to \mathbb{R} \text{ is increasing}, & \\ f(0) = 0, & \\ f(h) = 1, & \\ 0 < f(\alpha) \leq 1 & \text{if } 0 < \alpha < h, \\ \left. \dfrac{d^n f(\alpha)}{d\alpha^n} \right|_{\alpha = 0, h} = 0 & \forall \ n \geq 0. \end{cases}$$

Let $\{L_i\}$ be a covering of L. Then the sets

$$U_i = \{ \ z(\alpha; x) \mid 0 < \alpha < h, \quad x \in L_i \ \}$$

form an open covering of B. For each U_i we consider a local coordinate system

$$\phi_i : U_i \to \phi_i(U_i) \subset \mathbb{R}^2$$

such that ϕ_i maps the geodesic interval $L_x(0, h)$ to the x_2-axis in $\mathbb{R}^2$ and the curves $\Gamma_\alpha = \{ \ z(\alpha; x) \mid x \in L_i\}$ to lines parallel to the x_1-axis.

It is easy to see that the Hamiltonian $J\nabla H$ on B has the following form in the local coordinate system:

$$J\nabla H = -\frac{1}{\sqrt{g}} \frac{\partial H}{\partial x_2} \frac{\partial}{\partial x_1} + \frac{1}{\sqrt{g}} \frac{\partial H}{\partial x_1} \frac{\partial}{\partial x_2}.$$

We now define the vector field v required in the lemma by

$$v(z) = \begin{cases} 0 & \text{if } z \in M - B, \\ J\nabla H(z) & \text{if } z \in B. \end{cases}$$

Then it is easy to see that $v \in C^r(TM)$ and is a divergence-free vector field. Moreover, $v \neq 0$ in B, and B is a circle band of v, due to the fact that f is monotone.

The remaining part of the proof is trivial, and the proof is complete.

□

REMARK 2.1.10. If M is not orientable, then Assertions (1) and (2) in Lemma 2.1.9 are still true.

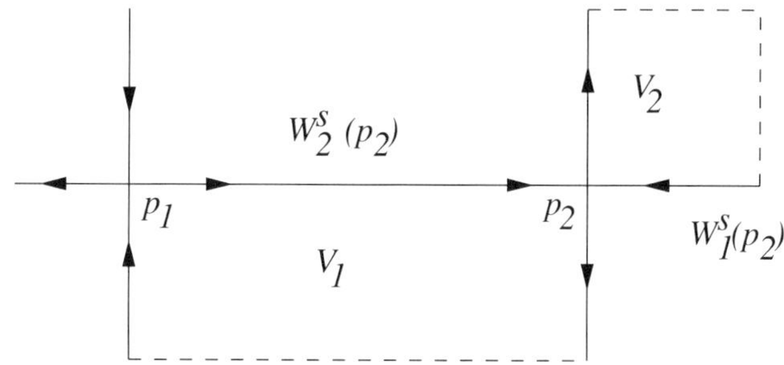

FIGURE 2.1.1. Saddle connection between two saddle points p_1 and p_2.

2.1.3. Breaking saddle connections. In this subsection, we demonstrate a typical method of breaking saddle connections of divergence-free vector fields. This method will useful in Sections 2.2, 3.1, 3.2 and 3.3.

Let $M \subset S^2$ be a two-dimensional compact manifold, let $v \in D^r(TM)$ be regular, and let $V \subset M$ be the open set consisting of closed orbits and centers of v. By the structural classification theorem, V consists precisely of circle cells and circle bands, and $\overline{V} = M$. The set $N = \overline{V} - V = M - V$ is called the saddle connection set of v, and V is called the closed orbit set of v. Obviously N has a finite number of connected components, i.e., $N = \cup_{k=1}^{K} N_k$.

DEFINITION 2.1.11. *Let $M \subset S^2$ be a two-dimensional compact manifold. A regular vector field $v \in D^r(TM)$ is called a self-connection vector field if the saddle connection set $N = \cup_{k=1}^{K} N_k$ of v satisfies that*

(1) *if $N_k \subset \overset{\circ}{M}$, then N_k contains exactly one interior saddle of v, and*
(2) *if $N_k \cap \partial M \neq \emptyset$, then $N_i \cap \partial M$ is exactly one connected component of ∂M, and all saddles on N_i are on ∂M.*

For simplicity, let

$$D_0^r(TM) = \{v \in D^r(TM) \quad | \quad v \text{ is regular }\},$$
$$D_1^r(TM) = \{v \in D_0^r(TM) \quad | \quad v \text{ is a self-connection field }\}.$$

Then Theorem 2.1.2 amounts to saying that $v \in D^r(TM)$ is structurally stable if and only if $v \in D_1^r(TM)$, and $D_1^r(TM)$ is open and dense in $D^r(TM)$.

LEMMA 2.1.12. *Let $M \subset S^2$ be a two-dimensional compact manifold. Then, $D_1^r(TM)$ is dense in $D_0^r(TM)$.*

PROOF. Let $v \in D^r(TM)$ be regular, and let V be its closed orbit set. Then $M = \overline{V}$ since $M \subset S^2$. We then only have to consider the case where $N = M - V \neq \emptyset$, and to prove that v can be approximated in $D^r(TM)$ topology by the self-connection vector fields.

Step 1. First we demonstrate a method to break saddle connections. To this end, we may assume that $N_1 \subset \overset{\circ}{M}$ contains $m \geq 2$ interior saddles. For simplicity, let $p_1,\ p_2 \in N_1$ be two such saddle points that are connected either as shown in Figure 2.1.1 or with V_2 being surrounded by V_1 (i.e., with V_2 being flipped entirely inside V_1). For simplicity, we consider here only the situation as shown in Figure 2.1.1; the other situation can be handled in exactly the same fashion.

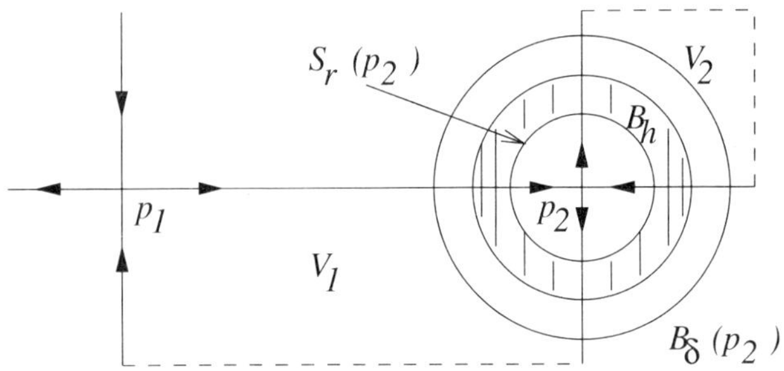

FIGURE 2.1.2. Open annulus B_h.

Observe that two parts $W_1^s(p_2)$ and $W_2^s(p_2)$ of the stable manifold $W^s(p_2)$ at p_2 satisfy (see Figure 2.1.1)

$$W_1^s(p_2) \subset \partial V_1, \quad W_2^s(p_2) \subset \partial V_2.$$

As V_1 and V_2 are two open sets consisting of closed orbits of v, there exists a real number $\delta > 0$ such that for any $x \in B_\delta(p_2) \cap [V_1 \cup V_2]$ with $x \neq p_2$, the orbit $\Phi(x,t)$ of v is closed. Here $B_\delta(p_2)$ is the geodesic ball with radius δ and center p_2.

Let r, $h > 0$ be two real numbers with $r + h < \delta$. Let $S_r(p_2)$ be the geodesic circle with radius r and center p_2. It is clear that, as a closed chain, $S_r(p_2)$ is homological to zero. By Lemma 2.1.10, there is an open annulus B_h with width h and with $S_r(p_2)$ as one connected component of its boundary, and there exists $v_1 \in D^r(TM)$ as in Lemma 2.1.10 (see Figure 2.1.2).

Obviously, for $\varepsilon > 0$ small, the vector field $v + \varepsilon v_1$ and v have the same saddles and the same saddle connections except the connected component N_1, which may be broken due to the perturbation εv_1.

Step 2. We now show that, for $\varepsilon > 0$ sufficiently small, the connected component N_1 of saddle connections set N of v is divided into two connected components $N_1^{(1)}$ and $N_1^{(2)}$ such that $N_1^{(1)}$ is a self-connection of p_2 and $N_1^{(2)}$ has $m-1$ saddle points.

As shown in Figure 2.1.3, the orbits born from p_2 intersect with $S_r(p_2)$ at a and a'. For simplicity, we discuss only trajectories in V_1. Under the perturbation εv_1, the orbit $\overline{p_2 a}$ will travel through B_h to the point $b \in V_1 \cap B_\delta(p_2)$. Since $b \in V_1 \cap B_\delta(p_2)$, $\Phi(t; b)$ of v is closed and contained in V_1, and hence $\Phi(t, b)$ will intersect with ∂B_h at a point c. Then, again under the action $y \in V_1$, the orbit $\overline{p_2 abc}$ travels through B_h and intersects with $S_r(p_2)$ at d.

We now show that the point d is on the interval $\overline{p_1 p_2}$, hence the orbit $\overline{p_2 abcdp_2}$ of $v + \varepsilon v_1$ is a self-connection of the saddle point p_2. In the same fashion, we can obtain that the new orbit $\overline{p_2 a'b'c'd'p_2}$ of $v + \varepsilon v_1$ is also a self-connection of the saddle p_2.

Take a point $x \in (B_\delta(p_2) \backslash B_{r+h}(p_2)) \cap V_1$. It is easy to see that $\Phi(t; x)$ is a closed orbit. If d is below the interval $\overline{p_1 p_2}$ (see Figure 2.1.4), then denote by A the domain enclosed by the *closed* orbit $\Phi(t; x)$ and the closed curve $\overline{abcdea}$.

Obviously, $\Phi(\partial A, t) \subset A$ for $t > 0$, i.e.,

$$|\Phi(t, A)| < |A| \quad \text{for } t > 0,$$

a contradiction to the fact that $v + \varepsilon v_1$ preserves the area.

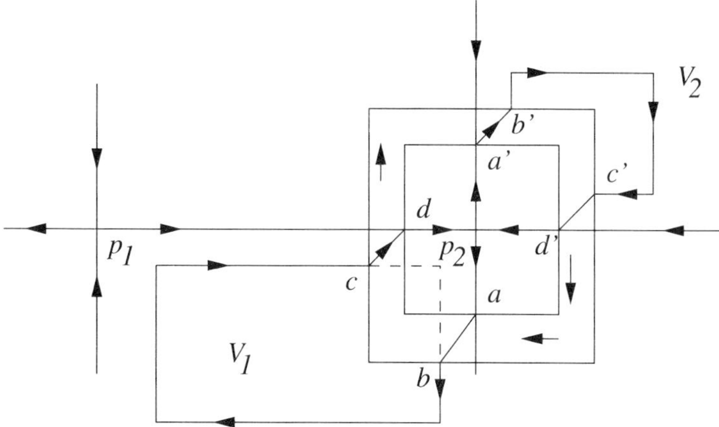

FIGURE 2.1.3. Schematic illustrating Step 2 in the proof of Lemma 2.1.12.

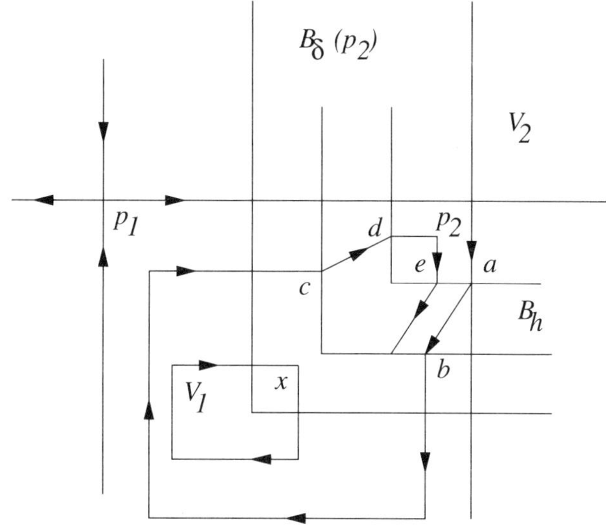

FIGURE 2.1.4. The point d is under the interval $\overline{p_1p_2}$.

Now if d is above the interval $\overline{p_1p_2}$ (see Figure 2.1.5), then we denote by A the domain *enclosed* by $\Phi(t;x)$ and the closed curve $\overline{fghijf}$. It is clear that $\partial A \subset \Phi(\partial A, t)$ for $t > 0$ and $\Phi(\partial A, t) \neq \partial A$. Therefore, we have

$$|A| < |\Phi(A, t)| \text{ for } t > 0,$$

also a contradiction. Hence d is on the interval $\overline{p_1p_2}$.

Then by induction, v can be C^r approximated by regular divergence-free vector fields with node number 1.

Step 3. Assume now that there is a connected component of N, say N_1 again, such that $N_1 \cap \partial M \neq \emptyset$. Let $\Sigma \subset N_1$ be a connected component of ∂M. Let p_1, p_2 be two saddle points on Σ such that p_1, p_2 are on the boundary of a circle cell/band

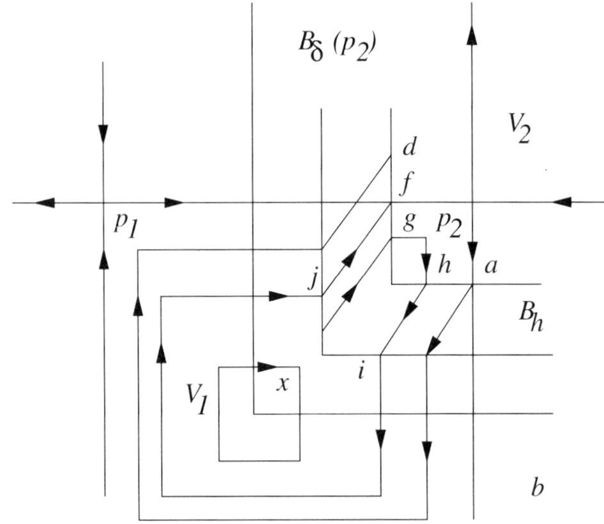

FIGURE 2.1.5. The point d is above the interval $\overline{p_1p_2}$.

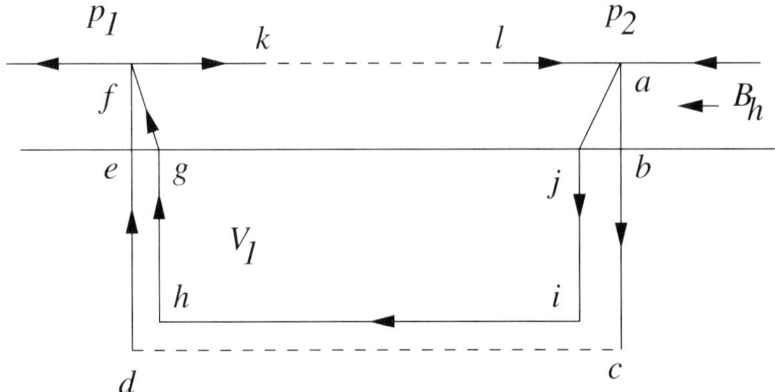

FIGURE 2.1.6. Breaking a saddle connection from a saddle on one component of the boundary to interior saddles or to saddles on other components of the boundary.

V_1 as shown in Figure 2.1.6. Here there might be saddle points of v on the dotted lines $\overline{cd}$ and $\overline{kl}$.

Since ∂M is $C^r(r \geq 1)$, by Lemma 2.1.10, for the closed curve $L \subset \partial M$ with $p_1, p_2 \in L$, we take an open annulus $B_h(h > 0)$ with L as one connected component of its boundary and with width h, and we take a vector field $v_1 \in D^r(TM)$ as in Lemma 2.1.10. By a similar analysis as in Steps 1–2, we can prove that for any $\varepsilon > 0$ sufficiently small, the vector field $v + \varepsilon v_1$ has a saddle connection in $\overset{\circ}{M}$ between p_1 and p_2 (i.e., $\overline{ajihgf}$ as shown in Figure 2.1.6).

In fact, this perturbation $v + \varepsilon v_1$ breaks all saddle connections from saddles on Σ to either interior saddles or saddles on other components of the boundary.

Namely, a connected component of the saddle connection set of $v + \varepsilon v_1$ containing Σ has all saddles on Σ.

Using the above procedure inductively to each connected component of ∂M, we can approximate v by divergence-free vector fields with a saddle connection set satisfying Condition (2) in Definition 2.1.11.

Step 4. We have the saddle connection set $N = \cup_{k=1}^{k} N_k$ satisfying Condition (2) in Definition 2.1.11. For each $N_k \subset \overset{\circ}{M}$, let m_i be the number of saddles on N_k. For simplicity, we call

$$m = \max\{m_k \quad | \quad N_k \subset \overset{\circ}{M}\}$$

the *node number* of v.

Assume that the node number $m \geq 2$. It suffices for us to show that v can be approximated by a divergence-free vector field with node number less than m. For simplicity, we assume $m_1 = m$; i.e., $N_1 \subset \overset{\circ}{M}$ and N_1 contains m saddles. Using then Steps 1 and 2, the vector field can be approximated by new divergence-free vector fields with connected component N_1 of saddle connections set N being divided into two connected components $N_1^{(1)}$ and $N_1^{(2)}$ such that $N_1^{(1)}$ is a self-connection of p_2 and $N_1^{(2)}$ has $m - 1$ saddle points. By induction, we can approximate the vector field by self-connection vector fields. The proof is complete.

□

2.1.4. Genericity and stability of extended orbits.

THEOREM 2.1.13. *Let $M \subset S^2$ be a two-dimensional compact manifold. Then $D_1^r(TM)$ is open and dense in $D^r(TM)$.*

PROOF. As in C. Robinson [**88**], using local Hamiltonians, the openness and density of $D_0^r(TM)$ in $D^r(TM)$ are obvious. Hence, it suffices to prove Lemma 2.1.14 below. □

LEMMA 2.1.14. *Let $M \subset S^2$ be a two-dimensional compact manifold. For any $v \in D_1^r(TM)$, there exists an open neighborhood $\mathcal{O}$ of v in $D^r(TM)$ such that $\mathcal{O} \subset D_1^r(TM)$.*

PROOF. *Step 1.* Since v is regular, we can choose $\mathcal{O}$ small enough such that $\mathcal{O} \subset D_0^r(TM)$. By Theorem 1.3.3, the (global) structure of each $u \in \mathcal{O}$ consists of circle cells, circle bands and saddle connections.

Since v is regular, there are only a finite number of saddles, denoted by $\{p_i \mid i = 1, \dots, \mu\}$, where p_i is either an interior saddle or a boundary saddle point. By the implicit function theorem, there exists a small $\varepsilon > 0$ such that

(a) $\{B(p_i, \varepsilon) \mid i = 1, \dots, \mu\}$ are disjoint open balls centered at the saddle points with radius ε;
(b) the neighborhood $\mathcal{O}$ of v can be chosen sufficiently small such that each $u \in \mathcal{O}$ has the same number of saddles as v, and each $B(p_i, \varepsilon)$ contains exactly one saddle of u; and
(c) if p_i is a boundary saddle of v, then the saddle point in $B(p_i, \varepsilon)$ of each $u \in \mathcal{O}$ is also a boundary saddle of u.

Step 2. We now prove that when $\mathcal{O}$ is sufficiently small, all $u \in \mathcal{O}$ are self-connected. Assuming otherwise, there exists a sequence v^n satisfying the following

(a) $v^n \to v$ as $n \to \infty$,

(b) the saddles of v^n are $\{p_i^n \mid i = 1, \ldots, \mu\}$, such that $p_i^n \in B(p_i, \varepsilon)$, $i = 1, \ldots, \mu$, and

(c) without loss of generality, p_1^n is connected by an orbit of v^n to p_2^n, but p_1 is not connected by an orbit of v to p_2.

However, by the stability theorem of extended orbits stated below, Theorem 2.1.15, this is a contradiction.

The proof is complete. □

In the above proof, we have used a stability theorem of extended orbits. Let $v \in C_n^r(TM)$ be a vector field. A curve $\Gamma \subset M$ is called an extended orbit of v connected by singular points p_i $(i = 1, \cdot, I)$, if γ is a finite union of curves

$$\Gamma = \bigcup_{i=1}^{I} \gamma_i$$

such that each γ_i is a complete orbit of v with starting point p_i and ending point p_{i+1}, i.e.,

$$\alpha(\gamma_1) = p_1, \quad \omega(\gamma_i) = \alpha(\gamma_{i+1}) = p_{i+1} \quad (i = 1, \cdots, I).$$

The point $p_1 = \alpha(\gamma_1)$ is called the starting point of the extended orbit Γ. Notice that there might be some index i such that $\gamma_i = p_i = p_{i+1}$.

Then we have the following theorem on stability of extended orbits; the proof of the theorem is trivial, and we omit the details.

THEOREM 2.1.15 (Stability of Extended Orbits). *Let $v^k \in C_n^r(TM)$ $(r \geq 1)$ be a consequence of vector fields with $\lim_{k\to\infty} v_k = v_0 \in C_n^r(TM)$. Let γ_k be an extended orbit of v_k with finite length uniform with respect to k, and let the starting points p_1^k of Γ_k converge to p_1. Then the extended orbits Γ_k of v_k converge, up to taking a subsequence, to an extended orbit Γ of v_0 with starting point p_1.*

2.1.5. Completion of the proof of the structural stability theorem. After the long preparation in the previous subsections, we are now in a position to complete the proof of the structural stability theorem.

1. Notice that degenerate singular points are unstable, and by Lemma 2.1.12, the structure of connections between different saddle points is also unstable. Hence if $v \in X$ is structurally stable, then v must satisfy stability conditions (1)–(3) in Theorem 2.1.2.

2. Let $v \in X$ satisfy stability conditions 1–3 in Theorem 2.1.2. As v is a self-connection vector field, by Lemma 2.1.14, there exists a real number $\varepsilon > 0$ such that when $\| u - v \| < \varepsilon$, u is also a self-connection vector field. Moreover, u and v have the same numbers of saddle points and centers.

Obviously, both u and v have the same isomorphic saddle connection diagrams. By Theorem 1.4.6 (Topological Classification Theorem), u and v are topologically equivalent.

The proof of Theorem 2.1.2 is complete. □

2.2. Structural Stability for Divergence-Free Vector Fields with Dirichlet Boundary Conditions

2.2.1. D-regular vector fields in $B_0^r(TM)$. In this subsection we study some properties of boundary singular points of a vector field $u \in B_0^r(TM)$.

DEFINITION 2.2.1. *Let $u \in B_0^r(TM)(r \geq 2)$.*

(1) *A point $p \in \partial M$ is called a ∂-regular point of u if $\frac{\partial u_\tau(p)}{\partial n} \neq 0$; otherwise, $p \in \partial M$ is called a ∂-singular point of u.*
(2) *A ∂-singular point $p \in \partial M$ of u is called non-degenerate if*

$$\det \begin{pmatrix} \dfrac{\partial^2 u_\tau(p)}{\partial \tau \partial n} & \dfrac{\partial^2 u_\tau(p)}{\partial n^2} \\ \\ \dfrac{\partial^2 u_n(p)}{\partial \tau \partial n} & \dfrac{\partial^2 u_n(p)}{\partial n^2} \end{pmatrix} \neq 0. \tag{2.2.1}$$

A non-degenerate ∂-singular point of u is also called a ∂-saddle point of u.

DEFINITION 2.2.2. *$u \in B_0^r(TM)$ $(r \geq 2)$ is called D-regular if*

(1) *u is regular in $\overset{\circ}{M}$, and*
(2) *all ∂-singular points of u on ∂M are non-degenerate.*

LEMMA 2.2.3. *Each non-degenerate ∂-singular point of $u \in B_0^r(TM)$ is isolated. Therefore, if all ∂-singular points of u on ∂M are non-degenerate, then the number of ∂-singular points of u is finite.*

PROOF. Since ∂M is $C^r (r \geq 2)$, there is a collar neighborhood $D \subset M$ of ∂M such that for any $x, y \in \partial M, x \neq y$, the normal lines λn_x and λn_y do not intersect in D. There is a vector field N defined in D with unit modulus $|N| = 1$ such that the orbits of N are the normal lines λn in D. We define the vector field $\frac{\partial u}{\partial n}$ as the N-direction derivative of u in D, i.e.,

$$\frac{\partial u}{\partial n} = (N \cdot \nabla) u. \tag{2.2.2}$$

Obviously, $\partial u / \partial n \in C^{r-1}(TD) (r \geq 2)$. Since $u_\tau|_{\partial\Omega} = 0$, we infer from the divergence-free condition of u that

$$\frac{\partial u}{\partial n} \cdot n|_{\partial M} = \frac{\partial u_n}{\partial n}|_{\partial M} = -\frac{\partial u_\tau}{\partial \tau}|_{\partial M} = 0.$$

Hence, the vector field $\partial u / \partial n$ is tangent to the boundary ∂M.

Thus, by (2.2.1), we see that $p \in \partial M$ is a non-degenerate ∂-singular point of u if and only if p is a non-degenerate singular point of $\partial u / \partial n$ on boundary; therefore, non-degenerate ∂-singular points of u are isolated. The proof is complete. □

Let $p \in \partial M$, and let (x_1, x_2) be an orthogonal coordinate system with origin at p, x_1-axis tangent to ∂M at p, and the orientation of the x_2-axis in the inward normal direction. We call (x_1, x_2) a canonical coordinate system at $p \in \partial M$.

THEOREM 2.2.4. *Let $u \in B_0^r(TM)(r \geq 2)$ and (x_1, x_2) be a canonical coordinate system at $p \in \partial M$. Then near p $(x = 0)$, $u = (u_1(x_1, x_2), u_2(x_1, x_2))$ has the following the Taylor expansion at p $(x = 0)$:*

$$\begin{cases} u_1(x_1, x_2) = c x_2 - \dfrac{c}{2} k(p) x_1^2 - 2a x_1 x_2 + b x_2^2 + o(|x|^2), \\ u_2(x_1, x_2) = c k(p) x_1 x_2 + a x_2^2 + o(|x|^2), \end{cases} \tag{2.2.3}$$

where $k(p)$ is the curvature of ∂M at p.

PROOF. Let ds be the arclength differential of ∂M at $p \in \partial M$. Since $u|_{\partial M} = 0$, we have

$$\frac{\partial u_\tau}{\partial s} = \frac{\partial u_\tau}{\partial x_1}\frac{\partial x_1}{\partial s} + \frac{\partial u_\tau}{\partial x_2}\cdot\frac{\partial x_2}{\partial s} = 0, \tag{2.2.4}$$

$$\frac{\partial u_n}{\partial s} = \frac{\partial u_n}{\partial x_1}\frac{\partial x_1}{\partial s} + \frac{\partial u_n}{\partial x_2}\cdot\frac{\partial x_2}{\partial s} = 0, \tag{2.2.5}$$

$$\begin{aligned}\frac{\partial^2 u_\tau}{\partial s^2} &= \frac{\partial^2 u_\tau}{\partial x_1^2}\left(\frac{\partial x_1}{\partial s}\right)^2 + 2\frac{\partial^2 u_\tau}{\partial x_1 \partial x_2}\frac{\partial x_1}{\partial s}\frac{\partial x_2}{\partial s} \\ &\quad + \frac{\partial^2 u_\tau}{\partial x_2^2}\left(\frac{\partial x_2}{\partial s}\right)^2 + \frac{\partial u_\tau}{\partial x_1}\frac{\partial^2 x_1}{\partial s^2} + \frac{\partial u_\tau}{\partial x_2}\frac{\partial^2 x_2}{\partial s^2} = 0,\end{aligned} \tag{2.2.6}$$

$$\begin{aligned}\frac{\partial^2 u_n}{\partial s^2} &= \frac{\partial^2 u_n}{\partial x_1^2}\left(\frac{\partial x_1}{\partial s}\right)^2 + 2\frac{\partial^2 u_n}{\partial x_1 \partial x_2}\frac{\partial x_1}{\partial s}\cdot\frac{\partial x_2}{\partial s} \\ &\quad + \frac{\partial^2 u_n}{\partial x_2^2}\left(\frac{\partial x_2}{\partial s}\right)^2 + \frac{\partial u_n}{\partial x_1}\frac{\partial^2 x_1}{\partial s^2} + \frac{\partial u_n}{\partial x_2}\frac{\partial^2 x_2}{\partial s^2} = 0.\end{aligned} \tag{2.2.7}$$

On the other hand, let α be the angle between τ and the x_1-axis; then

$$dx_1 = \cos\alpha ds, \quad dx_2 = \sin\alpha ds, \quad \alpha|_{x=0} = 0.$$

Hence, at $x = 0$,

$$\begin{aligned}\frac{\partial x_1}{\partial s} &= \cos\alpha = 1, \\ \frac{\partial x_2}{\partial s} &= \sin\alpha = 0, \\ \frac{\partial^2 x_1}{\partial s^2} &= -\sin\alpha\frac{d\alpha}{ds}, \\ \frac{\partial^2 x_2}{\partial s^2} &= \cos\alpha\frac{d\alpha}{ds} = k(p).\end{aligned}$$

We then infer from (2.2.4–2.2.7) that at $x = 0$,

$$\frac{\partial u}{\partial x_1} = 0, \tag{2.2.8}$$

$$\frac{\partial^2 u_\tau}{\partial x_1^2} = -k(p)\frac{\partial u_\tau}{\partial x_2}, \tag{2.2.9}$$

$$\frac{\partial^2 u_n}{\partial x_1^2} = -k(p)\frac{\partial u_n}{\partial x_2}. \tag{2.2.10}$$

Thanks to the divergence-free condition, (2.2.8) yields that at $x = 0$,

$$\begin{aligned}\frac{\partial u_1}{\partial x_1} &= \frac{\partial u_\tau}{\partial x_1} - u\cdot\frac{\partial \tau}{\partial x_1} = 0, \\ \frac{\partial u_2}{\partial x_1} &= \frac{\partial u_n}{\partial x_1} - u\frac{\partial n}{\partial x_1} = 0, \\ \frac{\partial u_2}{\partial x_2} &= -\frac{\partial u_1}{\partial x_1} = 0.\end{aligned}$$

Therefore, the Taylor expansion of u at p is

$$\begin{cases} u_1(x_1, x_2) = cx_2 + a_{11}x_1^2 - 2b_{22}x_1x_2 + a_{22}x_2^2 + o(|x|^2), \\ u_2(x_1, x_2) = \qquad b_{11}x_1^2 - 2a_{11}x_1x_2 + b_{22}x_2^2 + o(|x|^2). \end{cases} \tag{2.2.11}$$

Furthermore, we observe that on ∂M,

$$\frac{\partial u_\tau}{\partial x_1} = \frac{\partial u}{\partial x_1} \cdot \tau + u \cdot \frac{\partial \tau}{\partial x_1},$$

and at $x = 0$,

$$\begin{aligned}
\frac{\partial^2 u_\tau}{\partial x_1^2} &= \frac{\partial^2 u}{\partial x_1^2} \cdot \tau + 2\frac{\partial u}{\partial x_1} \cdot \frac{\partial \tau}{\partial x_1} + u \cdot \frac{\partial^2 \tau}{\partial x_1^2}, \\
\frac{\partial^2 u_n}{\partial x_1^2} &= \frac{\partial^2 u}{\partial x_1^2} \cdot n + 2\frac{\partial u}{\partial x_1} \cdot \frac{\partial n}{\partial x_1} + u \cdot \frac{\partial^2 n}{\partial x_1^2}, \\
\frac{\partial \tau}{\partial x_1}|_{x=0} &= k(p)n, \\
\frac{\partial n}{\partial x_1} &= -k(p)\tau, \\
\frac{\partial^2 u}{\partial x_1^2} \cdot \tau &= \frac{\partial^2 u_1(0)}{\partial x_1^2}, \\
\frac{\partial^2 u}{\partial x_1^2} \cdot n &= \frac{\partial^2 u_2(0)}{\partial x_1^2}.
\end{aligned}$$

By (2.2.9) and (2.2.10) we derive that at $x = 0$,

$$\begin{aligned}
a_{11} &= \frac{1}{2}\frac{\partial^2 u_1(0)}{\partial x_1^2} \\
&= -\frac{1}{2}\left[k\frac{\partial u_\tau}{\partial x_2} + 2k\frac{\partial u_2}{\partial x_1} + u \cdot \frac{\partial^2 \tau}{\partial x_1^2}\right] \\
&= -\frac{1}{2}k(p)\frac{\partial u_1(0)}{\partial x_2} \\
&= -\frac{1}{2}k(p)c, \\
b_{11} &= \frac{1}{2}\frac{\partial^2 u_2(0)}{\partial x_1^2} \\
&= -\frac{1}{2}\left[k\frac{\partial u_n}{\partial x_2} - 2k\frac{\partial u_2}{\partial x_1} + u \cdot \frac{\partial^2 n}{\partial x_1^2}\right] \\
&= 0.
\end{aligned}$$

Finally set $a = b_{22}, b = a_{22}$. Then (2.2.3) follows from (2.2.11), and the proof is complete. □

THEOREM 2.2.5. *Let $u \in B_0^r(TM)(r \geq 2)$, and let $p \in \partial M$ be a ∂-saddle point of u. Then u has the following Taylor expansion at p ($x = 0$):*

$$\begin{cases} u_1(x_1, x_2) = -2ax_1x_2 + bx_2^2 + o(|x|^2), \\ u_2(x_1, x_2) = \qquad\qquad\quad ax_2^2 + o(|x|^2). \end{cases} \tag{2.2.12}$$

PROOF. Since $z_0 \in \partial M$ is a ∂-saddle point of u, the coefficient c in (2.2.3) is

$$c = \frac{\partial u_1(0)}{\partial x_2} = \frac{\partial u_\tau}{\partial n}|_{x=0} = 0.$$

Then (2.2.12) follows from (2.2.3). □

The following theorem characterizes the local orbit structure of a D-regular vector field $u \in B_0^r(TM)$ near a boundary.

THEOREM 2.2.6. *Let $u \in B_0^r(TM)(r \geq 2)$ be D-regular, let $p \in \partial M$ be a ∂-saddle point of u, and let the constants a and b be as given in (2.2.12). Then the following assertions hold true:*

(1) *The number of interior singular points of u is finite.*
(2) *No orbit of u connects to a ∂-regular point of u.*
(3) *There is exactly one orbit of u in $\overset{\circ}{M}$, connected to p, whose ω-limit (resp. α-limit) is p if $a < 0$ (resp. if $a > 0$), which is called a stable orbit (resp. an unstable orbit) connected to p.*
(4) *The orbit γ connected to p is transversal to ∂M at p. Moreover, let α be the angle between γ and ∂M at p; then*

$$\cot \alpha = \frac{b}{3a}. \tag{2.2.13}$$

REMARK 2.2.7. Theorem 2.2.6 shows that the behavior of a D-regular vector field $u \in B_0^r(TM)$ is very similar to that of a regular vector field in $D^r(TM)$. This explains why a non-degenerate ∂-singular point of u is called ∂-saddle point. Moreover, Lemma 2.2.3 and Theorem 2.2.6 tell us that the number of ∂-saddle points of a D-regular vector field is even, and half of the ∂-saddle points are connected by stable orbits, the other half by unstable orbits.

To prove Theorem 2.2.6, we introduce a local coordinate transformation that is area-preserving. Let $z_0 \in \partial M$, and let (x_1, x_2) be a canonical coordinate system at z_0. The boundary ∂M can be locally expressed at z_0 by

$$x_2 = f(x_1) \text{ with } f(0) = 0, f'(0) = 0. \tag{2.2.14}$$

We take the local coordinate transformation

$$\widetilde{x}_1 = x_1, \qquad \widetilde{x}_2 = x_2 - f(x_1), \tag{2.2.15}$$

where $f(x_1)$ is defined as in (2.2.14).

Obviously, the transformation (2.2.15) takes a neighborhood $B \subset M$ of z_0 to a domain

$$D \subset \mathbb{R}_+^2 = \{(\widetilde{x}_1, \widetilde{x}_2) \in \mathbb{R}^2 \quad | \quad \widetilde{x}_2 > 0\},$$

and maps the boundary part $B \cap \partial M$ to the $\widetilde{x}_1$-axis. We denote the transformation (2.2.15) by $\Phi : B \to D$.

Since the boundary ∂M is C^{r+1}, the transformation Φ induces an isomorphism $\Phi^* : C^r(TM) \to C^r(TD)$ by

$$\Phi^*(u) = \widetilde{u} = D\Phi \circ u \circ \Phi^{-1}. \tag{2.2.16}$$

It is known that u and $\Phi^*(u)$ are topologically equivalent, i.e., Φ takes orbits of u to orbits of $\Phi^*(u)$, preserving the orientation.

LEMMA 2.2.8. *The following assertions hold true:*

(1) *If $u \in C^r(TB)$ is divergence-free, then $\widetilde{u} = \Phi^*(u)$ is also divergence-free;*
(2) *If $u \in B_0^r(TM)$ have the Taylor expansion at z_0 as (2.2.3) in B, then $\widetilde{u} = \Phi^*(u)$ can be Taylor expanded at $\widetilde{x} = 0$ as follows:*

$$\widetilde{u} = \begin{cases} c\widetilde{x}_2 + \widetilde{x}_2 g_1(\widetilde{x}_1, \widetilde{x}_2), \\ \qquad \widetilde{x}_2 g_2(\widetilde{x}_1, \widetilde{x}_2), \end{cases} \tag{2.2.17}$$

where

$$\lim_{\widetilde{x}\to 0} g_i(\widetilde{x}) = 0 \quad (i = 1, 2);$$

(3) *If z_0 is a ∂-saddle point of $u \in B_0^r(TM)$, then $\widetilde{u} = \Phi^*(u)$ has the Taylor expansion at $\widetilde{x} = 0$ as follows:*

$$\widetilde{u} = \begin{cases} -2a\widetilde{x}_1\widetilde{x}_2 + b\widetilde{x}_2^2 + \widetilde{x}_2 h_1(\widetilde{x}_1, \widetilde{x}_2), \\ a\widetilde{x}_2^2 + \widetilde{x}_2 h_2(\widetilde{x}_1, \widetilde{x}_2), \end{cases} \tag{2.2.18}$$

where

$$h_i(\widetilde{x}) = o(|\widetilde{x}|) \quad (i = 1, 2).$$

PROOF. First, we infer from Definition (2.2.16) that

$$\widetilde{u}(\widetilde{x}) = \begin{pmatrix} 1 & 0 \\ -f'(\widetilde{x}_1) & 1 \end{pmatrix} \begin{pmatrix} u_1 \\ u_2 \end{pmatrix} \circ \Phi^{-1}(\widetilde{x}) = \begin{pmatrix} u_1 \\ u_2 - f'u_1 \end{pmatrix} \circ \Phi^{-1}(\widetilde{x}). \tag{2.2.19}$$

By (2.2.15) and (2.2.19), we get

$$\begin{aligned} \frac{\partial \widetilde{u}_1}{\partial \widetilde{x}_1} &= \frac{\partial \widetilde{u}_1}{\partial x_1} \cdot \frac{\partial x_1}{\partial \widetilde{x}_1} + \frac{\partial \widetilde{u}_1}{\partial x_2}\frac{\partial x_2}{\partial \widetilde{x}_1} \\ &= \frac{\partial u_1}{\partial x_1} + \frac{\partial u_1}{\partial x_2} f'(x_1), \\ \frac{\partial \widetilde{u}_2}{\partial \widetilde{x}_2} &= \frac{\partial \widetilde{u}_2}{\partial x_1}\frac{\partial x_1}{\partial \widetilde{x}_2} + \frac{\partial \widetilde{u}_2}{\partial x_2}\frac{\partial x_2}{\partial \widetilde{x}_2} \\ &= \frac{\partial}{\partial x_2}[u_2 - f'u_1] = \frac{\partial u_2}{\partial x_2} - \frac{\partial u_1}{\partial x_2} f'(x_1). \end{aligned}$$

Hence, we have

$$\operatorname{div} \widetilde{u} = \frac{\partial \widetilde{u}_1}{\partial \widetilde{x}_1} + \frac{\partial \widetilde{u}_2}{\partial \widetilde{x}_2} = \frac{\partial u_1}{\partial x_1} + \frac{\partial u_2}{\partial x_2} = 0.$$

Substituting (2.2.15) into (2.2.3) and by (2.2.14) and noticing that $f(\widetilde{x}_1) = o(|\widetilde{x}_1|)$, we have

$$\begin{aligned} \widetilde{u} &= \begin{pmatrix} u_1 \\ u_2 - f'u_1 \end{pmatrix} \circ \Phi^{-1}(\widetilde{x}) \\ &= \begin{pmatrix} c(\widetilde{x}_2 + f(\widetilde{x}_1)) + o(|\widetilde{x}|) \\ o(|\widetilde{x}|) - c(\widetilde{x}_2 + f(\widetilde{x}_1)) f'(\widetilde{x}_1) + o(|\widetilde{x}|) \end{pmatrix} \\ &= \begin{pmatrix} c\widetilde{x}_2 + \widetilde{g}_1(\widetilde{x}_1, \widetilde{x}_2) \\ \widetilde{g}_2(\widetilde{x}_1, \widetilde{x}_2) \end{pmatrix}, \end{aligned}$$

where

$$\widetilde{g}_i(\widetilde{x}_1, \widetilde{x}_2) = o(|\widetilde{x}|) \quad (i = 1, 2).$$

Since $\widetilde{u}(\widetilde{x}_1, 0) = 0$ (by $u|_{\partial M} = 0$), we get that

$$\widetilde{g}_i(\widetilde{x}_1, \widetilde{x}_2) = \widetilde{x}_2 g_i(\widetilde{x}_1, \widetilde{x}_2) \quad (i = 1, 2).$$

Then (2.2.17) follows, and (2.2.18) can be derived in the same fashion from (2.2.12). The proof is complete. □

PROOF OF THEOREM 2.2.6. If u has an infinite number of singular points $\{z_n\} \subset \overset{\circ}{M}$, as u regular in $\overset{\circ}{M}$, $\{z_n\}$ must converge to some points on ∂M. Hence if we can prove that, for any point $z \in \partial M$, there is a neighborhood $B \subset M$ of z such that u has no singular point in $\overset{\circ}{B}$, then Assertion 1 follows.

For $z \in \partial M$, by Lemma 2.2.8, we assume, without loss of generality, that the neighborhood of z on ∂M is a straight line, and the x_1-axis of the canonical coordinate system (x_1, x_2) at z is on the straight line.

If $z \in \partial M$ is a ∂-regular point of u, then, in a neighborhood B of z in M, u has the expansion as in (2.2.17). Namely, $u = x_2 v$ in B, where

$$v = \begin{cases} c + g_1(x), \\ \qquad g_2(x). \end{cases} \tag{2.2.20}$$

Here $c \neq 0$, and $g_i(0) = 0, i = 1, 2$. Obviously, the singular points of u in $\overset{\circ}{M}$ are the same as those of v, and by (2.2.20), v has no singular points in an interior neighborhood $\mathcal{O} \subset \overset{\circ}{B}$ of z; therefore, u has no singular points in $\mathcal{O}$.

If $z \in \partial M$ is a ∂-saddle point of u, then by (2.2.18), $u = x_2 v$ in B, where

$$v = \begin{cases} -2ax_1 + bx_2 + h_1(x), \\ \qquad\qquad ax_2 + h_2(x). \end{cases} \tag{2.2.21}$$

Here $a \neq 0$, and $h_i(x) = o(|x|)(i = 1, 2)$. From (2.2.21) we see that $x = 0$ $(z \in \partial M)$ is a non-degenerate saddle point of v; hence $u = x_2 v$ has no singular point in an interior neighborhood of z in M.

It is known that for any function $g(x)$ in a domain $D \subset R^2$ with $g(x) \neq 0, x \in D$, the orbits of a vector field v defined in D coincide with those of $g \cdot v$. Assertions 2 and 3 then follow from $u = x_2 v$, (2.2.20) and (2.2.21).

Finally, let γ be the orbit of u connected to a ∂-saddle point $z_0 \in \partial M$ of u. Then γ can be locally expressed at z_0 by

$$x_1 = F(x_2),$$

and let

$$k = \cot\, \alpha = \lim_{x_2 \to 0} \frac{F(x_2)}{x_2}, \qquad 0 < \alpha < \pi.$$

Then

$$\lim_{x_2 \to 0} \frac{F(x_2)}{x_2} = \lim_{x_2 \to 0} \frac{u_1(F(x_2), x_2)}{u_2(F(x_2), x_2)}.$$

We then infer from (2.2.12) that

$$k = \frac{-2ak + b}{a}. \tag{2.2.22}$$

Therefore, Assertion 4 follows from (2.2.22), and the proof of Theorem 2.2.6 is complete. □

2.2.2. Structural stability theorem. In this section, we shall prove the following structural stability theorem for divergence-free vector fields with the Dirichlet boundary conditions.

Thanks to the characterization given in the previous sections, we generalize Theorem 2.1.2 in this section to divergence-free vector fields with the Dirichlet boundary conditions.

THEOREM 2.2.9. *Let $u \in B_0^r(TM)(r \geq 2)$. Then u is structurally stable in $B_0^r(TM)$ if and only if*

(1) *u is D-regular;*
(2) *all interior saddle points of u are self-connected; and*

(3) *each ∂-saddle point of u on ∂M is connected to a ∂-saddle point on the same connected component of ∂M.*

Moreover, the set of all structurally stable vector fields is open and dense in $B_0^r(TM)$.

PROOF. *Step 1.* Let $u \in B_0^r(TM)$ satisfy Conditions (1)–(3) above. Then u is structurally stable in $B_0^r(TM)$.

In the proof of Lemma 2.2.3, we can see that the number of the ∂-singular points of u is invariant under small perturbations in $B_0^r(TM)$, and the ∂-singular points remain nondegenerate.

The proof of Theorem 2.2.6 shows that there is a neighborhood $U \subset B_0^r(TM)$ of u such that for any $v \in U, v$ is D-regular, and the number of interior singular points of v equals that of u. Then with the same arguments as in the proof of Theorem 2.1.2, we can prove that u is structurally stable.

Step 2. If u does not satisfy either Conditions (2) or (3), we can prove, in the same fashion as in the proof of Theorem 2.1.3, that u is structurally unstable. Hence it suffices to prove that if u is not D-regular, then u is unstable. We proceed only for the case of degenerate ∂-singular points.

Let $z_0 \in \partial M$ be a degenerate ∂-singular point of u. If there are $m \geq 0$ $(m \neq 1)$ orbits connected to z_0, we take a vector field $v \in B_0^r(TM)$ such that $z_0 \in \partial M$ be a non-degenerate ∂-singular point. Obviously, for any $\lambda \neq 0$ sufficiently small, $z_0 \in \partial M$ is a ∂-saddle point of $u + \lambda v$. By Theorem 2.2.6, $u + \lambda v$ has only one orbit connected to z_0. It implies that $u + \lambda v$ and u are not locally topologically equivalent. Therefore, u is unstable. If there is one orbit γ of u connected to z_0, and γ is stable (resp. unstable) connected to z_0, then we take $v \in B_0^r(TM)$ such that $z_0 \in \partial M$ is a ∂-saddle point of v, and the orbit of v connected to z_0 is unstable (resp. stable). Thus, for any $\lambda > 0$ sufficiently small, the orbit of $u + \lambda v$ connected to z_0 is unstable (resp. stable), which means that $u + \lambda v$ and u are not locally topologically equivalent; therefore, u is unstable.

Step 3. The openness and density of all structurally stable vector fields in $B_0^r(TM)$ follow from the same argument as in the proof of Theorem 2.1.2.

The proof is complete. □

2.2.3. Non-homogeneous boundary condition case. In this subsection, we study structural stability of vector fields in $B_\phi^r(TM)$. Obviously, $B_\phi^r(TM)$ is not a linear space, and for any given vector field $u_0 \in B_\phi^r(TM)$,

$$B_\phi^r(TM) = \{u_0 + v | v \in B_0^r(TM)\}.$$

By definition, the given boundary value $\phi \in C_n^r(TM)$ satisfies the no-normal flow condition $\phi \cdot n|_{\partial M} = 0$. We set

$$\begin{aligned}
L_1 &= \left\{ x \in \partial M \quad | \quad |\phi(x)| + |\frac{\partial \phi_\tau(x)}{\partial \tau}| \neq 0 \right\}, \\
L &= \left\{ x \in \partial M \quad | \quad |\phi(x)| + |\frac{\partial \phi_\tau(x)}{\partial \tau}| = 0 \right\}, \\
L_2 &= \{x \in L \quad | \quad x \text{ is an isolated singular point of } \phi\}, \\
L_3 &= L - L_2,
\end{aligned}$$

where ϕ_τ is the tangential component of ϕ. Obviously,

$$L_i \cap L_j = \emptyset \qquad \forall i \neq j,\ 1 \leq i,\ j \leq 3,$$
$$\partial M = L_1 + L_2 + L_3.$$

Hereafter, we always assume that the given boundary vector field ϕ satisfies the following basic assumption.

BASIC ASSUMPTION. *The set $L_2 \subset \partial M$ contains only a finite number of points, and $L_3 \subset \partial M$ consists of a finite number of closed arc segments on ∂M.*

DEFINITION 2.2.10. *Let $z_0 \in L_2$, and let (x_1, x_2) be the canonical coordinate system at z_0.*

(1) *z_0 is called an odd singular point of ϕ, if*

$$\text{sign } \phi_\tau(x_1, x_2) = -\text{sign } \phi_\tau(-x_1, x_2')$$

for $(x_1, x_2), (-x_1, x_2') \in \partial M$ and $x_1 \neq 0$ sufficiently small.

(2) *z_0 is called an even singular point of ϕ, if*

$$\text{sign } \phi_\tau(x_1, x_2) = \text{sign } \phi_\tau(-x_1, x_2').$$

REMARK 2.2.11. If $z_0 \in L_2 \subset \partial M$ is an odd singular point (resp. even singular point) of ϕ, then for any $u \in B^r_\phi(TM)$ with z_0 being an isolated singular point of u in M, there must be odd number of orbits (resp. even number of orbits) of u connected to z_0. □

DEFINITION 2.2.12. *A vector field $u \in B^r_\phi(TM)$ $(r \geq 2)$ is called ∂_ϕ-regular on the boundary, if*

(1) *all ∂-singular points of u on $L_3 \subset \partial M$ are nondegenerate,*
(2) *when $z_0 \in L_2$ is an even singular point of ϕ, then*

$$\frac{\partial u_\tau(z_0)}{\partial n} = c \neq 0, \tag{2.2.23}$$

(3) *when $z_0 \in L_2$ is an odd singular point of ϕ, then either (2.2.23) holds, or*

$$\frac{\partial^2 u_n(z_0)}{\partial n^2} = a \begin{cases} > 0 & \text{if sign } \phi_\tau(x_1, x_2) = -1, \\ < 0 & \text{if sign } \phi_\tau(x_1, x_2) = 1, \end{cases} \tag{2.2.24}$$

where $(x_1, x_2) \in \partial M$ with $x_1 > 0$ sufficiently small.

For $u \in B^r_\phi(TM)$, each boundary point $z \in L_1$ is a singular point of ϕ; each $z \in L_3$ is a nondegenerate ∂-singular point of u; and each $z \in L_2$ that satisfies (2.2.23) and (2.2.24) is called a ∂_ϕ–saddle point of u.

DEFINITION 2.2.13. *$u \in B^r_\phi(TM)(r \geq 2)$ is called D_ϕ-regular if u is regular in $\overset{\circ}{M}$ and ∂_ϕ-regular on ∂M.*

LEMMA 2.2.14. *Let $u \in B^r_\phi(TM)(r \geq 2)$ be D_ϕ-regular. Then the following assertions hold true.*

(1) *all ∂_ϕ-saddle points of u on ∂M are isolated in M;*
(2) *the number of interior singular points of u is finite;*
(3) *for an odd singular point $z_0 \in L_2$ of ϕ, there is only one orbit of u connected to z_0;*

(4) *for an even singular point* $z_0 \in L_2$ *of* ϕ, *if for any* $x_1 > 0$ *small,*

$$\frac{\partial u_\tau(z_0)}{\partial n} \begin{cases} > 0 & \text{if sign } \phi_\tau(x_1, x_2) = 1, \\ < 0 & \text{if sign } \phi_\tau(x_1, x_2) = -1, \end{cases} \tag{2.2.25}$$

then either there is no orbit of u *connected to* z_0, *or there are exactly two orbits of* u *connected to* z_0, *one of which is stable and the other one is unstable.*

PROOF. *Step 1.* Assertion (1) follows from Assertion (2). By Lemma 2.2.8, for $z_0 \in \partial M$, we may assume that the neighborhood of z_0 in ∂M is a straight line, and the x_1–axis is on the straight line.

Step 2. If u has an infinite number of singular points $\{x_n\} \subset \overset{\circ}{M}$, then $\{x_n\}$ must converge to points on L_2. Let $x_n \to z_0 \in L_2$ and let z_0 satisfy (2.2.23). Then u has the following Taylor expansion at z_0:

$$u = (cx_2 + \phi_\tau(x_1) + x_2 f_1(x_1, x_2), \ -x_2\phi'_\tau(x_1) + x_2^2 f_2(x_1, x_2)), \tag{2.2.26}$$

where $c \neq 0$ is given as in (2.2.23), $f_1(0) = 0$, and ϕ is the given boundary condition field, which satisfies that

$$\begin{aligned} &\phi_\tau(x_1, x_2) \neq 0 \ \text{ if } \ x_1 \neq 0, \\ &\phi_\tau(0) = \phi'_\tau(0) = 0. \end{aligned}$$

Let $x_n = \{x_1^n, x_2^n\}$. Then by $u(x_n) = 0$, we have

$$-\phi'_\tau(x_1^n) - \frac{f_2(0)}{c}\phi_\tau(x_1^n) + o(\phi_\tau(x_1^n)) = 0,$$

as $x_1^n \to 0 (n \to \infty)$, a contradiction to the following:

$$\frac{\phi_\tau(x)}{\phi'_\tau(x)} \to 0, \qquad \text{as } \phi'(x) \neq 0, \ x \neq 0, \ x \to 0.$$

If $z_0 \in L_2$ satisfies $\frac{\partial u_\tau(z_0)}{\partial n} = 0$ and (2.2.24), then

$$u = \begin{cases} -2ax_1x_2 + bx_2^2 + \phi(x_1) + x_2 f_1(x_1, x_2), \\ ax_2^2 - x_2\phi'(x_1) + x_2^2 f_2(x_1, x_2), \end{cases} \tag{2.2.27}$$

where $a \neq 0, f_2(0) = 0$, and for $x \neq 0$,

$$\phi'(x) \begin{cases} > 0 & \text{if } a < 0, \\ < 0 & \text{if } a > 0. \end{cases} \tag{2.2.28}$$

By $u(x_n) = 0$ we have

$$\begin{aligned} &(a + f_2(x_n))x_2^n - \phi'(x_1^n) = 0, \\ &x_2^n > 0, \\ &\lim_{n \to \infty} f_2(x_n) \to 0, \end{aligned}$$

contrary to (2.2.28). Hence Assertion (2) is proved.

Step 3. If there are $2n (n \geq 1)$ orbits of u connecting to $z_0 \in L_2$, then there must be at least two curves Γ_1 and Γ_2 (resp. Γ'_1 and Γ'_2) connecting to z_0 such that $u_1(x) = 0$ on Γ_1 and Γ_2 (resp. $u_2(x) = 0$ on Γ'_1 and Γ'_2), and the angle between Γ_1 and Γ_2 (resp. between Γ'_1 and Γ'_2) at z_0 satisfies $0 \leq \alpha < \pi$.

When z_0 satisfies (2.2.23), by (2.2.26), $\frac{\partial u_1(0)}{\partial x_2} = c \neq 0$, and by the implicit function theorem, there is only one curve Γ passing through z_0 such that $u_1(x) = 0$ on Γ, a contradiction. When z_0 satisfies (2.2.24), by (2.2.27), $u_2 = x_2 v_2$, and $\frac{\partial v_2(0)}{\partial x_2} = a \neq 0$; hence, there is only one curve Γ passing through z_0 such that $u_2(x) = 0$ on Γ, a contradiction. The proof of Assertion (3) is complete.

Step 4. If z_0 satisfies (2.2.25), by (2.2.26), one can follow that $u_1(x) \neq 0 (x \neq 0)$ in a neighborhood of z_0 in $M(x_2 \geq 0)$; hence, there is no orbit of u connecting to z_0. If otherwise,

$$\text{sign } c = -\text{sign } \phi(x_1), \quad x_1 \neq 0,$$

by (2.2.26) there is exactly one curve Γ passing through z_0 (or, equivalently, there are exactly two curves Γ_1 and Γ_2 connecting to z_0, and the angle between Γ_1 and Γ_2 at z_0 equals π, i.e., $\Gamma = \Gamma_1 + \Gamma_2$), such that $u_1(x) = 0$ on Γ. It implies that there are exactly two orbits of u connected to z_0.

The proof of Lemma 2.2.14 is complete. □

REMARK 2.2.15. In the proof of Claim 3 of Lemma 2.2.14, we have used the fact that if an odd singular point z_0 satisfies (2.2.24) and $\partial u_\tau(z_0)/\partial n = 0$, then there is only one orbit connecting to z_0 which is transversal to the x_1-axis; moreover, the orbit is stable as $a < 0$, and is unstable as $a > 0$. □

With the same proof as that of Lemma 2.2.14, we obtain

LEMMA 2.2.16. *Let $z_0 \in L_2$ be an odd singular point of $u \in B^r_\phi(TM)(r \geq 2)$, and let z_0 satisfy that for $x_1 > 0$ small,*

$$\frac{\partial u_\tau(z_0)}{\partial n} = 0; \quad \frac{\partial^2 u_n(z_0)}{\partial n^2} \begin{cases} < 0 & \text{if sign } \phi_\tau(x_1, x_2) = -1, \\ > 0 & \text{if sign } \phi_\tau(x_1, x_2) = 1. \end{cases}$$

Then there are exactly three orbits of u connected to z_0.

We are now in a position to state and prove the structural stability theorem for divergence-free vector fields with non–homogeneous Dirichlet boundary conditions.

THEOREM 2.2.17. *Let $u \in B^r_\phi(TM)(r \geq 2)$. Then u is structurally stable in $B^r_\phi(TM)$ if and only if*

1. *u is D_ϕ-regular;*
2. *all interior saddle points of u are self-connected; and*
3. *each ∂_ϕ-saddle point of u on ∂M is connected to a ∂_ϕ-saddle point on the same connected component of ∂M.*

Moreover, the set of all structurally stable vector fields is open and dense in $B^r_\phi(TM)$.

PROOF. First, for a neighborhood $U \subset B^r_\phi(TM)$ of u,

$$U = \{u + v | v \in O \subset B^r_0(TM)\},$$

where $O \subset B^r_0(TM)$ is a neighborhood of $v = 0$.

In the same fashion as in the proof of Theorem 1.1.3, it suffices to prove the stability of orbits of u connected to singular points on L_2.

Let $z_0 \in L_2$ be an even singular point of ϕ. For any $v \in B^r_0(TM)$ with $\|v\|_{C^r}$ sufficiently small, the real number c_1 in (2.2.3) is sufficiently small. Therefore,

$$\frac{\partial(u_\tau(z_0) + v_\tau(z_0))}{\partial n} = c + c_1 \neq 0, \text{ and } \text{sign}(c + c_1) = \text{sign } c.$$

By Lemma 2.2.14, the orbits of $u+v$ connected to z_0 are topologically the same as the orbits of u connected to z_0.

Let $z_0 \in L_2$ be an odd singular point of ϕ. If z_0 satisfies (2.2.23), we can prove in the same fashion that the orbit of u connected to z_0 is stable. If z_0 satisfies (2.2.24) and $\frac{\partial u_\tau(z_0)}{\partial n} = 0$, then for any $v \in B_0^r(TM)$ with $\|v\|_{C^r}$ sufficiently small, when $\frac{\partial v_\tau(z_0)}{\partial n} \neq 0$, then $u+v$ at z_0 satisfies (2.2.23), and when $\frac{\partial v_\tau(z_0)}{\partial n} = 0$, then $u+v$ at z_0 satisfies (2.2.24). Then, by Lemma 1.1.6, the orbit of u connected to z_0 is stable.

Second, by Lemmas 1.1.6 and 1.1.7, it is easy to see that if $z_0 \in L_2$ is not a ∂_ϕ–saddle point, then the orbits of u connected to z_0 are unstable; i.e., for any $\varepsilon > 0$ sufficiently small, there is a $v \in B_0^r(TM)$ with $\|v\|_{C^r} < \varepsilon$ such that the orbits of $u+v$ connected to z_0 are topologically different from those of u.

The proof of the genericity is the same as that of Theorem 2.2.9.

The proof of Theorem 2.2.17 is complete. □

2.3. Two Dimensional Hamiltonian Structural Stability

Let M be a two-dimensional compact orientable Riemannian manifold, let $\{g_{ij}\}$ be the Riemannian metric under a local coordinate system on M and let $g = \det(g_{ij})$. There is a global two-form on M defined by

$$\omega = \sqrt{g}dx_1 \wedge dx_2.$$

Let $J : T^*M \to TM$ be the symplectic isomorphism induced by the symplectic form ω. A vector field v is called a Hamiltonian vector field if $v = J\nabla H$ for some Hamitonian function H defined on M.

Let $\mathcal{H}^r(TM)$ be the set of all C^r Hamiltonian vector fields on M, i.e.,

$$\mathcal{H}^r(TM) = \{v \in C^r(TM) \mid v \text{ is Hamiltonian }\}.$$

2.3.1. Connection Lemma I.

LEMMA 2.3.1 (Connection Lemma I). *Let $M \subset \mathbb{R}^2$ be a C^r manifold, let $u, v \in D^r(TM)$ with v sufficiently small, and let $L \subset M$ be an extended orbit of u starting at p. Then an extended orbit γ of $u+v$ starting at p passes through a point $q \in L$ if and only if*

$$\int_{L[p,q]} v \times d\ell = 0, \tag{2.3.1}$$

where $L[p,q]$ is the curve segment on L from p to q.

PROOF. We divide the proof into two steps.

Step 1. We first discuss the geometric meaning of condition (2.3.1). By the Green formula, we know that for a curve Γ with end points p and q, $v \in D^r(TM)$ satisfies

$$\int_\Gamma v \times d\ell = \int_\Gamma v \cdot n\, ds = 0$$

if and only if there is an orbit line Γ_1 of v passing through p and q, i.e., $\Gamma[p,q] \cup \Gamma_1[p,q]$ consists of simple closed curves; see Figure 2.3.1.

Hence, it suffices to prove that the orbit line γ of $u+v$ starting at p will pass through $q \in L$ if and only if the orbit line L_1 of v starting at p will connect to q.

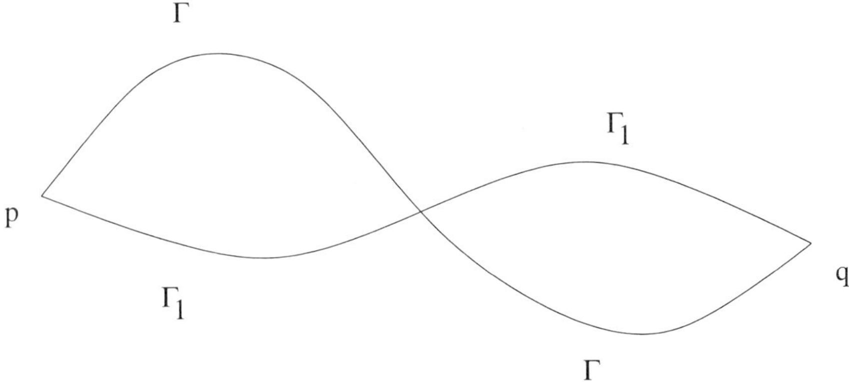

FIGURE 2.3.1. Schematic illustrating (2.3.1).

Step 2. Because $M \subset \mathbb{R}^2$, there exist two Hamiltonian functions $H_1, H_2 \in C^1(M)$ such that

$$\begin{aligned} u &= J\nabla H_1 = \left(\frac{\partial H_1}{\partial x_2}, -\frac{\partial H_1}{\partial x_1}\right), \\ v &= J\nabla H_2 = \left(\frac{\partial H_2}{\partial x_2}, -\frac{\partial H_2}{\partial x_1}\right). \end{aligned}$$

Namely, the level-set lines of H_1 and H_2 are respectively the orbit lines of u and v. Hence, the orbit line v of $u+v$ starting at p will pass through $q \in L$ if and only if p and q are on the same level-set line of $H_1 + H_2$, i.e., L and L_1 intersect at p and q. The proof is complete. □

Likewise, for the Hamiltonian vector fields on a general orientable two-dimensional manifold M, the connection lemma is also valid.

LEMMA 2.3.2. *Let $u, v \in H^r(TM)$ with v sufficiently small, and let L be the extended orbit of u starting at p. Then the extended orbit γ of $u+v$ starting at p will connect to a point $q \in L$ if and only if (2.3.1) holds true.*

2.3.2. Connection Lemma II. We now introduce a connection lemma, which is crucial in classifying the structure of stable Hamiltonian structures of vector fields in $\mathcal{H}^r(TM)$.

Let $u, v \in \mathcal{H}^r(TM)$ $(r \geq 1)$, and let x_0 and $y_0 \in M$ be two nondegenerate saddle points of u. Let $t > 0$ be sufficiently small, and let the vector field $u + tv$ have the singular points $x(t)$ and $y(t)$ with

$$\lim_{t\to 0} x(t) = x_0, \quad \lim_{t\to 0} y(t) = y_0. \tag{2.3.2}$$

LEMMA 2.3.3 (Connection Lemma II). *Let $h \in C^{r+1}(M)$ be the Hamiltonian function of v. If $h(x_0) \neq h(y_0)$, then there exists a number $t_0 > 0$ such that for any $0 < t < t_0$, the orbit of $u + tv$ starting at $x(t)$ will not connect to $y(t)$.*

PROOF. Let $H(x)$ be the Hamiltonian function of u. If the orbit of $u + tv$ starting at $x(t)$ connects to $y(t)$, then the Hamiltonian function $H + th$ must have the same values at $x(t)$ and $y(t)$, namely

$$H(x(t)) - H(y(t)) = t(h(y(t)) - h(x(t))). \tag{2.3.3}$$

We make the Taylor expansions near x_0 and y_0

$$\begin{aligned}
H(x(t)) &= H(x_0) + D^2H(x_0)(x-x_0)^2 + o(|x-x_0|^2),\\
H(y(t)) &= H(y_0) + D^2H(y_0)(y-y_0)^2 + o(|y-y_0|^2),\\
h(x(t)) &= h(x_0) + Dh(x_0)(x-x_0) + o(|x-x_0|),\\
h(y(t)) &= h(y_0) + Dh(y_0)(y-y_0) + o(|y-y_0|).
\end{aligned}$$

If $H(x_0) \neq H(y_0)$, then the lemma follows directly from (2.3.3). Hence, we assume that $H(x_0) = H(y_0)$. Thus, it follows from (2.3.3) that

$$\begin{aligned}
t(h(x_0) - h(y_0)) + o(t) =& D^2H(x_0)(x-x_0)^2\\
&- D^2H(y_0)(y-y_0)^2 + o(|x-x_0|^2, |y-y_0|^2).
\end{aligned} \tag{2.3.4}$$

On the other hand, since $x(t)$, $y(t)$ are the singular points of $u + tv$, we have

$$\begin{cases}
\dfrac{\partial H(x(t))}{\partial x_i} = -t\dfrac{\partial h(x(t))}{\partial x_i},\\
\dfrac{\partial H(y(t))}{\partial y_i} = -t\dfrac{\partial h(y(t))}{\partial y_i},
\end{cases} \tag{2.3.5}$$

for $i = 1, 2$. From the Taylor expansions above and (2.3.5), we derive that

$$\begin{cases}
D^2H(x_0)(x-x_0) = -\frac{1}{2}tDh(x_0) + o(t) + o(|x-x_0|),\\
D^2H(y_0)(y-y_0) = -\frac{1}{2}tDh(y_0) + o(t) + o(|y-y_0|).
\end{cases} \tag{2.3.6}$$

By assumption, x_0 and y_0 are nondegenerate singular points of u; therefore, the Hessians $D^2H(x_0)$ and $D^2H(y_0)$ are nondegenerate. Hence, we infer from (2.3.6) that

$$\begin{cases}
x(t) - x_0 = \alpha t + o(t),\\
y(t) - y_0 = \beta t + o(t),
\end{cases} \tag{2.3.7}$$

where α, β are constants. If $h(x_0) \neq h(y_0)$, (2.3.1) and (2.3.7) lead to a contradiction. The proof is complete. □

REMARK 2.3.4. In Lemma 2.3.3, the condition of $h(x_0) \neq h(y_0)$ can be equivalently replaced by the following: for any line segment L with x_0 and y_0 as its end points, we have

$$\int_L v \times d\ell = \int_L v \cdot n ds \neq 0, \tag{2.3.8}$$

where n is the normal vector on L.

To see this, we note that

$$\int_L v \times d\ell = \int_L \operatorname{curl} h \times d\ell = \int_\ell \nabla h \cdot d\ell = h(x_0) - h(y_0).$$

2.3.3. Hamiltonian structural stability. We first introduce the well-known Liouville-Arnold invariant tori theorem; see [**1, 42, 71**].

DEFINITION 2.3.5. *A Hamiltonian H on M is completely integrable if there exists n functions $f_1 = H,\ f_2,\ \cdots, f_n \in C^\infty(M)$ such that*

1) (Poisson bracket free condition) $\{f_i, f_j\} = 0$ *for* $1 \leq i, j \leq n$, *and*
2) $df_1(x) \wedge \cdots \wedge df_n(x) \neq 0$ *at some point* $x \in M$.

THEOREM 2.3.6 (Liouville-Arnold Theorem). *Let $dH^\# \in \mathcal{H}(TM)$ be completely integrable, and*

$$M_z = \{x \in M \quad | \quad f_i(x) = z_i,\ Df_i(x) \text{ are linearly independent, } 1 \leq i \leq n\}.$$

Then

1) *M_z is an invariant manifold under the Hamiltonian flow Φ_H of H, where each connected component is diffeomorphic to the n-torus $\mathbb{T}^n$; and*
2) *Φ_H is conjugate to a linear flow on $\mathbb{T}^n$.*

When M is a two-dimensional symplectic compact manifold, M_z is given by

$$M_z = \{x \in M | H(x) = z, DH \neq 0\},$$

which is diffeomorphic to the circle S^1. If the Hamiltonian vector field $dH^\# \in \mathcal{H}(TM)$ is regular, then dH has only a finite number of zeroes on M. Because the set of all regular Hamiltonian vector fields is open and dense in $\mathcal{H}(TM)$, we immediately obtain the following structural classification theorem for a regular Hamiltonian vector field.

THEOREM 2.3.7. *Let M be a two-dimensional symplectic compact manifold and let $dH^\# \in \mathcal{H}(TM)$ be regular. Then the topological set of orbits of H consists of finite connected components, which are of the following types:*

1) *circle cells and circle bands, and*
2) *saddle connections.*

Moreover, due to the non-existence of ergodic sets, with the same proof as that of Theorem 2.1.2, we have the following structural stability theorem for Hamiltonian vector fields on *general* two-dimensional compact manifolds.

THEOREM 2.3.8. *Let M be a two-dimensional compact symplectic manifold and $dH^\# \in \mathcal{H}(TM)$. Then the Hamiltonian H is structurally stable under perturbations of Hamiltonian fields if and only if*

1) *$dH^\#$ is regular, and*
2) *all saddle points are self-connected.*

Moreover, the set of structurally stable Hamiltonian vector fields is open and dense in $\mathcal{H}(TM)$.

PROOF. It is clear that if $dH^\#$ is not a self-connection, then $dH^\#$ must be structurally unstable. It suffices then to consider self-connected Hamiltonian vector fields.

Let $p \in M$ be a saddle point of $dH^\#$, and let Γ be a saddle connection. Then Γ is a simple closed curve. If we can prove that, under a Hamiltonian perturbation $dh^\#$, the saddle self-connection Γ' of p' of $dH^\# + dh^\#$ is homological to Γ and Γ' can be deformed into Γ on M, then the saddle connection diagram of $dH^\#$ is isomorphic to that of $dH^\# + dh^\#$. Therefore, as in the proof of Theorem 1.4.6, we infer from Theorem 2.3.8 and Lemma 1.4.10 that $dH^\#$ and $dH^\# + dh^\#$ are topologically equivalent.

By Lemma 2.3.2, it suffices then to prove that

$$\int_\Gamma dh^\# \times dl = 0. \tag{2.3.9}$$

By definition,

$$dh^{\#} = \text{curl } h = \left(\frac{\partial h}{\partial x_2}, -\frac{\partial h}{\partial x_1}\right),$$
$$dl = (dx_1, dx_2).$$

Hence,

$$dh^{\#} \times dl = dh.$$

Then (2.3.9) follows from the Stokes formula.

The proof is complete. □

REMARK 2.3.9. The above stability result was obtained for general two-dimensional compact symplectic manifolds. We prove in the next chapter, however, that when M is a two-dimensional orientable manifold with nonzero genus (e.g., a torus), no C^1 divergence-free vector field is structurally stable under either the perturbations of general C^1 vector fields or C^1 divergence-free vector fields. In view of these results, it is easy to see that the instability of a divergence-free vector field on a general two-dimensional manifold is due to the presence of ergodic sets caused by the harmonic perturbations.

LEMMA 2.3.10. *Let $u \in \mathcal{H}^r(TM)$ $(r \geq 1)$ be a regular vector field, and let $\{x_i \mid 1 \leq i \leq n\}$ be all saddle points of u. If $v = curl\ h \in \mathcal{H}^r(TM)$ satisfies $h(x_i) \neq h(x_j)$ $\forall i \neq j$, then there is a $t_0 > 0$ such that for all $0 < t < t_0$, the vector fields $u + tv$ are Hamiltonian structurally stable.*

PROOF. Since u is regular, for $t > 0$ sufficiently small, the vector fields $u + tv$ are regular. By Lemma 2.3.1, we can deduce that all saddle points of $u + tv$ are self-connected. It follows from Theorem 2.3.8 that the vector fields $u + tv$ are Hamiltonian structurally stable. The proof is complete. □

2.4. Block Structure of Hamiltonian Vector Fields

By the Hamiltonian structural classification and stability theorems, we see that each saddle point of a stable Hamiltonian vector field has exactly two saddle self-connections. Obviously, each saddle self-connection is a simple closed curve. We restrict ourselves in this section to the case where $M = \mathbb{T}^2$, and we interchangeably use M and $\mathbb{T}^2$.

DEFINITION 2.4.1. *Let v be a Hamiltonian vector field on a torus $\mathbb{T}^2$, and let $Q \subset \mathbb{T}^2$ be an invariant set of v.*

1. *Q is called a D-block if Q is homeomorphic to an open disk such that ∂Q is a saddle self-connection of a saddle point $p \in \partial Q$, and the other saddle self-connection of p, as a closed chain, is not homological to zero in $H_1(\mathbb{T}^2)$;*
2. *Q is called an S-block if Q is homeomorphic to an open disk such that ∂Q consists of two saddle self-connections of a saddle point $p \in \partial Q$;*
3. *Q is called a T-block if Q is a closed domain such that the interior $\mathring{Q}$ does not contain singular points of v, and $\mathbb{T}^2 - Q$ consists of D and S-blocks of v.*

In the above definition, the letters D, S and T stand for disk, sphere and torus respectively; see Figure 2.4.1 for schematic pictures of the D-, S- and T-blocks.

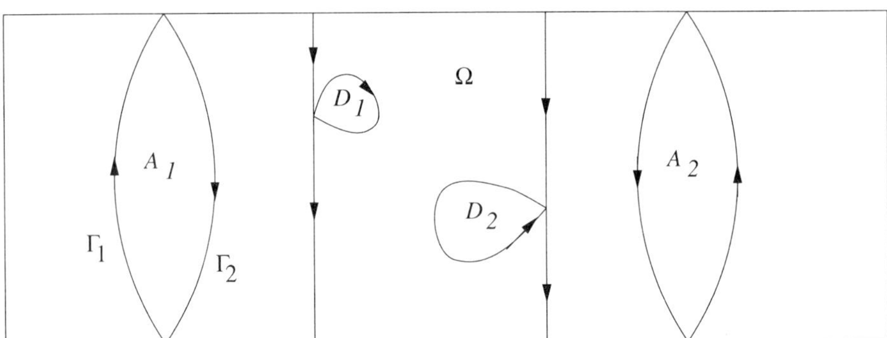

FIGURE 2.4.1. A_1 and A_2 are two S-blocks with left-hand and right-hand orientations respectively, D_1 and D_2 are D-blocks with left-hand and right-hand orientations, and Ω is a T-block.

DEFINITION 2.4.2. *Let $Q \subset M$ be a circle cell of $v \in D^r(TM)$. We say that Q has a right-hand orientation (resp. a left-hand orientation) if the interior of Q is on the left side of ∂Q (resp. on the right side of ∂Q), traveling in the direction of the orbits of v on ∂Q.*

The right-hand and left-hand orientations for the D and S-blocks of a stable Hamiltonian vector field v on $\mathbb{T}^2$ can be defined in the same fashion.

The Hamiltonian structural stability theorem tells us that a stable Hamiltonian vector field v on $\mathbb{T}^2$ has a block decomposition as follows:

$$\mathbb{T}^2 = \Omega + \cup_{i=0}^{I} D_i + \cup_{k=1}^{K} A_k \text{ with } D_0 = \emptyset. \tag{2.4.1}$$

Here Ω is a unique T-block, $D_i (0 \leq i \leq I)$ are D-blocks and $A_k (1 \leq k \leq K)$ are S-blocks of v, which are topologically equivalent to a block structure as shown in Figure 2.4.1.

The following theorem shows that each stable Hamiltonian vector field on $\mathbb{T}^2$ has exactly one T-block, at least two S-blocks and possibly zero D-blocks.

THEOREM 2.4.3. *A vector field $v \in \mathcal{H}^r(TM)$ $(r \geq 1)$ is Hamiltonian structurally stable if and only if*

(1) *v has a block decomposition as*
$$\mathbb{T}^2 = \Omega + \cup_{i=0}^{I} D_i + \cup_{k=1}^{k} A_k \quad (D_0 = \phi),$$
where Ω is a T-block, D_i are D-blocks and A_k the S-blocks, and all saddle points in D- and S-blocks are self-connected;

(2) *the T-block Ω of v is unique; and*

(3) *the number K of S-blocks of v is even, and $K = 2m \geq 2$ $(m \geq 1)$, half of which have right-hand orientation and the other half have left-hand orientation.*

PROOF. Notice that the boundary of each S-block A_k has two saddle self-connections, say Γ_1 and Γ_2 as shown in Figure 2.4.1, and the orientation of closed orbits in Ω near Γ_1 are reverse to the orientation of the closed orbits in Ω near Γ_2. Obviously, a D-block does not reverse the orientation of the orbits in Ω. Therefore, the number of S-blocks of v must be even, i.e., $K = 2m$, and m of these S-blocks have right-hand orientation, and the other m S-blocks have left-hand orientation.

Now we show that $m \geq 1$. If we assume otherwise, i.e., $m = 0$, then the T-block Ω of v is a compact manifold with boundary, which consists of closed orbits and saddle self-connections of v. Let $v = dH^{\#}$, where $H \in C^{r+1}(M)$ is a Hamiltonian function. Then all orbits in $\overset{\circ}{\Omega}$ are level lines, and there is an orbit $\gamma \subset \overset{\circ}{\Omega}$ such that H takes the maximum value on γ.

For any point $q \in \gamma$, we take an orthogonal coordinate system (x_1, x_2) on the tangent space $T_q(\mathbb{T}^2)$ with the origin at q and the x_1-axis tangent to γ at q. Since γ is a level line of H, and H takes the maximum value at $q \in \gamma$ along the x_2-axis, we have

$$\frac{\partial H(q)}{\partial x_1} = 0, \qquad \frac{\partial H(q)}{\partial x_2} = 0,$$

which implies that $v = 0$ on Γ, contrary to v being regular.

The proof is complete. □

COROLLARY 2.4.4. *Let $v \in \mathcal{H}^r(TM)$ be a regular Hamiltonian vector field on $M = \mathbb{T}^2$. Then v has at least four singular points, two of which are centers and the other two being saddle. If v has exactly four singular points, then v must be Hamiltonian structurally stable.*

PROOF. By the Poincaré–Hopf index theorem, the number of centers of v equals the number of saddle points. Obviously, by Theorem 2.4.3, a regular Hamiltonian has at least four singular points, and if v has exactly four singular points, then the structure of v can only be topologically equivalent to the Hamiltonian vector field that has the block decomposition as follows:

$$\mathbb{T}^2 = \Omega + A_1 + A_2,$$

where Ω and A_i are as in (2.4.1). It implies that v is Hamiltonian structurally stable. The proof is complete.

□

2.5. Local Structural Stability

The main objectives of this section are:

(1) to generalize the classical Hartman-Grobman theorem to divergence-free vector fields on both the two- and three-dimensional compact manifolds with or without boundary, and
(2) to study the local structural stability for divergence-free vector fields.

2.5.1. Hartman-Grobman theorem for divergence-free vector fields.

DEFINITION 2.5.1. *Let (φ, U) be a local coordinate of $p \in M$. We say that a vector field v is locally topologically equivalent to its linear part at p if the vector field $u = D\varphi \circ U \circ \varphi^{-1}$ is locally topologically equivalent to its linear part $J(u(0))x$, where $x \in \mathbb{R}^n$ and $\varphi(p) = 0$.*

Then we have the following theorem, which is a version of the classical Hartman-Grobman theorem for divergence-free vector fields. The classical Hartman-Grobman theorem amounts to saying that given a general (not necessarily divergence-free) vector field v and a *hyperbolic* singularity p of v, v is locally topologically equivalent to its linear part at p. Namely, the hyperbolic condition of p is crucial in the classical theorem. In our new theorem below, we do not assume hyperbolicity of p.

THEOREM 2.5.2. *Let M be a two- or three-dimensional manifold with or without boundary and let $v \in D^r(TM)(r \geq 1)$. If $p \in M$ is a non-degenerate singular point of v, then v is locally topologically equivalent to its linear part Dv at $p \in M$.*

PROOF. By the Hartman-Grobman theorem, we only have to prove this theorem for the cases where p is either on the boundary ∂M or a center ($n = 2$).

1. Let $p \in \partial M$, and let $n = 2$ or 3. It is easy to see that p is hyperbolic. Consider a local coordinate $\{\varphi, U\}$ of p such that

$$\begin{cases} \varphi : U \to \mathbb{R}^n_+ = \{x \in \mathbb{R}^n \mid x_n \geq 0\}, \\ \varphi(U \cap \partial M) \subset \partial \mathbb{R}^n_+, \\ \varphi(p) = 0. \end{cases} \tag{2.5.1}$$

Under the local coordinate (2.5.1), we have

$$\begin{cases} u = D\varphi \circ v \circ \varphi^{-1} = (u_1, \ldots, u_n) = Ax + g(x), \\ u_n(x) = 0, \quad \forall \ x \in \partial \mathbb{R}^n_+, \\ g(x) = o(\| x \|), \\ Ax = J(u(0))x. \end{cases} \tag{2.5.2}$$

Here $A = J(u(0))$ is the Jacobian matrix of u at $x = 0$, having the form

$$J(u(0)) = \begin{pmatrix} \frac{\partial u_1(0)}{\partial x_1} & \cdots & \frac{\partial u_1(0)}{\partial x_{nj-1}} & \frac{\partial u_1(0)}{\partial x_n} \\ \vdots & & & \\ \frac{\partial u_{n-1}(0)}{\partial x_1} & \cdots & \frac{\partial u_{n_1}(0)}{\partial x_{nj-1}} & \frac{\partial u_{n-1}(0)}{\partial x_n} \\ 0 & \cdots & 0 & \frac{\partial u_n(0)}{\partial x_n} \end{pmatrix}.$$

Since $u_n(x) = 0$ for any $x \in \partial \mathbb{R}^n_+$, we make the following reflective extension of u:

$$\widetilde{u}(x) = \begin{cases} u(x) & \text{if } x \in \mathbb{R}^n_+, \\ u^-(x^*) & \text{if } x \in \mathbb{R}^n_- = \{x \in \mathbb{R}^n \mid x_n \leq 0\}, \end{cases} \tag{2.5.3}$$

where $u^-(x) = (u_1^-(x), \ldots, u_n^-(x))$, $u_i^-(x) = 2u_i(\widetilde{x}) - u_i(x)$ $(1 \leq i \leq n)$, and $x^* = (x_1, \ldots, x_{n-1}, -x_n)$, $\widetilde{x} = (x_1, \ldots, x_{n-1}, 0)$ if $x = (x_1, \ldots, x_{n-1}, x_n)$. It is easy to see that $\widetilde{u} \in C^1(T\mathbb{R}^n)$, and $J(\widetilde{u}(0)) = J(u(0))$. Since $p \in \partial M$ is hyperbolic, i.e., there is no pure imaginary eigenvalue of $J(u(0))$, $x = 0$ is a hyperbolic point of $\widetilde{u}(x)$. By a slightly modified version of the classical Hartman-Grobman theorem, $\widetilde{u}(x)$ is locally topologically equivalent to $J(\widetilde{u}(0))x$. Therefore, u is locally topologically equivalent to $J(u(0))x$.

2. Let $n = 2$ and let $p \in \overset{\circ}{M}$ be a center. Using the notations in Definition (2.5.2), it is easy to see that $x = 0$ is a center for both u and $J(u(0))x$. Then, applying Lemma 1.4.10, u is locally topologically equivalent to $J(u(0))x$ near $x = 0$.

The proof is complete. □

2.5.2. Local structural stability of divergence-free vector fields. We now state and prove the local structural stability theorem for divergence-free vector fields. Here again, as in the previous Hartman-Grobman type theorem, we *do not assume* the hyperbolicity of the singularity point p.

THEOREM 2.5.3. *Let M be a two- or three-dimensional manifold with or without boundary, and let $v \in D^r(TM)(r \geq 1)$. If $p \in M$ is a non-degenerate singularity of v, then v is locally incompressibly structurally stable at $p \in M$.*

PROOF. 1. Assume $p \in \overset{\circ}{M}$. If $n = 3$, then p is hyperbolic. Hence, v is locally structurally stable at $p \in M$.

If $n = 2$, then p is either a saddle, which is hyperbolic, or a center. When p is a saddle, v is locally structurally stable. Hence we only have to consider the case of p being a center. It is then easy to see that there is an open neighborhood $N(v) \subset D^r(TM)$ of v and an open neighborhood $U \subset M$ of p such that every vector field $v_1 \in N(v)$ has a unique center p_1 in U. As v is locally equivalent to $Dv(p)$ and v_1 is locally equivalent to $Dv_1(p_1)$, this proves that v is locally equivalent to v_1. Hence, v is locally structurally stable at $p \in M$.

2. Assume $p \in \partial M$. For $\varepsilon > 0$ sufficiently small, we consider $v_1 \in D^r(TM)$ such that

$$\| v^1 - v \|_{C^r} < \varepsilon. \tag{2.5.4}$$

Under the local coordinate system (2.5.1), v^1 is expressed by

$$\begin{cases} u^1(x) = D\varphi \circ v^1 \circ \varphi^{-1} = (u_1^1(x), u_2^1(x), \dots, u_n^1(x)) \\ u_n^1(x) = 0, \quad \forall \ x \in \partial\mathbb{R}_+^n. \end{cases} \tag{2.5.5}$$

For simplicity, let

$$w = u|_{\partial\mathbb{R}_+^n}, \qquad w^1 = u^1|_{\partial\mathbb{R}_+^n}.$$

We infer from (2.5.2) and (2.5.5) that

$$\begin{cases} w, w^1 : \mathbb{R}^{n-1} \longrightarrow \mathbb{R}^{n-1}, \\ w(0) = 0, \\ J(w(0)) = J(u(0))|_{\partial\mathbb{R}_+^n} : \mathbb{R}^{n-1} \to \mathbb{R}^{n-1}. \end{cases} \tag{2.5.6}$$

By (2.5.6), we have

$$J(w(0)) = \begin{pmatrix} \frac{\partial u_1(0)}{\partial x_1} & \cdots & \frac{\partial u_1(0)}{\partial x_{n-1}} \\ \vdots & & \vdots \\ \frac{\partial u_{n-1}(0)}{\partial x_1} & \cdots & \frac{\partial u_{n-1}(0)}{\partial x_{n-1}} \end{pmatrix}. \tag{2.5.7}$$

Since the matrix $J(u(0))$ is non-singular, the matrix (2.5.7) is also non-singular. Hence, $x^1 = (x_1, \dots, x_{n-1}) = 0$ is a non-degenerate singular point of w. We deduce from (2.5.4) that

$$\| w^1 - w \|_{C^r} < \varepsilon. \tag{2.5.8}$$

When $\varepsilon > 0$ is sufficiently small, by Proposition 3.1, p. 55 in [**77**], there exists a unique singular point x_0 of w_1 in a neighborhood of $x = 0$ such that $J(w_1(x_0))$ and $J(w(0))$ have the same index. Hence, x_0 is an isolated singular point of $u^1(x)$, and $J(u^1(x_0))$ and $J(u^1(0))$ have the same index. Moreover, $J(u^1(x_0))x$ is topologically equivalent to $J(u(0))x$. By Theorem 2.5.2, v is locally topologically equivalent to v_1 near $p \in \partial M$.

The proof is complete. □

Notes for Chapter 2

1. The study of structural stability has been one of the main driving forces behind much of the development of dynamical systems theory; see, among many others, D. V. Anosov and V. Arnold [**3**], V. Arnold [**5**], J. Guckenheimer and P. J. Holmes [**33**], J. Moser [**74**], S. Smale [**96**], J. Palis and de Melo [**77**], M. Peixoto [**82**], C. Pugh [**85**], C. Robinson [**88**], M. Shub [**94**], and the references therein.

2. For the proof of Peixoto's classical theorem, see J. Palis and de Melo [**77**] and M. Peixoto [**82**]. Structural stability for divergence-free vector fields with either the free boundary condition or the Dirichlet boundary condition is obtained by the authors in [**51, 57, 58**]. The block structure for Hamiltonian vector fields in Section 2.3 is based on [**54, 62, 68**].

CHAPTER 3

Block Stability of Divergence-Free Vector Fields on Manifolds with Nonzero Genus

In this chapter, we continue to develop a kinematic theory for incompressible flows on a two-dimensional manifold of nonzero genus. In particular, the theory developed in this chapter will be useful for understanding the periodic structure and its evolutions of two-dimensional incompressible fluid flows with double boundary periodic conditions; see Chapter 4.

3.1. Instability on Manifolds with Nonzero Genus

3.1.1. An instability theorem on tori. We start with a structural instability theorem for the case where $M = \mathbb{T}^2$.

THEOREM 3.1.1. *Let $M = \mathbb{T}^2$ be the torus. Then no divergence-free vector field is structurally stable in $D^r(TM)$ $(r \geq 1)$.*

The proof of this instability theorem occupies the remainder of this section. We need to start with a few lemmas.

First, by Lemma 2.1.12, we only have to prove the instability of a self-connected vector field. Therefore, hereafter we always assume that $v \in D^r(TM)$ $(r \geq 1)$ is a self-connected vector field.

Let $p \in \mathbb{T}^2$ be a saddle point of v and let Γ be a saddle connection of p. Then Γ is a simple closed curve since v is a self-connection. Γ is called retractable if Γ is homological to zero in $H_1(\mathbb{T}^2)$, and Γ is called irretractable if Γ is not homological to zero.

Thanks to the properties of ergodic sets stated in Theorem 1.3.14, it is easy to see that:

(1) Each saddle point $p \in \mathbb{T}^2$ has at least one saddle connection Γ of p;
(2) If Γ is the unique saddle connection of $p \in \mathbb{T}^2$, then Γ is retractable.

It is clear that a retractable saddle connection cannot be broken under small perturbations of divergence-free vector fields. For irretractable saddle connections, we have the following lemma.

LEMMA 3.1.2. *Let $p \in \mathbb{T}^2$ be a saddle point of v. If p has two irretractable saddle connections, then the vector field v is structurally unstable in $D^r(TM)$ $(r \geq 1)$.*

PROOF. We divide the proof into several steps.

Step 1. Let $p \in M$ be a saddle point of v, having two irretractable saddle connections γ_1 and γ_2. First of all, we shall show that as two closed chains in $H_1(\mathbb{T}^2)$, γ_1 is homological to $-\gamma_2$.

To this end, we first consider the characterization of a nonzero element in a one-dimensional integer coefficient homology group $H_1(\mathbb{T}^2)$. We know that a torus

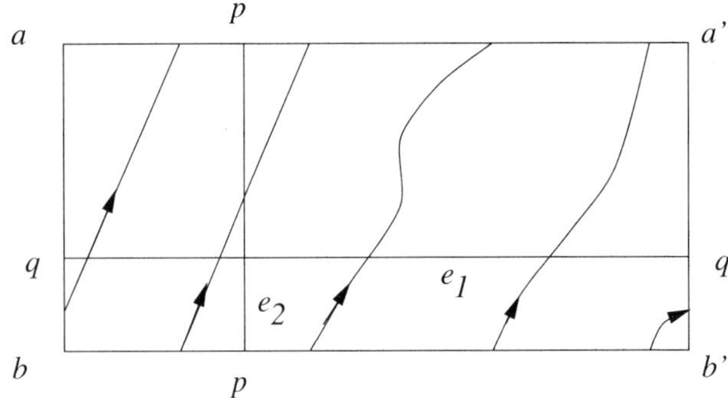

FIGURE 3.1.1. The edges ab and $a'b'$, aa' and bb' are identified.

is homeomorphic to a rectangle with each pair of opposite edges identified as shown in Figure 3.1.1. We take two closed curves e_1 and e_2 with orientations (see Figure 3.1.1) in M as two basic elements in $H_1(\mathbb{T}^2)$.

Then for any $\gamma \in H_1(\mathbb{T}^2)$ with $\gamma \neq 0$, we have $\gamma = n\ e_1 + m\ e_2,\ n, m \in \mathbb{Z}$. If $\gamma \subset M$ is a closed curve with orientation, then

$$n = \text{ the intersection number of } \gamma \text{ with } e_2,$$
$$m = \text{ the intersection number of } \gamma \text{ with } e_1.$$

Let $\widetilde{M}$ be homeomorphic to M, represented by an annulus with the pair of edges identified. Given a closed curve $\gamma \subset M$ with $\gamma \neq 0$ in $H_1(\mathbb{T}^2)$, we can define a homeomorphism

$$\phi = \phi_\gamma\ :\ M \to \widetilde{M} \tag{3.1.1}$$

such that

1) $\phi(\gamma) = \beta$, which is one of the edges identified on $\widetilde{M}$, and which is a closed curve; and
2) ϕ maps each closed curve, retractable to γ in M, to a closed curve retractable to β in $\widetilde{M}$.

Now we return to the two irretractable saddle connections γ_1 and γ_2 of p. Let $\gamma_1 = n\ e_1 + m\ e_2$; we need to prove that $\gamma_2 = -(n\ e_1 + m\ e_2)$. For the closed curve γ_1 with orbit orientation, we take a homeomorphism $\varphi = \varphi_{\gamma_1} : M \to \widetilde{M}$ as defined by (3.1.1); see Figure 3.1.2. By Theorem 1.1.1 in ([**36**], page 132), for any closed curve $\alpha = n_1 e_1 + m_1 e_2$ retractable to γ_1 on M, $|n_1| = |n|, |m_1| = |m|$. If $\gamma_2 \neq -\gamma_1$ and $\gamma_2 = k\ e_1 + r\ e_2$, then it is easy to see that $|k| \neq |n|$ or $|r| \neq |m|$. It implies that γ_2 cannot be retracted to γ_1 in M, and hence the image $\beta_2 = \varphi(\gamma_2)$ cannot be retracted to $\beta_1 = \varphi(\gamma_1)$ in $\widetilde{M}$; see Figure 3.1.2.

Obviously, the intersection $x = \varphi(p)$ in $\widetilde{M}$ of β_1 and β_2 is not a saddle point. And it is impossible that, under a homeomorphism, a saddle point could be mapped to a non-saddle point. Therefore we have that $\gamma_2 = -\gamma_1$. It implies that the two closed curves γ_1 and γ_2 enclose a domain $D \subset M$, homeomorphic to a sub-manifold of S^2; see Figure 3.1.3.

We take a C^r simple closed curve L in M such that L transversally intersects each of the irretractable saddle connections γ_1 and γ_2 at exactly one point; see Figure 3.1.3. For the closed curve L, let B be an open annulus as in Lemma 2.1.10

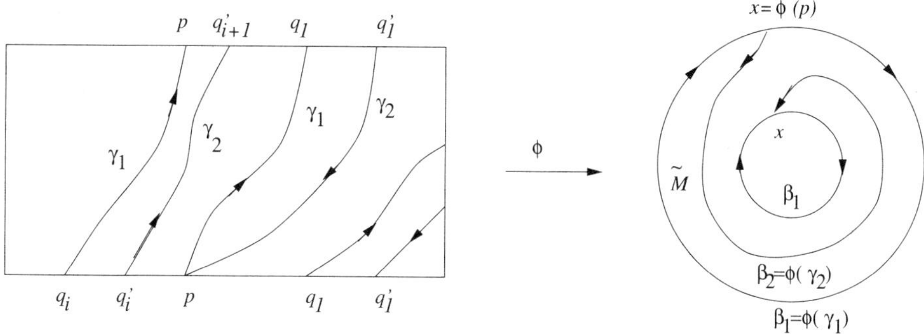

FIGURE 3.1.2. Schematic showing the definition of the mapping ϕ in Step 1 of the proof of Lemma 3.1.2.

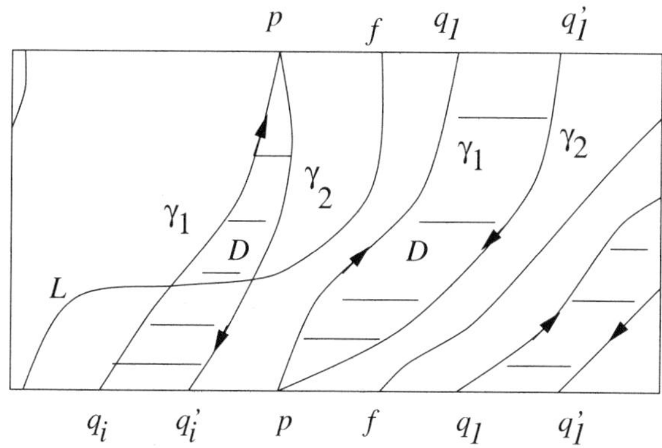

FIGURE 3.1.3. Two closed curves γ_1 and γ_2 enclose a domain $D \subset M$.

with width $h > 0$ sufficiently small. By Lemma 2.1.10, there exists a tubular divergence-free vector field w such that

$$w\Big|_{M-B} = 0, \quad w\Big|_B \neq 0$$

and B is a circle band of w. We will show that the vector field $v + \varepsilon w$ is not topologically equivalent to the vector field v for any $\varepsilon > 0$ sufficiently small.

To this end, we consider the topological structure of the vector field v in the domain D enclosed by the two irretractable saddle connections of p; see Figure 3.1.3. Since D is an invariant set of v and $\overset{\circ}{D}$ is homeomorphic to an open sub-manifold in S^2, the Poincaré-Bendixson Theorem of divergence free vector fields holds true in the domain D, which says that the α (resp. ω) limit set must be either a closed orbit or a saddle point. By Theorem 1.1.6, the domain D consists of circle cells, circle bands and saddle connections of v. Therefore, by Lemma 1.1.3, there exists a neighborhood Q of the two irretractable saddle connections in D such that for any $x \in Q$, the orbit $\Phi(x, \cdot)$ is closed. Then, as in the proof of Lemma 2.1.12, we can show that for any $\varepsilon > 0$ sufficiently small, there is a saddle connection of p of $v + \varepsilon w$, which is retractable on M; see Figure 3.1.4.

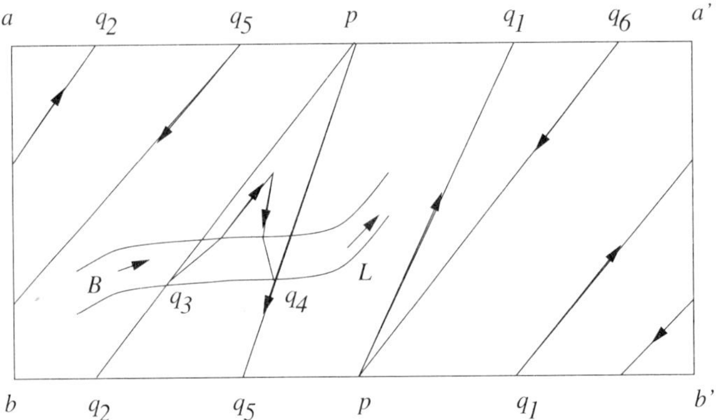

FIGURE 3.1.4. Saddle connection $\overline{pq_1q_2q_3q_4q_5q_6p}$ is retractable.

Obviously, the saddle connections of v are not homeomorphic to the saddle connections of $v + \varepsilon w$ for any $\varepsilon > 0$ sufficiently small. Hence, v and $v + \varepsilon w$ are not topologically equivalent.

The proof of Lemma 3.1.2 is complete. □

By Lemma 3.1.2, we only have to consider a self-connected vector field v such that each of its saddle points has at most one irretractable saddle connection.

For a saddle point with two retractable saddle connections Γ_1 and Γ_2, by Lemma 1.1.10, there is a circle band $B \subset M - D_1 - D_2$ such that $\Gamma_1 \cup \Gamma_2 \subset \partial B$, where $D_1, D_2 \subset M$ are enclosed by Γ_1 and Γ_2.

Obviously, there is a largest domain $A \subset M$ with $B \cup D_1 \cup D_2 \subset A$ such that ∂A consists of one retractable saddle connection of a saddle point $q \in \mathbb{T}^2$, and either q is connected by two nontrivially recurrent orbits or q has another irretractable saddle connection; see Figure 3.1.5.

A self-connected vector field $v \in D^r(TM)$ $(r \geq 1)$ is called a *basic vector field* if each saddle point of v has at most one irretractable saddle connection. Then Theorem 3.1.1 follows from the following lemma.

LEMMA 3.1.3. *Any basic vector field v is structurally unstable in $D^r(TM)$.*

PROOF. It is easy to see that for a basic vector field v, $\mathbb{T}^2$ can be decomposed into the following invariant sets of v:

$$\mathbb{T}^2 = \Omega \cup_{i=0}^{K} A_i,$$

where $K \geq 0$, $A_0 = \emptyset$, A_i $(1 \leq i \leq K)$ are open disks as shown in Figure 3.1.6, and Ω is a flow-invariant submanifold of $\mathbb{T}^2$ with genus one. Namely, a basic vector field v is topologically equivalent to vector fields given in the schematic picture in Figure 3.1.6.

We take a simple closed curve $\alpha \subset \Omega \subset \mathbb{T}^2$ such that α is transversal to v; see Figure 3.1.7. Let B be an open annulus given by Lemma 2.1.10 corresponding to the closed curve α with $h > 0$ sufficiently small, and let ω be the tubular divergence-free vector field in B.

We set an x-coordinate on α as follows. Choose arbitrarily a point $q \in \alpha$, and let $x = 0$ at the point q. Then for any point $q_1 \in \alpha$, we take x to be equal to the arc length from q to q_1 along the direction of the flow of w in B. Without loss of

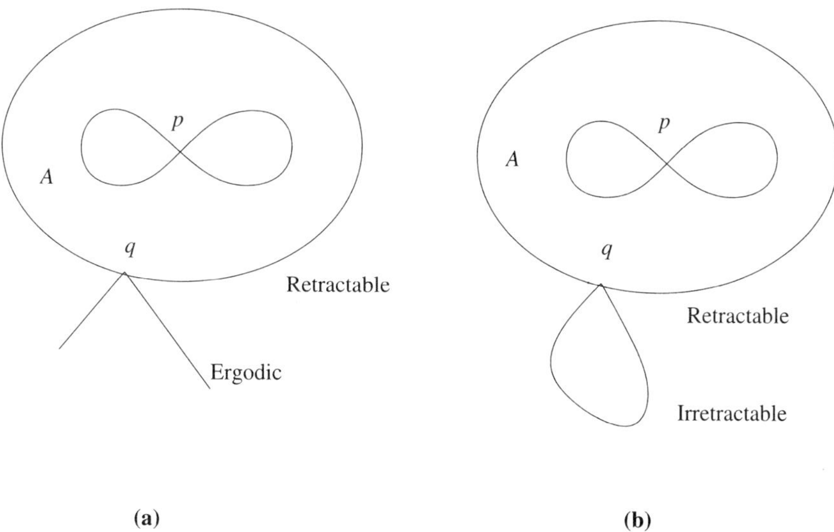

FIGURE 3.1.5. A retractable saddle connection is attached either to two nontrivially recurrent orbits or to an irretractable saddle connection.

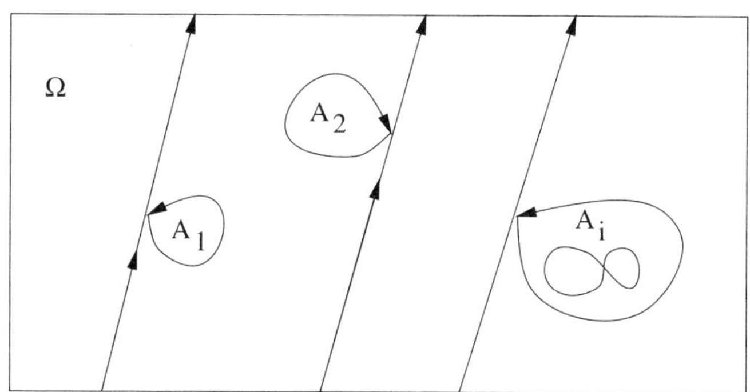

FIGURE 3.1.6. Schematic showing a basic vector field.

generality, we assume that the length of α equals one, i.e., $0 \leq x \leq 1$; see Figure 3.1.7.

Let $D_i \subset M$ be the domain enclosed by the retractable saddle connection of p_i, and let $D = \cup_i D_i$. We take $\bar{B} \cap D = \emptyset$. Hence, for any $x \in \partial B = \alpha \cup L$, x is not a singular point of v. Since α is transversal to the vector field v, the vector field w in domain B can be chosen to be transversal to v.

For any point $x \in \alpha$, the flow from x of v crosses $M - B$ and intersects the other boundary L of ∂B at y. Under the perturbation of $\epsilon w(\epsilon > 0)$, the flow from y of $v + \epsilon w$ crosses B and intersects with α at x_1 (see Figure 3.1.7). We denote by $g(x, \epsilon)$ the arc length from x to x_1. Obviously, the function $g(x, \epsilon)$ is continuous

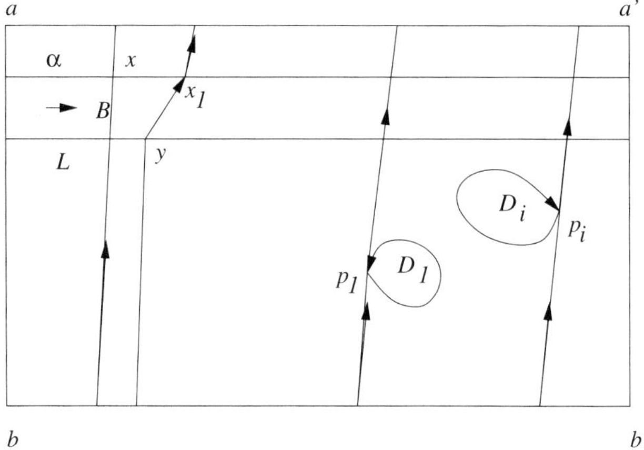

FIGURE 3.1.7. Perturbation under a non-Hamiltonian divergence-free vector field.

and satisfies

$$\begin{cases} g(x+1,\epsilon) = g(x,\epsilon) & \text{for } \epsilon > 0, \forall x \in \alpha, \\ 0 < b \le g(x,\epsilon) & \text{for } \epsilon > 0, \forall x \in \alpha, \\ g(x,0) \ge 0, \\ g(x,\epsilon) \text{ is an increasing function of } \epsilon \text{ for } \epsilon > 0 \text{ small.} \end{cases} \tag{3.1.2}$$

Since $g(x,\epsilon)$ is a periodic function of x, we define a sequence of real numbers as follows:

$$h_1(\epsilon) = g(0,\epsilon), \cdots, h_n(\epsilon) = g\left(\sum_{i=1}^{n-1} h_i, \epsilon\right), \cdots. \tag{3.1.3}$$

We can see that the orbit $\Phi(0,t)$ of $v+\epsilon w$ is closed if and only if there is an integer $m \ge 1$ such that

$$\sum_{i=1}^{m} h_i(\epsilon) = \text{integer}. \tag{3.1.4}$$

The set $M - D$ consists of either circle bands or an ergodic set of $v+\epsilon w$, depending entirely on whether the orbit $\Phi(0,t)$ of $v+\epsilon w$ is closed.

By (3.1.2), $h_n(\epsilon) \ge c > 0$ for a fixed $\epsilon > 0$ and all $n \ge 1$; moreover, $h_n(\epsilon)(n \ge 1)$ are continuous and exactly increasing on ϵ. Therefore, in an interval $[0,\delta](\delta > 0)$, there are countably infinite many real numbers $\{\epsilon_i\} \subset [0,\delta]$ with $\{\epsilon_i\}$ dense in $[0,\delta]$, such that (3.1.4) holds. Namely, the vector fields $v+\epsilon_i w (i = 1,2,\cdots)$ have no ergodic set in M, and the vector fields $v+\epsilon w$ with $\epsilon \in [0,\delta]$ and $\epsilon \ne \epsilon_i (i = 1,\cdots)$ have an ergodic set in M. It implies that the vector field v is unstable. □

3.1.2. Structural instability on higher genus manifolds.

THEOREM 3.1.4. *Let M be a 2D compact orientable manifold with genus $k \ge 1$ and $\partial M = \emptyset$. Then no divergence-free vector field $v \in D^r(TM)(r \ge 1)$ is structurally stable in $D^r(TM)(r \ge 1)$.*

We omit the proof of Theorem 3.1.4. In fact, in the next section, we shall see that a divergence-free vector field on M can be approximated by basic vector fields (see Definition 3.2.1), and the structural instability will become transparent.

3.2. Block Structure and Block Stability

3.2.1. Main theorem. In Section 3.1, we proved that if M is orientable with nonzero genus, then no divergence-free vector field $v \in D^r(TM)$ is structurally stable under the perturbation of C^r divergence-free vector fields. Hence, we introduce a new concept called block stability of incompressible vector fields on two-dimensional compact manifolds.

Let N be a two-dimensional compact manifold with boundary. It is well known that the genus of N is defined by

$$g(N) = 1 - \frac{1}{2}(\chi(N) + r),$$

where $\chi(N)$ is the Euler characteristics of N, and r is the number of connected components of ∂N. Obviously, the manifold N must be homeomorphic to a submanifold $\widetilde{N}$ of M, where M is a compact manifold without boundary with genus $g(M) = g(N)$, such that each connected component of $\partial\widetilde{N}$ is retractable to a point in M.

Let M be a two-dimensional compact orientable manifold with genus $g \geq 1$. We now introduce a new class of divergence-free vector fields, called basic vector fields on M, which are those having the block structure.

DEFINITION 3.2.1. *A regular divergence-free vector field $v \in D^r(TM)$ is called a basic vector field if M can be decomposed into blocks as $M = \cup_{j=1}^R \Omega_j + \cup_{i=0}^K A_i$ with empty intersections between them, and with $A_0 = \emptyset$ such that*

(1) *each A_i is an open flow-invariant sub-manifold with genus zero, and Ω_j is a compact invariant sub-manifold with genus $g(\Omega_j) \geq 1$;*
(2) *v is a self-connected field in A_i with only one saddle point on each connected component Γ of $\partial A_i (1 \leq i \leq k)$, and Γ, as a closed chain, is homological to zero in $H_1(M)$;*
(3) *$\overset{\circ}{\Omega}_j$ does not contain any invariant closed curve that is homological to zero in $H_1(M)$.*

For simplicity, let $D_B^r(TM)$ be all C^r basic vector fields.

DEFINITION 3.2.2. *Let $v_1, v_2 \in D_B^r(TM)$ $(r \geq 1)$ with block decompositions $M = \cup_{j=1}^R \Omega_j^{(\alpha)} + \cup_{i=0}^K A_i^{(\alpha)}$ $(\alpha = 1, 2)$. We say that v_1, v_2 have isomorphic block decompositions if there exists a homeomorphism $\varphi : M \to M$, which takes the blocks $\Omega_j^{(1)}$ and $A_i^{(1)}$ of v_1 to the corresponding blocks $\Omega_j^{(2)}$ and $A_i^{(2)}$ of v_2, preserving orbit orientations.*

The main result of this section is

THEOREM 3.2.3. (Block Stability Theorem) *Let M be a two-dimensional compact orientable manifold with genus $g \geq 1$. Then $D_B^r(TM)$ is open and dense in $D^r(TM)$. Furthermore, let $v \in D_B^r(TM)$ be a basic vector field with block decomposition $M = \cup_{j=1}^R \Omega_j + \cup_{i=0}^K A_i$. Then there exists a neighborhood $\mathcal{O} \subset D^r(TM)$ of v such that*

(1) *each* $v_1 \subset \mathcal{O}$ *has a block decomposition* $M = \cup_{j=1}^{R} \Omega_j^{(1)} + \cup_{i=0}^{K} A_i^{(1)}$, *which is isomorphic to the block decomposition of* v;
(2) $v_1|_{A_i^{(1)}}$ *is topologically equivalent to* $v|_{A_i}$;
(3) *there is a dense set* $\widetilde{\mathcal{O}} \subset \mathcal{O}$ *such that for any* $v_1 \subset \widetilde{\mathcal{O}}$, *each* $\Omega_j^{(1)}$ *is an ergodic set of* v_1;
(4) $\sum_{j=1}^{R} g(\Omega_j) = g(M)$, *where* $g(\Omega_j)$ *is the genus of* Ω_j.

A few remarks are now in order.

REMARK 3.2.4. By Definition 1.3.1, $\Omega_j \cap \Omega_\ell = \emptyset$ for $j \neq \ell$, which implies that

$$\cup_{j=1}^{R} \partial\Omega_j = \cup_{i=0}^{K} \partial A_i. \tag{3.2.1}$$

Hence, each connected component of $\partial\Omega_j$ is a self-connection to a unique saddle point, and is homological to zero.

Each Ω_j may contain a few flow invariant subsets of nonzero measure. But in any small neighborhood of v, there exists a basic vector field v_1 such that the corresponding $\Omega_j^{(1)}$ does not contain any invariant sets of nonzero measure, i.e., $\Omega_j^{(1)}$ is a minimal invariant set of nonzero measure.

REMARK 3.2.5. When M is non-orientable, the block stability problem is much more complicated. The reason is that if M is orientable and M has no invariant closed curve which is homological to zero in $H_1(M)$, then under a perturbation M can become an ergodic set, but if M is non-orientable, we don't know whether the property still holds for $g(M) \geq 4$. □

We remark also that when $M = \mathbb{T}^2$ is a torus, the above block structure and block stability can be simplified as follows; see Figure 3.2.1.

DEFINITION 3.2.6. *Let* $M = \mathbb{T}^2$. $v \in D_0^r(TM)$ *is called a basic vector field if* M *can be decomposed into blocks as* $M = \Omega \cup_{i=0}^{K} A_i$ *with* $A_0 = \emptyset$, $\Omega \cap A_i = \emptyset$, $A_i \cap A_j = \emptyset (i \neq j)$ *such that*
(1) Ω *and* A_i *are invariant sets of* v,
(2) *each* A_i *is homeomorphic to an open disk and retractable to a point in* M, *and* Ω *is a compact sub-manifold of* M *with genus one,*
(3) *the restriction of* v *in* A_i *are self-connections,*
(4) v *has only one saddle point in each* $\partial A_i (1 \leq i \leq K)$, *and*
(5) v *has no singular point in* $\overset{\circ}{\Omega}$.

THEOREM 3.2.7. *Let* $M = \mathbb{T}^2$. *Then the following assertions hold true:*
(1) $D_B^r(TM)$ *is open and dense in* $D^r(TM)$;
(2) *Each* $v \in D_B^r(TM)$ *has a block decomposition* $M = \Omega \cup_{i=0}^{K} A_i$ *for some* $K \geq 0$ *with* $A_0 = \emptyset$;
(3) *For any* $v \in D_B^r(TM)$, *there exists a neighborhood* $\mathcal{O} \subset D^r(TM)$ *of* v *such that*
   (a) *each* $v_1 \in \mathcal{O}$ *has block decomposition* $M = \Omega^{(1)} \cup_{i=0}^{K} A_i^{(1)}$, *which is isomorphic to the block decomposition of* v;
   (b) $v|_{A_i}$ *is topologically equivalent to* $v_1|_{A_i^{(1)}}$ *for* $v_1 \in \mathcal{O}$; *and*
   (c) *there is a dense set* $\widetilde{\mathcal{O}} \subset \mathcal{O}$ *such that for any* $v_1 \in \widetilde{\mathcal{O}}$, $\Omega^{(1)}$ *is an ergodic set of* v_1.

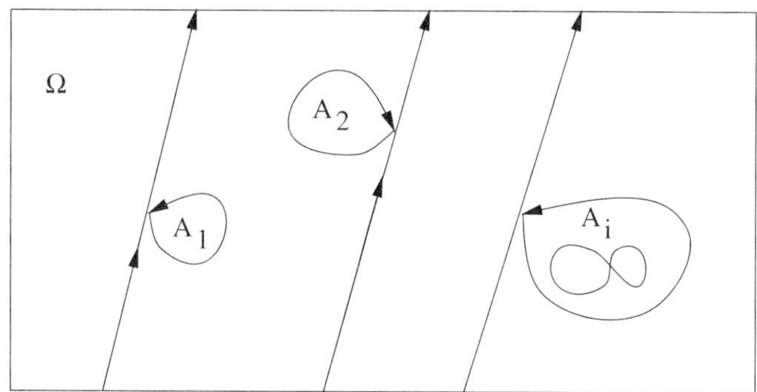

FIGURE 3.2.1. Schematic of a basic vector field on $\mathbb{T}^2$ with $\mathbb{T}^2 = \Omega + \cup_{i=0}^{K} A_i$. Here Ω contains either periodic orbits (with nonzero homology) or nontrivially recurrent orbits, and each A_i is isomorphic to an open disk such that $v|_{A_i}$ is a self-connected vector field.

3.2.2. A generalized lemma of breaking saddle connections. In Section 2.1.3, we introduced a lemma for breaking saddle connections of a divergence-free vector field v such that $M = \bar{V}$, where V is the set of all closed orbits. We now study the case without this restriction. For simplicity, let

$$D_1^r(TM) = \{v \in D^r(TM) \quad | \quad v \text{ is a self-connection } \}.$$

LEMMA 3.2.8. *$D_1^r(TM)$ $(r \geq 1)$ is dense in $D^r(TM)$.*

REMARK 3.2.9. If $M = S^2$ (or $M = P^2$, or $M = K^2$), then $D_1^r(TM)$ is also open in $D_0^r(TM)$. For general 2–manifolds, however, $D_1^r(TM)$ is not open in $D^r(TM)$.

PROOF OF LEMMA 3.2.8. Let $v \in D^r(TM)$ be regular, and let $p_1, p_2 \in M$ be two saddle points of v, which are connected as shown in Figure 3.2.2.

Let S_1 and S_2 be two concentric circles with the same center p_1, let B_h be an annulus with width $h > 0$ and with $\partial B_h = S_1 \cup S_2$, and let r be the radius of S_2 (see Figure 3.2.2). Assume that r and h be sufficiently small, and let w be a tubular incompressible vector field in B_h.

Obviously, the vector fields $v+\epsilon w$ and v have the same saddle points. It suffices then to prove that for almost all $\epsilon > 0$ sufficiently small, the saddle point p_1 of $v+\epsilon w$ is self-connected.

As shown in Figure 3.2.2, the trajectories born from p_1 intersect with S_1 at a_1 and b_1. Under the perturbation ϵw, the trajectories $\overline{p_1 a_1}$ and $\overline{p_1 b_1}$ will travel through B_h to a_2 and b_2, respectively. The trajectory γ_1 of v intersects with S_2 at c_1. Under the perturbation ϵw, γ_1 travels through B_h to c_2. The orbit $\Phi(c_2, t)$ of v from c_2 will intersect S_1 at c_3, then travel through B_h to c_4.

It is easy to see that for almost all points $x \in S_2$, the orbits $\Phi(x, t)$ of v will not connect to the orbits γ_1 and γ_2 of v. Hence, if for any $\epsilon > 0$ sufficiently small, the point c_4 on the trajectory $\overline{c_1 c_2 c_3 c_4}$ of $v + \epsilon w$ lies on the trajectory of v connecting to p_1 and p_2, then for almost all $\epsilon > 0$ sufficiently small, the trajectories of $v + \epsilon w$ connected to p_1 will be either self–connected or ergodic, i.e., p_1 is self–connected.

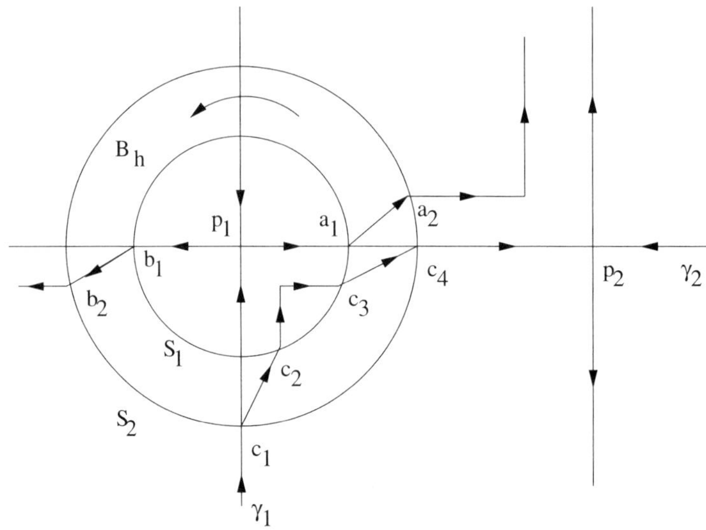

FIGURE 3.2.2. Breaking the saddle connections connecting two saddles $p_1, p_2 \in M$.

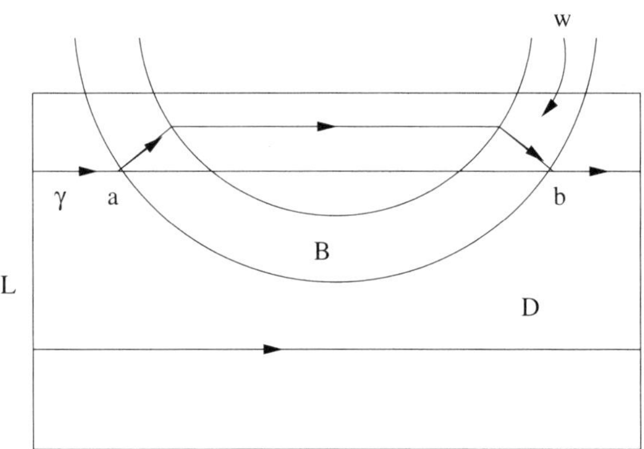

FIGURE 3.2.3. The points a and b are still connected with the perturbation.

We now show that for any $\epsilon > 0$ sufficiently small, the point c_4 will lie in the trajectory of v connecting to p_1 and p_2. In fact, the question can be equivalently stated as follows.

Let v_0 be an incompressible vector field defined in a local tubular region D, let $v_0 \neq 0$ in D, and let B be a U-tubular region with $B \cap D \neq \emptyset$. Also, let w be an incompressible vector field in B defined as in Lemma 2.1.10; see Figure 3.2.3.

A trajectory γ of v_0 in D, under the perturbation ϵw, will travel from point a through B twice to b. We need to prove that point b lies on the trajectory γ.

We extend the region D to an annulus A, and we extend the vector field v_0 to a C^0 vector field v in A whose flow is area-preserving. The existence of the extension v is obvious; this is equivalent to finding an incompressible vector field u in $A - D$ such that $u|_{\partial(A-D)}$ equals a given boundary condition. Let L be an arc transversal to v_0 (see Figure 3.2.3). By the Poincaré-Bendixson Theorem for incompressible

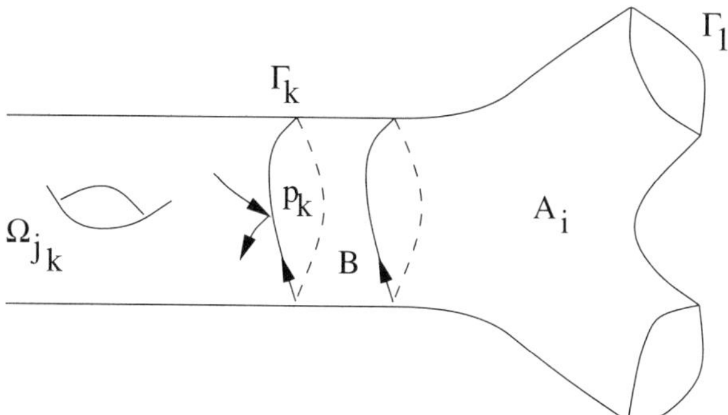

FIGURE 3.2.4. Existence of a circle band B.

flows, the set

$$\{x \in L|\ \text{the orbit } \Phi(x,t) \text{ of } v\ in\ A \text{ is closed}\}$$

is open and dense in L. Without loss of generality, we assume that the orbits of v in A are closed. As in the proof of Lemma 2.1.12, it is easy to see that if b does not lie on the trajectory γ, then there is a limit cycle of $v+\epsilon w$ in A for $\epsilon \neq 0$ sufficiently small, a contradiction to $v \in \epsilon w$ being area-preserving.

The proof is complete. □

3.2.3. Proof of the main theorem. We proceed in a few steps as follows.

Step 1. Let $v \in D^r_B(TM)$ be a basic vector field with block decomposition $M = \cup_{j=1}^R \Omega_j + \cup_{i=0}^K A_i$. The genus zero of A_i means that each invariant set A_i is homeomorphic to an open sub-manifold of S^2. By definition, each connected component Γ of ∂A_i, as a closed chain, is homological to zero. Hence, we infer from (3.2.1) that if ∂A_i has $m(\geq 1)$ connected components, $\partial A_i = \sum_{k=1}^m \Gamma_k$, then there are m different invariant sub-manifolds $\Omega_{j_1}, \cdots, \Omega_{j_m}$ such that each Ω_{j_k} has exactly one Γ_k $(1 \leq k \leq m)$ as one of the connected components of its boundary. It is easy to see that each boundary component Γ_k is a single saddle self-connection. Let $p_k \in \Gamma_k$ be the saddle point on Γ_k, and let

$$\Omega_{j_k} \cap \bar{A}_i = \Gamma_k. \tag{3.2.2}$$

Now we prove that the invariant open sets A_i are stable. More precisely, under a small perturbation in $D^r(TM)$, the blocks A_i cannot be destroyed but are merely displaced slightly in different positions in M. To this end, it suffices to show that the boundary components Γ_k $(1 \leq k \leq m)$ are stable.

Let $w \in D^r(TM)$ with $\|w\| < \epsilon$ sufficiently small. In a neighborhood of each saddle point $p_k \subset \Gamma_k$ $(1 \leq k \leq m)$, there exists a unique saddle point q_k of $v+w$. By definition, $\overset{\circ}{\Omega}_{j_k}$ does not contain any invariant closed cycles homological to zero. Since Γ_k contains only one saddle point p_k, both of the other two orbits connected to p_k must be in $\overset{\circ}{\Omega}_{j_k}$. By Lemma 2.1.10, there is a collar neighborhood B of Γ_k in A_i such that B is a circle band of v; see Figure 3.2.4.

If, under the perturbation of w, the self-connection orbit Γ_k of v does break, then the flows of $v+w$ in B will either enter into or leave $\overset{\circ}{\Omega}_{j_k}$; see Figure 3.2.5. By

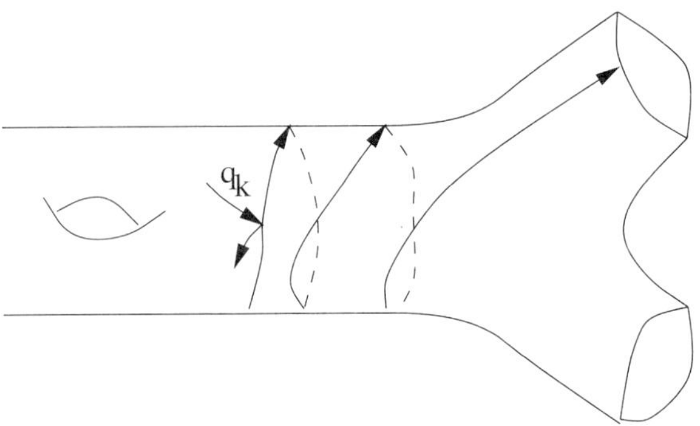

FIGURE 3.2.5. Orbit structure after the perturbation.

(3.2.2), in either case, the flows of $v+w$ in B will not return again, a contradiction to the Poincaré Recurrence Theorem.

If the saddle self-connection Γ_k does not break under the perturbation of w, then it is easy (as in the proof of Lemma 2.1.14) to see that the saddle point q_k of $v+w$ still has a self-connection orbit $\widetilde{\Gamma}_k$ homological to zero.

Therefore, the perturbated vector field $v+w$ has a decomposition of invariant sets $M = \cup_{j=1}^{R}\Omega_j^{(1)} + \cup_{i=0}^{K} A_i^{(1)}$ with $\Omega_j^{(1)}$ and $A_i^{(1)}$ homeomorphic to Ω_j and A_i, respectively, and each connected component $\Gamma^{(1)}$ of $\partial A_i^{(1)}$ is homological to zero and contains only one saddle point.

Step 2. Notice that $A_i \subset S^2$ and $v|_{A_i}$ is a self-connection. Theorem 2.1.2 yields that, under a perturbation of $D^r(TN)$, a self-connection vector field on a submanifold $N \subset S^2$ remains a self-connection vector field. Therefore the restriction of $v+w$ vector field on $A_i^{(1)}$ is also a self-connection vector field. Moreover, $v|_{A_i}$ is topologically equivalent to $(v+w)|_{A_i^{(1)}}$.

Step 3. We show that $v+w \in D_B^r(TM)$ for $\|w\|$ sufficiently small, i.e., $D_B^r(TM)$ is open in $D^r(TM)$.

By Steps 1–2, the perturbated vector field $v+w$ has a decomposition of invariant sets $M = \cup_{j=1}^{R}\Omega_j^{(1)} + \cup_{i=0}^{K} A_i^{(1)}$ with $\Omega_j^{(1)}$ and $A_i^{(1)}$ homeomorphic to Ω_j and A_i, respectively. It suffices then to prove that each invariant closed set $\Omega = \Omega_j^{(1)}$ of $v+w$ does not contain any invariant closed curves of homology zero.

Assuming otherwise, there exist $v_n \in D^r(TM)$ with $\lim_{n\to\infty} v_n = v$ such that each v_n has an invariant closed curve $L^{(n)} \subset \Omega^{(n)}$ of homology zero. Here, each $\Omega^{(n)}$ is an invariant closed set of v_n with $\lim_{n\to\infty}\Omega^{(n)} = \Omega$. As in the proof of Lemma 2.1.14, we can show that there exists an invariant closed curve $L \subset \Omega$ of v such that $\lim_{n\to\infty} L^{(n)} = L$. We notice that a closed curve $\Gamma \subset M$ is homological to zero if and only if Γ consists of boundaries of $k(\geq 2)$ disjoint open sets $N_1, \cdots, N_k \subset M$, i.e., $N_i \cap N_j \neq \emptyset$ $(i \neq j)$ and $\Gamma = \sum_{i\neq j} \bar{N}_i \cap \bar{N}_j$. From this property, we deduce that the closed curves homological to zero must converge to a closed curve homological to zero. Therefore, $L \subset \Omega$ is a closed curve homological to zero, a contradiction to $v \in D_B^r(TM)$. Hence, $D^r(TM)$ is open in $D^r(TM)$.

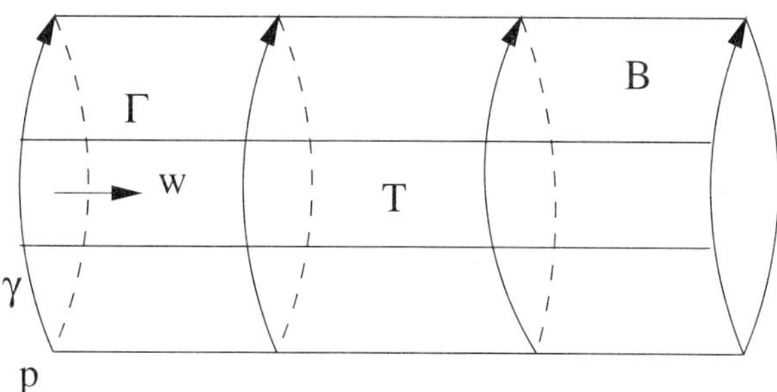

FIGURE 3.2.6. Construction of tubular flows.

Step 4. Proof of Assertion 3. It suffices to prove that v can be arbitrarily approximated by v_1 with block decomposition $M = \cup_{j=1}^{R} \Omega_j^{(1)} + \cup_{i=0}^{K} A_i^{(1)}$ such that $\Omega_j^{(1)}$ $(j = 1, \cdots, K)$ are ergodic sets of v_1. By Lemma 3.2.8, the set of self-connected vector fields is dense in $D^r(TM)$. Hence, we may assume that $v \in \mathcal{O}$ is a self-connected basic vector field.

Let $\Omega \subset M$ be a compact, v-flow invariant sub-manifold with genus $g(\Omega) \geq 1$ such that there are no zero homology closed orbits of v_1 in Ω. Then there are no centers of v in Ω. Notice that on each connected component Γ of $\partial\Omega$, there is only one saddle point $p \in \Gamma$ and the four orbits connected to p are all in Ω, two of which on Γ and the other two in $\overset{\circ}{\Omega}$. If we regard each Γ as a point, then Ω is equivalent to a compact manifold N without boundary and with genus $g = g(\Omega)$, and v is topologically equivalent to a basic incompressible vector field on N with exactly $s = 2g - 2$ saddle points in N. Therefore, without loss of generality we assume that $\partial\Omega = \emptyset$.

By the Structural Classification Theorem, Ω is an ergodic set of v if and only if Ω has no closed orbits, or equivalently, $\overset{\circ}{\Omega}$ has no saddle self-connections. Assume that Ω is not an ergodic set of v; then Ω has $k(\geq 1)$ circle bands $\widetilde{B}_i \subset \Omega (1 \leq i \leq k)$, and $k \leq 2s - 1$ where $s = 2g - 2$ is the number of saddle points of v.

Hereafter, we introduce a general procedure to approximate v arbitrarily by v_1 such that all corresponding $\Omega_j^{(1)}$ contain no circle bands, and, therefore, are ergodic sets. Namely, we will break the circle bands $\widetilde{B}_i$ with arbitrarily small perturbations of v.

To this end, we first introduce a general procedure to construct a tubular flow associated with each circle band, which will be used to break the circle band.

Construction of tubular flows:

(1) For each such circle band, denoted by B for simplicity, the invariant closed orbits of v in B are not homological to zero. Hence, there is a simple closed curve $\Gamma \subset \Omega$ such that Γ transversally intersects with each invariant cycle in $\bar{B}$ at exactly one point.
(2) We take a tubular domain T with width $r > 0$ sufficiently small and $\partial T = \Gamma + \widetilde{\Gamma}$. Furthermore, let w be a tubular vector field on T defined as in Lemma 2.1.9. Then for any $\lambda \neq 0$, the flows of $v + \lambda w$ will enter into B from one boundary component of ∂B and will leave B from the other boundary component of ∂B; see Figure 3.2.6.

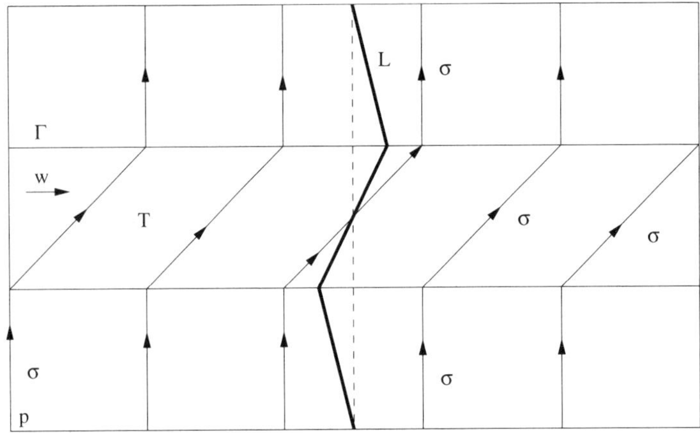

FIGURE 3.2.7. Construction of a simple closed curve L.

(3) Then we can construct a simple closed curve L as shown in Figure 3.2.7 such that L is transversal to $v+\lambda w$. Furthermore, we take a collar neighborhood C with $\partial C = L + \widetilde{L}$ such that the width $h > 0$ is sufficiently small and $\widetilde{L}$ is transversal to $v + \lambda w$.

Notice that the original circle band B has a self-connection γ of the saddle point p as one component of its boundary. With the perturbation of λw, this orbit becomes σ in Figure 3.2.7, which intersects with L transversally at exactly one point. With the above procedure, two situations could happen near the orbit σ for the new vector field $v + \lambda w$: either a) σ becomes non-trivially recurrent, or b) σ is one component of the boundary of a circle band B' of $v + \lambda w$. In the second situation, L will intersect each orbit transversally in B' at exactly one point, and $\widetilde{L}$ will as well with the width $h > 0$ small. Moreover, further small perturbations of the vector field $v + \lambda w$ preserve either property a) or b) of the orbit σ connected to the saddle point p.

Now we apply the above procedure to $\widetilde{B}_1$ and v. Let γ_1 be the self-connection of a saddle p_1, and be one connected component of the boundary of $\widetilde{B}_1$. We obtain

1) the pertubated vector field $v + \lambda_1 w_1$ with w_1 being a tubular vector field,
2) a simple closed curve L_1 and a tubular neighborhood C_1 with $\partial C_1 = L_1 + \widetilde{L}_1$ and with the width $h_1 > 0$ sufficiently small such that both L_1 and $\widetilde{L}_1$ are transversal to $v + \lambda_1 w_1$,
3) a new orbit σ_1 connected to p_1 obtained from γ_1 under the perturbation satisfying properties a) or b) above.

Then we apply the same procedure to the new vector field $v+\lambda_1 w_1$ and a circle band of $v+\lambda_1 w_1$ different from the one described in the property b), preserving the properties of L_1 and C_1 obtained previously because of transversality. Repeating this procedure a finite number of times (at most twice the number of saddle points in Ω), we obtain that for any $\epsilon > 0$ sufficiently small, there is a vector field $u \in D^r(TM)$ with $\|v - u\| < \epsilon$ verifying the following properties:

1° $\Omega = E \cup \sum_{i=1}^{m} B_i$, where E consists of the ergodic sets of u, and B_i are the circle bands of u;

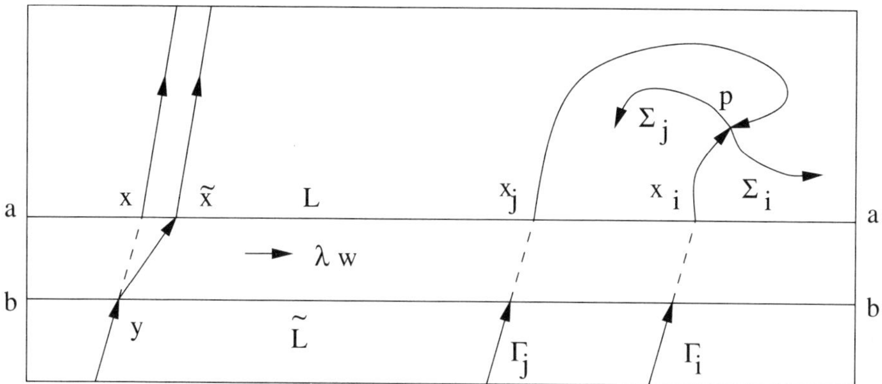

FIGURE 3.2.8. Schematic showing the definition of f_λ.

2° for each $i = 1, \cdots, m$, there exists a simple closed curve L_i such that L_i and its associated tubular neighborhood C_i with $\partial C_i = L_i + \widetilde{L}_i$, satisfying
  (a) both L_i and $\widetilde{L}_i$ are transversal to u and each intersects with each orbit in B_i at exactly one point;
  (b) the flow of u enters into C_i on the side $\widetilde{L}_i$ and leaves C_i on the other side L_i;

3° for each $i = 1, \cdots, m$, by Lemma 2.1.9, there exists a tubular incompressible flow, denoted again by w_i, corresponding to C_i.

Let $B = B_1, C = C_1, L = L_1, w = w_1$ and let the length of L be one, i.e., $|L| = 1$. For each $\lambda \in (-1, 1)$, we define a map

$$f_\lambda : L \to L, \ \lambda \in (-1, 1), \tag{3.2.3}$$

on L as follows. Let $x \in L$ and let $\Phi(x, t)$ be the orbit of $u + \lambda w$ with $\Phi(x, 0) = x$. By definition, the orbit $\Phi(x, t)$ starting from $x \in L$ crosses $\Omega - B$ and intersects with $\widetilde{L}$ at y; under the action of the vector field λw, the orbit $\Phi(x, t)$ from y crosses T and intersects with L at $\widetilde{x}$. Then we define that $f_\lambda(x) = \widetilde{x}$; see Figure 3.2.8.

Let Γ_i be the orbits of u whose ω-limits are saddles of u. Since the number of saddle points of u is finite, there are only a finite number of such orbits, denoted by Γ_i $(i = 1, \cdots, N)$. Let $x_i \in L$ be the last intersection point of Γ_k with L. We can see that, for each $\lambda \in (-1, 1)$, the map (3.2.3) is continuous except at the finite number of points x_i. In fact, the orbit Γ_i after x_i ends at a saddle point p (see Figure 3.2.8), and the orbit on the left side of Γ_i travels near (along) Σ_j. The orbit on the right side of Γ_i will travel near (along) Σ_i. Then we set $f_\lambda(x_i) =$ the point that is the intersection of $\sum_i$ with L; see Figure 3.2.8. With such a definition, $f_\lambda(x)$ is a right continuous function on L.

We define a function $g : L \times (-1, 1) \to \mathbb{R}$ by

$$g(x, \lambda) = \rho(x, f_\lambda(x)), \qquad \forall (x, \lambda) \in L \times (-1, 1),$$

where $f_\lambda(x)$ is as (3.2.3) and $\rho(x, y)$ is the arc length from x to y on L along the direction of the flow of w. It is easy to see that the finite number of discontinuous points $x_i (1 \leq i \leq N)$ of $f_\lambda(x)$ is independent of λ, and for any $x \neq x_i (1 \leq i \leq N)$,

$g(x,\lambda)$ is an increasing continuous function on $\lambda \in (-1,1)$ when the vector field w is taken sufficiently small. Furthermore, $g(x,\lambda)$ is also right continuous at $x_i \in L$.

Taking an arbitrary continuous point $x_0 \in L$ of f_λ, we define a sequence of real numbers:

$$h_1(\lambda) = g(x_0,\lambda), h_2(\lambda) = g(f_\lambda(x_0),\lambda), \cdots, h_n(\lambda) = g(f_\lambda^{n-1}(x_0),\lambda), \cdots.$$

Obviously, the orbit $\Phi(x_0,t)$ of $u + \lambda w$, with the ω-limit not being a saddle point, is closed if and only if there is an integer $m \geq 1$ such that

$$\sum_{i=1}^{m} h_i(\lambda) = \text{ integer } = \text{ integer } \times |L|. \tag{3.2.4}$$

We now show that for each $m \geq 1$, there are only finite real numbers $\lambda \in (-1,1)$ such that (3.2.4) holds true.

Let $\lambda_0 \in (-1,1)$ be a real number satisfying (3.2.4); namely, there is an integer $m \geq 1$ such that $x_0 = f_{\lambda_0}^{m-1}(x_0)$. If there is an $x_i (1 \leq i \leq N)$ such that $x_i = f_{\lambda_0}^k(x_0)(1 \leq k \leq m-1)$, then the invariant cycle containing x_0 is a boundary of a circle band. Without loss of generality, we assume that each point $f_{\lambda_0}^k(x_0) \neq x_i$ for any $1 \leq k \leq m-1$ and for any $1 \leq i \leq N$. Therefore, each function $h_k(\lambda)(1 \leq k \leq m-1)$ is continuous at λ_0. Let the discontinuous points $x_i (1 \leq i \leq N)$ of f_λ be arranged on L in order such that the open intervals $(x_j, x_{j+1}) \subset L$ contain no discontinuous points of f_λ. Then there is a real number $\delta > 0$ sufficiently small such that if $0 < \epsilon < \delta$, $f_{\lambda_0 \pm \epsilon}^k(x_0)$ and $f_{\lambda_0}^k(x_0)$ are in the same open interval $(x_j, x_{j+1}) \subset L$. Since M is orientable and L is transversal to u, for each $k(0 \leq k \leq m-1)$, $f_{\lambda_0+\epsilon}^k(x_0)$ is on the right side of $f_{\lambda_0}^k(x_0)$ (resp. $f_{\lambda_0-\epsilon}^k(x_0)$ is on the left side of $f_{\lambda_0}^k(x_0)$) under the direction of w and the meaning of minimum distance. Hence $f_{\lambda_0\pm\epsilon}^{m-1}(x_0) \neq x_0$, i.e., $\sum_{k=1}^{m} h_k(\lambda_0 \pm \epsilon) \neq$ integer for any $0 < \epsilon < \delta$. It implies that, for any point $x_0 \in L$, there is only a countable number of real numbers $\lambda \in (-1,1)$ such that (3.2.4) holds. Thus it follows that there exists a dense subset $I \subset (-1,1)$, such that for each $\lambda \in I$, the vector field $u + \lambda w$ has no closed orbits in $E + \bar{B}$.

Inductively ,we can prove that there is an open subset $I_1 \times \cdots \times I_m \subset (-1,1) \times \cdots \times (-1,1)$ such that for each $(\lambda_1, \cdots, \lambda_m) \in I_1 \times \cdots \times I_m$, the vector field $u + \sum_{i=1}^{m} \lambda_i w_i$ has no closed orbits in $\Omega = E + \cup_{i=1}^{m} \bar{B}_i$. Assertion 3 is proved.

Step 5. Proof of Assertion 4. By (3.2.1), each connected component of $\partial\Omega_j(1 \leq j \leq R)$ is homological to zero. Hence, for each $A_i(1 \leq i \leq K)$, if A_i is homeomorphic to an annulus, then there must be two Ω_{j_1} and Ω_{j_2} such that $\partial A_i = \Omega_{j_1} \cap \bar{A}_i + \Omega_{j_2} \cap \bar{A}_i$. Let $\Omega_{j_1} \# \Omega_{j_2} = \Omega_{j_1} + \Omega_{j_2} + A_i$. It suffices then to prove that

$$g(\Omega_{j_1} \# \Omega_{j_2}) = g(\Omega_{j_1}) + g(\Omega_{j_2}). \tag{3.2.5}$$

We have

$$\begin{aligned} g(\Omega_{j_i}) &= 1 - \frac{1}{2}(\chi(\Omega_{j_i}) + r_{j_i}), \ i = 1,2 \\ g(\Omega_{j_1} \# \Omega_{j_2}) &= 1 - \frac{1}{2}(\chi(\Omega_{j_1} \# \Omega_{j_2}) + \widetilde{r}). \end{aligned} \tag{3.2.6}$$

Obviously, the connected component number $\widetilde{r}$ of $\partial(\Omega_{j_1} \# \Omega_{j_2})$ equals $r_{j_1} + r_{j_2} - 2$. On the other hand, it is well known that $\chi(\Omega_{j_1} \# \Omega_{j_2}) = \chi(\Omega_{j_1}) + \chi(\Omega_{j_2})$. Therefore (3.2.5) follows from (3.2.6).

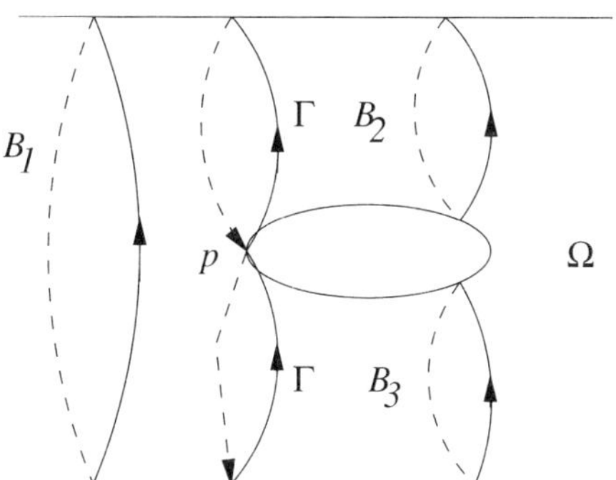

FIGURE 3.2.9. Intersection of circle bands B_1, B_2 and B_3.

Step 6. We have proved in Step 3 above that $D^r_B(TM)$ is open in $D^r(TM)$. Now we show that $D^r_B(TM)$ is dense in $D^r(TM)$. By Lemma 3.2.8, we only need to prove that any self-connected vector field $v \in D^r_1(TM)$ can be approximated by basic vector fields.

Let $v \in D^r_1(TM)$, and let $\Omega \subset M$ be an invariant closed domain whose interior $\overset{\circ}{\Omega}$ contains no flow-invariant closed curves of zero homology. Obviously, Ω consists of ergodic sets of v and the closure of a finite number of circle bands where each closed orbit is not homological to zero. Hence, $A = \overline{M - \Omega}$ is the closure of all closed orbits, each of which is homological to zero. It is easy to see that $\partial A = \partial\Omega$. If each connected component of ∂A is homeomorphic to S^1, then Ω and $\overset{\circ}{A}$ consist of invariant sub-manifolds as required in the definition of a basic vector field. Namely, v is a basic vector field.

Let v be not a basic vector field. Then there is a connected component $A_1 \subset A$, one of whose boundary components $\Gamma \subset \partial A_1$ is not homeomorphic to S^1. Because each boundary component of A is a saddle self-connection, Γ is homeomorphic to the topological set $S[x, y]$, where $S[x, y]$ is the quotient space

$$S[x, y] = S^1/\{x = y\},$$

and $x, y \in S^1$ are two different points on the cycle S^1. Hence, the invariant closed curve Γ is a common boundary of three circle bounds $B_i (1 \leq i \leq 3)$ with $B_1 \subseteq A_1, B_2, B_3 \subset \Omega$; see Figure 3.2.9.

Obviously, we can take a simple closed curve L that transversally intersects with each closed orbit in B_2 and B_3 at exactly one point. We take the tubular domain T and vector field w defined as in Lemma 2.1.10, then for any $\lambda \neq 0$ sufficiently small, the saddle point p of v on Γ is also a saddle point of $v + \lambda w$ and the saddle self-connection $\widetilde{\Gamma}$ of $v + \lambda w$ at p is homeomorphic to S^1; see Figure 3.2.10.

The invariant set $\widetilde{\Omega}$, which approximates Ω as $\lambda \to 0$, has no interior invariant closed curves homological to zero. The procedure shows that v can be approximated by basic vector fields.

The proof of the main theorem, Theorem 3.2.3, is complete. □

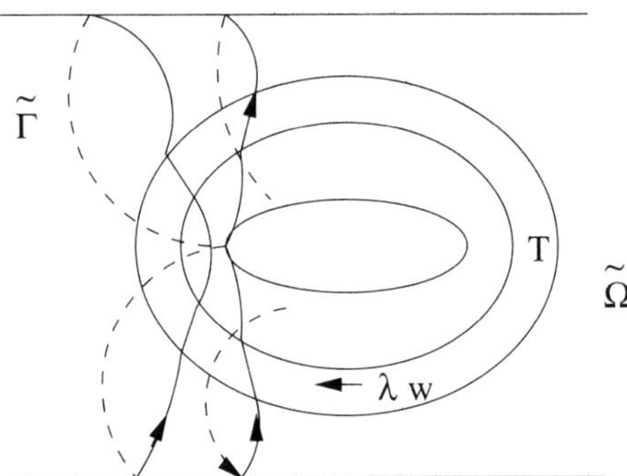

FIGURE 3.2.10. Flow structure with the small perturbation.

3.3. Structural Evolution of the Taylor Vortices

3.3.1. The Taylor fields on tori. Let

$$M = \mathbb{R}^2/(2\pi\mathbb{Z})^2.$$

The Taylor fields are defined by

$$v_{nm} = (m \cos nx_1 \cos mx_2, n \sin nx_1 \sin mx_2), \tag{3.3.1}$$

where $n, m \geq 1$ are integers. The Hamiltonian functions (or the stream functions) of (3.3.1) are given by

$$H_{nm}(x_1, x_2) = \cos nx_1 \sin mx_2. \tag{3.3.2}$$

From (3.3.1), it is easy to see that each Taylor field v_{nm} has $4nm$ saddle points and $4nm$ centers. The saddle points of v_{nm} are

$$\left(\frac{2k_1+1}{2n}\pi, \frac{k_2}{m}\pi\right) \quad \text{for } k_1 = 0, 1, \cdots, 2n-1, \quad k_2 = 0, 1, \cdots, 2m-1;$$

and the centers of v_{nm} are

$$\left(\frac{k_1}{n}\pi, \frac{2k_2+1}{2m}\pi\right) \quad \text{for } k_1 = 0, 1, \cdots, 2n-1, \quad k_2 = 0, 1, \cdots, 2m-1.$$

For instance, the phase diagram of the Taylor field v_{22} is shown in Figure 3.3.1.

By the structural stability theorem, the Taylor fields (3.3.1) are unstable because they are not self-connection fields. Since the set of all stable Hamiltonian vector fields $H_s^r(TM)$ is open and dense in $H^r(TM)$, under a perturbation in $H^r(TM)$, a Taylor field v_{nm} will be transformed into a stable Hamiltonian vector field. We consider the problem of what and how many types of stable Hamiltonian vector fields one can obtain by perturbing slightly the Taylor field v_{nm} with Hamiltonian fields. The procedure is to break saddle connections of the Taylor vortices. In the next section, we shall introduce some methods of breaking saddle connections, and then we will study the questions raised for the Taylor fields.

In this section, we introduce some methods to break saddle connections of the Taylor vortices. The main tool is the tubular flow introduced in Lemma 2.1.9.

Hamiltonian Breaking Method. We know the local structure of the flow pattern of a Taylor field v_{nm} is as shown in Figure 3.3.2(a).

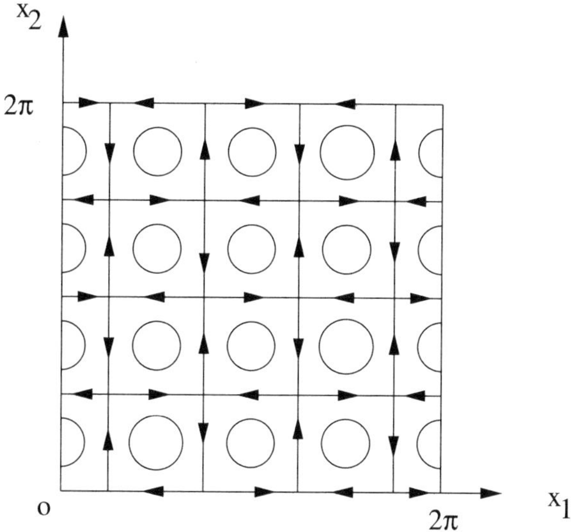

FIGURE 3.3.1. Phase diagram of the Taylor field v_{22}.

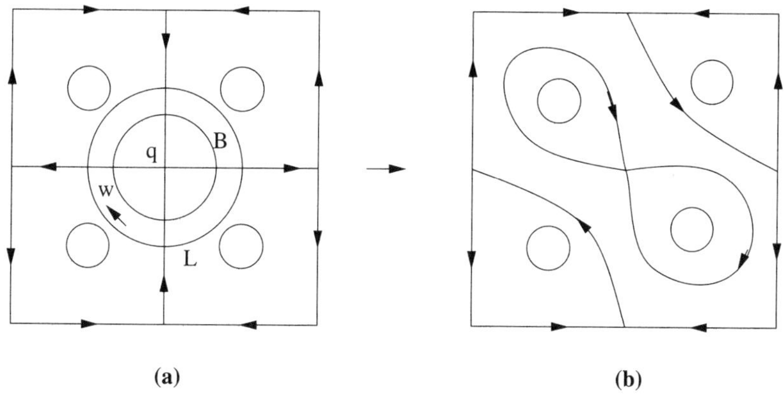

FIGURE 3.3.2. Local structure of a Taylor field v_{nm}, and a Hamiltonian perturbation.

For a saddle point q that is connected with other saddle points, we set an open annulus B with ∂B retractable to q in M, and we take a Hamiltonian vector field w as in Lemma 2.1.9 (see Figure 3.3.2(a)). Then for any $\lambda > 0$ sufficiently small, the pattern of v in Figure 3.3.2(a) transforms to the pattern in Figure 3.3.2(b) of $v + \lambda w$. We call this method the *Type I Local Breaking Method at the saddle point* q.

Notice that by reversing the orientation of w, the saddle points are connected slightly differently as illustrated by Figures 3.3.2 and 3.3.3.

REMARK 3.3.1. The perturbation field w can be constructed analytically as follows. Let $0 < r_1 < r_2$ be two given small numbers and let $\psi \in C^\infty[0, \infty)$ such

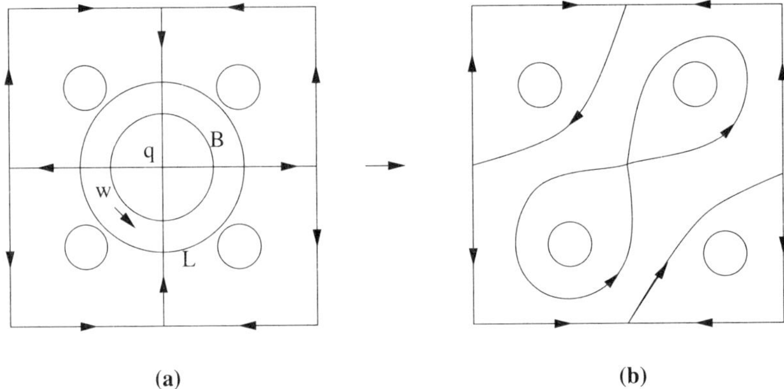

FIGURE 3.3.3. Another Hamiltonian perturbation.

that

$$\begin{cases} \psi(r) = 1 & \text{if } r \geq r_2, \\ \psi(r) = 0 & \text{if } r \leq r_1, \\ \psi(r) \in (0,1) \quad \text{and increasing} & \text{if } r \in (r_1, r_2). \end{cases} \tag{3.3.3}$$

Then for $0 < r_1 < r_2$ small, the divergence-free vector fields

$$\begin{cases} w_1 = \text{ curl } \psi(|(x_1, x_2) - (x_1^0, x_2^0)|), \\ w_2 = \text{ curl } [1 - \psi(|(x_1, x_2) - (x_1^0, x_2^0)|)], \end{cases} \tag{3.3.4}$$

are typical examples of the perturbation fields in Figures 3.3.2(a) and 3.3.3(a) respectively. Here, (x_1^0, x_2^0) is the coordinate of q, and r_1 and r_2 are the radii of the inner and outer circles of the circle band B.

Non-Hamiltonian Breaking Method. As we know, the Hamiltonian vector fields do not exhaust all divergence-free vector fields. We now construct two typical families of non-Hamiltonian divergence-free vector fields, and apply them to the flow structure of the Taylor vortices.

Consider again the saddle point q, which is connected by four saddle connections. In the Hamiltonian case, the closed curve L with zero homology has to be chosen to intersect each of the four saddle connections at least one time. In the non-Hamiltonian case, we can choose L to intersect either one or three of the four saddle connections as shown in Figures 3.3.4(a) and 3.3.5(a). Then, for any $\lambda > 0$ sufficiently small, the pattern in Figure 3.3.4(a) (resp. Figure 3.3.5(a)) of v transforms to the pattern in Figure 3.3.4(b) (resp. Figure 3.3.5(b)) of $v + \lambda w$. We call the method illustrated in Figure 3.3.4 the *Type II Local Breaking Method at the saddle point* q, and the method corresponding to Figure 3.3.5 the *Type III Local Breaking Method at the saddle point* q.

3.3.2. Structural evolution of the Taylor fields with Hamiltonian perturbations. Our first main result is the following theorem on the structural evolution of the general Taylor field $v_{nm}(1 \leq m \leq n)$. In the next subsection, we shall study the detailed structural evolution of v_{11}.

THEOREM 3.3.2. *Let* $v_{nm}(1 \leq m \leq n)$ *be a Taylor field. Then, for any sufficiently small neighborhood* $\mathcal{O} \subset H^r(TM)$ *of* v_{nm} *and* $v \in \mathcal{O} \cap H_s^r(TM)$*, we have:*

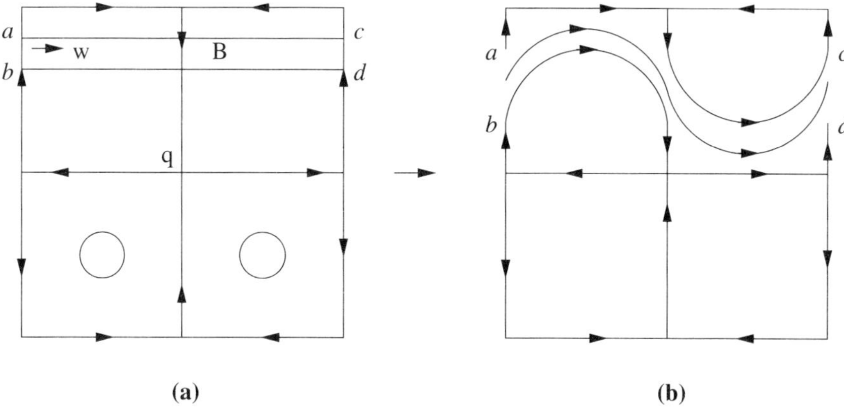

FIGURE 3.3.4. *Type II Local Breaking Method.* The segment ab (resp. cd) in (b) is the same as ab (resp. cd) in (a). In (b), any orbit passing through a point in the open segment ab is periodic.

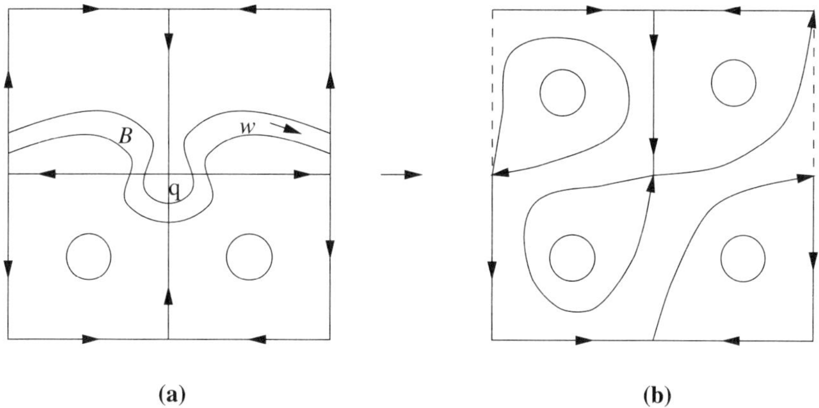

FIGURE 3.3.5. Type III Local Breaking Method.

(1) *the number $K = 2R$ of the S-blocks of v satisfies*

$$1 \leq R \leq n;$$

(2) *for each $R(1 \leq R \leq n)$, there are $v \in \mathcal{O} \cap H^r_s(TM)$ such that v have exactly $2R$ S-blocks;*

(3) *in each S-block of $v \in \mathcal{O} \cap H^r_s(TM)$, there are at least $2m$ centers (or $2m-1$ saddle points) of v; and*

(4) *when $m = 1$, all circle cells in right-hand D- and S-blocks (resp. in left-hand D- and S-blocks) are right-hand orientation (resp. left-hand orientation).*

PROOF. *Step 1.* Let $L \subset M$ be an orbit line of a vector field v. If a closed orbit line L, as a closed chain, is homological to zero, then there must be an open set $A \subset M$ such that $\partial A = L$, and we say that L is retractable if the closure $\bar{A}$ is retractable to a point in M.

The following lemma can be derived from Theorem 2.1.15.

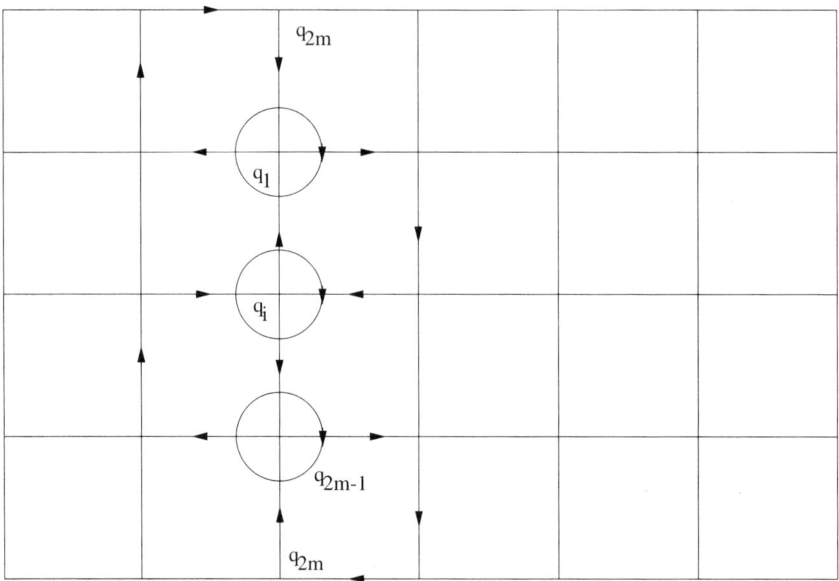

FIGURE 3.3.6. Single circles are used to denote the tubular flows near saddle points q_j here as well as in figures hereafter.

LEMMA 3.3.3. *Let $v_n, v \in D^r(TM)$ be regular with $v_n \to v(n \to \infty)$, and let L_n be a closed orbit line of v_n. Then there are a closed orbit line L of v and a subsequence still denoted by L_n such that $L_n \to L(n \to \infty)$, and*

(1) *L is homological to zero if and only if L_n are homological to zero.*
(2) *If L is retractable, then L_n are retractable.*
(3) *Let $\partial A_n = L_n$, $\partial A = L$ and $\bar{A}_n \to \bar{A}$. Then, for any n sufficiently large, the number of centers of v_n in A_n equals the number of centers of v in A.*

Now we come back to prove Assertions 1–4. A Taylor field $v_{nm}(1 \leq m \leq n)$ has $4nm$ centers, and from the flow pattern of v_{nm} (see Figure 3.3.1), we can see that each open set A enclosed by a closed non-retractable orbit curve must contain at least $2m$ centers. Hence Assertions 1 and 3 follow from Lemma 3.3.3.

Step 2. Proof of Assertion 2. We proceed in a few cases as follows.

Case $R = n$: We first construct vector fields in $\mathcal{O} \cap H^r_s(TM)$ with exactly $2R = 2n$ S-blocks. Consider a non-retractable closed orbit line L enclosing exactly $2m$ circle cells, and having exactly $2m$ self-intersecting points $q_1, \cdots, q_{2m}$.

Then we apply the Type I Local Breaking Method with clockwise tubular flows described at the end of the previous section to the Taylor field at the saddle points $q_i(1 \leq i \leq 2m-1)$; see Figure 3.3.6. The flow pattern in Figure 3.3.6 transforms to the flow pattern in Figure 3.3.7. Next, we apply again the Type I Local Breaking Method near the saddle point q_{2m} still with clockwise tubular flow; the flow pattern transforms into that in Figure 3.3.8, which contains one S-block A_1.

Next we consider a non–retractable closed orbit curve $\widetilde{L}$ enclosing exactly $2m$ circle cells, and with $2m$ self-intersecting saddle points $p_1, \cdots, p_{2m-1}, p_{2m}$ as shown in Figure 3.3.8. Applying the Type I Local Breaking Method with counter-clockwise

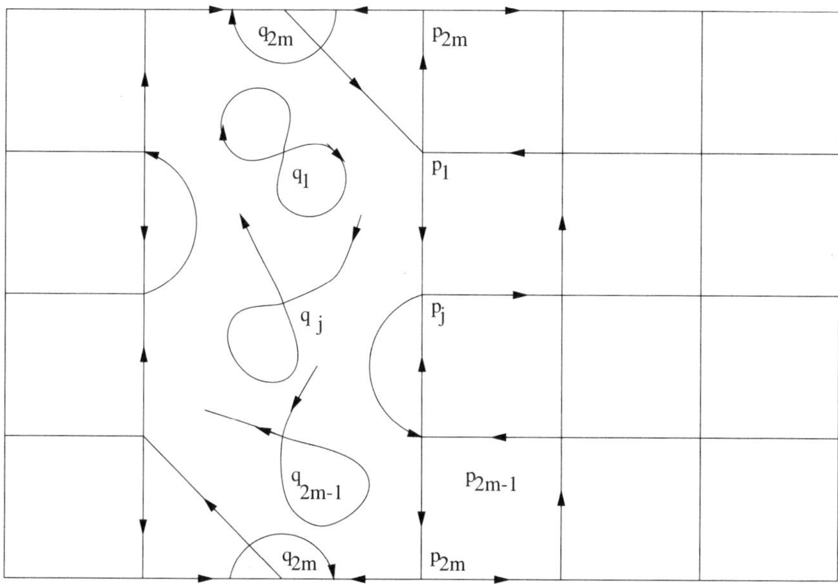

FIGURE 3.3.7. Flow structure after successive Type I perturbations.

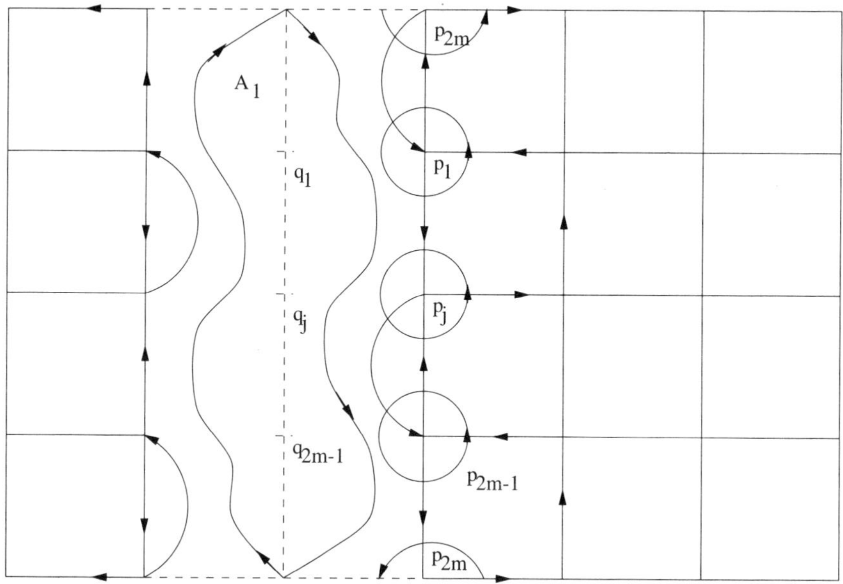

FIGURE 3.3.8. An S-block is derived after successive Type I perturbations.

tubular flows near these saddle points, we obtain another S-block, which we call A_2, enclosing the saddle points $p_1, \cdots, p_{2m-1}$; see Figure 3.3.9. Inductively, in exactly $2n$ steps, the flow pattern in Figure 3.3.7 will be transformed to the flow pattern having exactly $2n$ S-blocks.

Case $R = 1$: We construct vector fields in $\mathcal{O} \cap H^r_s(TM)$ with exactly $2R = 2$ S-blocks.

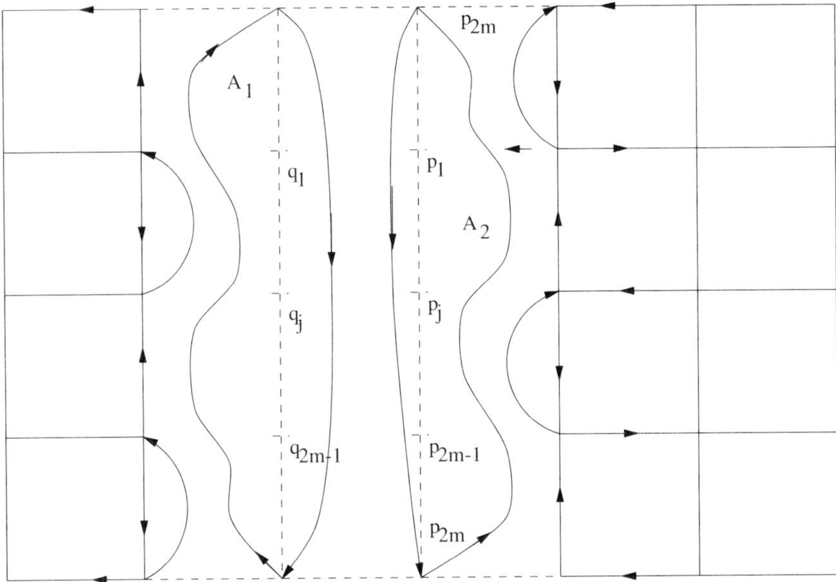

FIGURE 3.3.9. An S-block is derived after successive counterclockwise Type I perturbations.

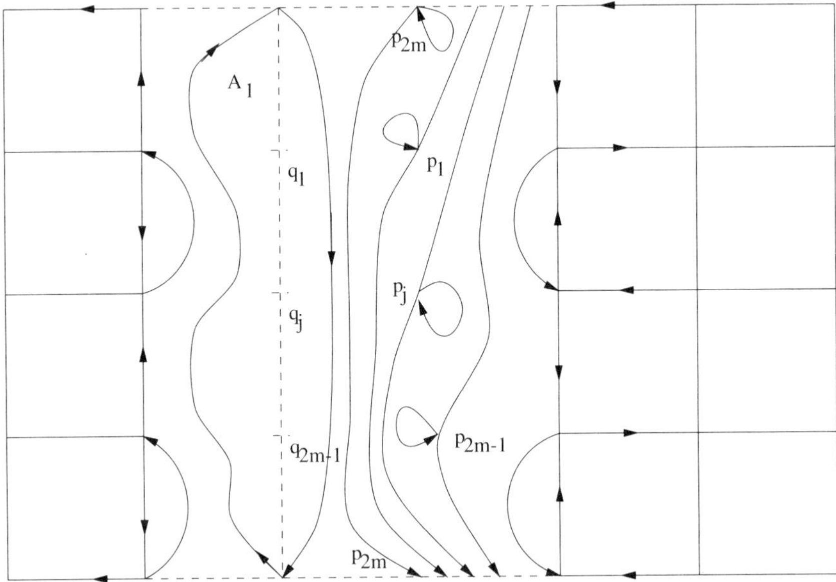

FIGURE 3.3.10. Derivation of $2m$ D-blocks.

As in the previous case, we obtain an $S-$block as shown in Figure 3.3.8. Then, instead of the counterclockwise tubular flows, we apply the Type I Local Breaking Method with clockwise tubular flows near the saddle points $p_1, \cdots, p_{2m-1}, p_{2m}$. We then obtain $2m$ D-blocks as shown in Figure 3.3.10.

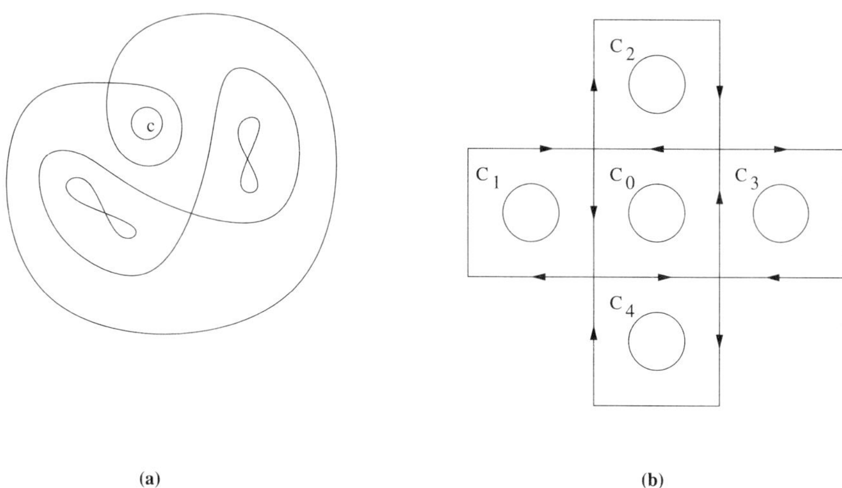

FIGURE 3.3.11. Schematic illustrating Step 3 in the proof of Theorem 3.3.2.

By repeating this procedure $2n-2$ times and then applying the Type I Local Breaking Method to the last non-retractable orbit curve on the right of the S-block A_1 with clockwise tubular flows, we obtain another S-block and all other saddles are connected to a D-block as shown in Figure 3.3.10.

Case $1 < R < n$: In this case, we construct the fields in $\mathcal{O} \cap H_s^r(TM)$ having exactly $2R$ S-blocks in two steps. First, we obtain $2R$ S-blocks as in the case where $R = n$; then we break all other saddles into D-blocks.

Step 3. Proof of Assertion 4. For any $v \in \mathcal{O} \cap H_s^r(TM)$, v is a self-connection field. Note that each closed orbit curve of v_{1n} contains at least four saddle points. Therefore if a left-hand S-block A^- contains a right-hand circle cell C, then the flow pattern adjacent to C must be as illustrated by Figure 3.3.11(a).

The flow pattern in Figure 3.3.11(a) can only be obtained by perturbing the flow pattern in Figure 3.3.11(b) in the Taylor field. In Figure 3.3.11(b), there are four different left-hand circle cells C_1, C_2, C_3 and C_4 adjacent to the right-hand circle cell C_0, corresponding to C in Figure 3.3.11(a). Obviously there is no such structure in the flow pattern of v_{1n}. Assertion 4) is proved.

The proof is complete. □

Uniqueness of the structural evolution of v_{11}. When $m = n = 1$, we have the following uniqueness result of the structural evolution of the Taylor field v_{11}. In the next subsection, we shall analytically construct the perturbation fields, and will make connections between the solutions of the Navier-Stokes equations.

THEOREM 3.3.4. *There is a neighborhood $\mathcal{O} \subset H^r(TM)$ of v_{11} such that $\mathcal{O} \cap H_s^r(TM)$ consists of exactly one topologically equivalent class. The standard phase portrait of the topologically equivalent class $\mathcal{O} \cap H_s^r(TM)$ is as shown in Figure 3.3.12.*

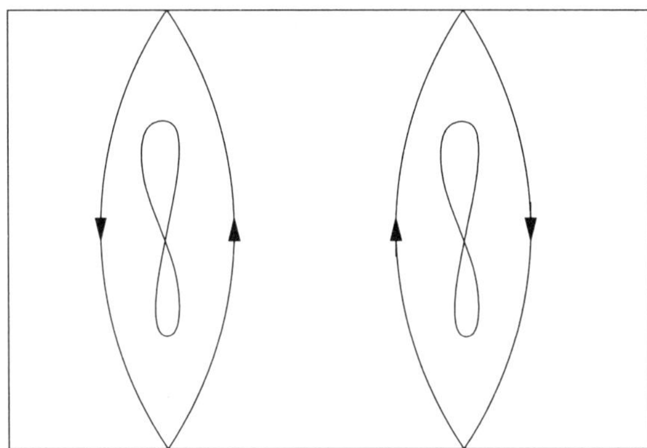

FIGURE 3.3.12. The standard phase portrait of the topologically equivalent Hamiltonian stable class near v_{11}.

PROOF. By Assertions (1), (3) and (4) in Theorem 3.3.2, each $v \in \mathcal{O} \cap H_s^r(TM)$ has a block decomposition isomorphic to the standard decomposition

$$M = \Omega + A^+ + A^-,$$

where Ω is a T-block, and A^+ and A^- are S-blocks with right-hand and left-hand orientations, respectively. Therefore, v has a saddle connection diagram isomorphic to that as shown in Figure 3.3.12. By the Topological Classification Theorem in Section 1.4, Theorem 3.3.3 follows. □

3.3.3. Structural evolution of the Taylor fields with divergence-free vector field perturbations. The structural evolution of the Taylor fields under the perturbation of divergence-free vector fields is quite different from that under perturbation of Hamiltonian fields. In this section, we only give an example to show the difference.

In Section 3.1, we proved that no divergence-free vector fields on the torus are structurally stable under the perturbation in $D^r(TM)$, due essentially to the existence of ergodic sets.

We denote by $D_B^r(TM)$ the set of all C^r basic vector fields.

The differences between the block structures of the divergence-free vector fields and the Hamiltonian vector fields are as follows: 1) there are no S-blocks in the block structure of basic divergence-free fields; 2) the T-block Ω of a basic divergence-free vector field is a manifold, which may be an ergodic set.

Due mainly to these differences, under a perturbation in $D^r(TM)$, a Taylor field v_{nm} may be transformed into a vector field that has $4nm$ D-blocks, i.e., each center corresponds to a D-block. More precisely, for a Taylor field v_{nm} and any sufficiently small neighborhood $\mathcal{O} \subset D^r(TM)$ of v_{nm}, the number K of the D-blocks of $v \in \mathcal{O} \cap D_B^r(TM)$ satisfies

$$2 \leq K \leq 4mn,$$

and for each $K (2 \leq K \leq 4mn)$, there are $v \in \mathcal{O} \cap D_B^r(TM)$ such that v have K D-blocks.

We now illustrate this point with the following example. Consider the Taylor field $v_{11} = \{\cos x_1 \cos x_2, \ \sin x_1 \sin x_2\}$, which has the flow pattern as shown in Figure 3.3.13(a). By Lemma 2.1.9, we take a tubular incompressible vector field

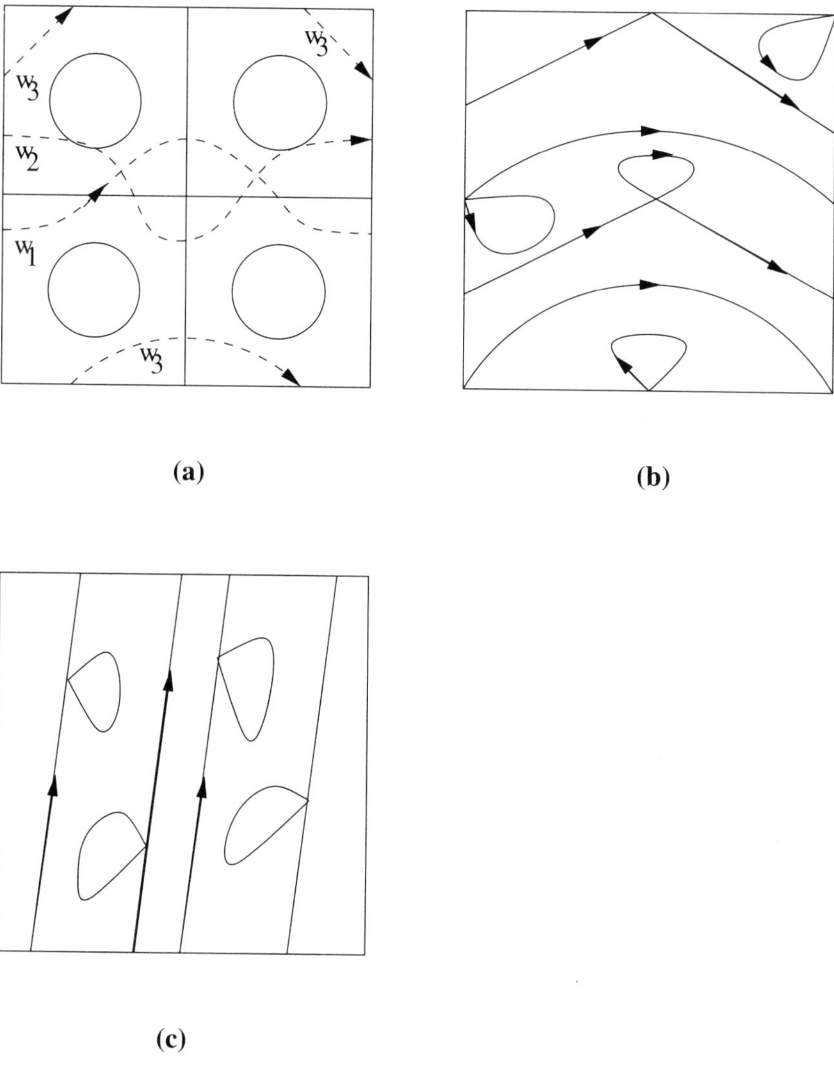

FIGURE 3.3.13. The dotted lines in (a) represent small tubular flows w_1, w_2 and w_3. With perturbation of $w_1 + w_2 + w_3$, the Taylor field v_{11} becomes the flow structure as shown in (b), or equivalently the block structures shown in (c).

w_1 as in Figure 3.3.13(a). Under the perturbation $\lambda_1 w_1$, the flow pattern in Figure 3.3.13(a) of v_{11} becomes the pattern in Figure 3.3.13(b) of $v_{11} + \lambda_1 w_1 (\lambda > 0)$. Then we take two tubular incompressible vector fields w_2 and w_3 as in Figure 3.3.13(b). Under the perturbation of $\lambda_2 w_2$ and $\lambda_3 w_3 (\lambda_2 > 0, \lambda_3 > 0)$, the flow pattern (b) will be transformed into the flow pattern (c) of $v_{11} + \lambda_1 w_1 + \lambda_2 w_2 + \lambda_3 w_3$, which is as desired.

Of course, there are other ways to break the saddle connections of v_{11}. For instance, by using the Type I Local Breaking Method (with Hamiltonian perturbations), we obtain the flow pattern as shown in Figure 3.3.12. Then by applying the

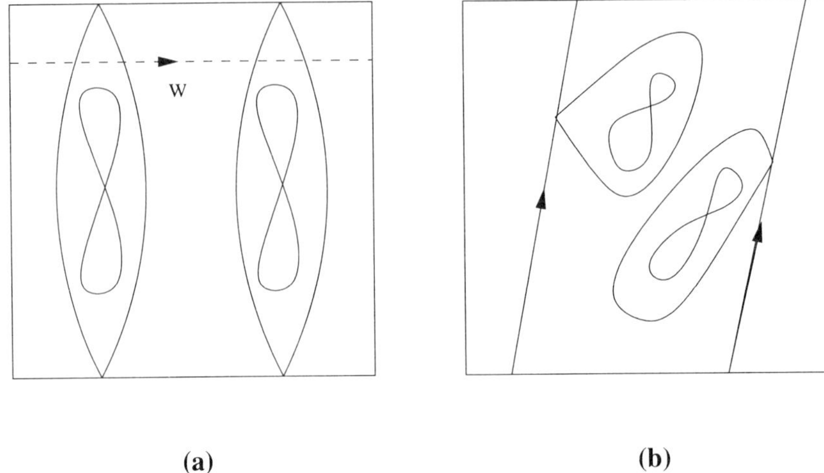

(a) **(b)**

FIGURE 3.3.14. A block structure near v_{11}.

Type II Local Breaking Method as shown in Figure 3.3.14(a), we obtain the block structure Figure 3.3.14(b). In this case, the final flow pattern Figure 3.3.14(b) has a T-block and two D-blocks, each of which contains two centers and an eight-shaped saddle connection.

Notes for Chapter 3

The material in this section is based on the authors [**53, 54, 67**]. We acknowledge that the study of the evolution of the Taylor vortices was bought to our attention by G. Papanicolaou.

CHAPTER 4

Structural Stability of Solutions of Navier-Stokes Equations

This chapter is oriented toward linking the kinematic theory to dynamics of incompressible fluid flows as described in Area B in the Introduction.

4.1. Genericity of Stable Steady States

4.1.1. Main theorem. The main objective of this section is to study the generic properties of structurally stable solutions of the steady-state Navier-Stokes equations with different sets of boundary conditions. The problems read as follows:

$$-\mu\Delta u + (u\cdot\nabla)u + \nabla p = f \qquad \text{in } \overset{\circ}{M}\subset \mathbb{R}^2, \tag{4.1.1}$$

$$\text{div } u = 0 \qquad \text{in } \overset{\circ}{M}, \tag{4.1.2}$$

with either the free boundary conditions

$$u_n = 0, \quad \frac{\partial u_\tau}{\partial n} = 0, \qquad \text{on } \partial M, \tag{4.1.3}$$

or the homogeneous Dirichlet boundary conditions

$$u = 0, \qquad \text{on } \partial M, \tag{4.1.4}$$

or the non-homogeneous Dirichlet boundary conditions

$$u = \varphi, \qquad \text{on } \partial M. \tag{4.1.5}$$

Notice that if $u \in B^{2+\alpha}(TM)$ (or $u \in B_0^{2+\alpha}(TM)$, or $u \in B_\varphi^{2+\alpha}(TM)$) satisfies (4.1.1), then u is a solution of the problem (4.1.1–4.1.3) (or the problem (4.1.1), (4.1.2) and (4.1.4), or the problem (4.1.1), (4.1.2) and (4.1.5)).
The main theorem in this section is

THEOREM 4.1.1. *Let X be either $B^{2+\alpha}(TM)$ or $B_0^{2+\alpha}(TM)$ or $B_\varphi^{2+\alpha}(TM)$. For any $\mu > 0$, there is an open and dense set $\mathcal{F} \subset C^\alpha(TM)(0 < \alpha < 1)$ such that for each $f \in \mathcal{F}$, the solutions $u \in X$ of (4.1.1) and (4.1.2) with corresponding boundary conditions are structurally stable in X.*

Let $S(f, \mu; X)$ be the set of all solutions $u \in X$ of (4.1.1) and (4.1.2) with corresponding boundary conditions. Then the above theorem can be restated as follows:

THEOREM 4.1.2. *For any $\mu > 0$, there is an open and dense set $F \subset C^\alpha(TM)$ such that for each $f \in F$, we have*

(1) (Foias and Temam [**23**]). *The set $S(f, \mu)$ is non-empty and finite.*

(2) *For each solution $u \in S(f,\mu;X)$, there exist a neighborhood $U \subset X$ of u and a neighborhood $\mathcal{O} \subset F$ of f such that, for any $f_1 \in \mathcal{O}$, there is a unique solution $u_1 \in S(f_1,\mu)$ in U, and u_1 and u are topologically equivalent.*

4.1.2. Some *a priori* estimates. The proof of Theorem 4.1.1 is based on some *a priori* estimates of the steady state solutions of the Navier-Stokes equations, and an infinite dimensional version of the Sard theorem due to S. Smale [**95**].

We start with some *a priori* estimates of the solutions of the following 2D Stokes equations:

$$\begin{cases} -\mu\Delta u + \nabla p = g & \text{in } \overset{\circ}{M}, \\ \operatorname{div} u = 0 & \text{in } \overset{\circ}{M}, \end{cases} \tag{4.1.6}$$

with one of the boundary conditions (4.1.3), or (4.1.4), or (4.1.5).

We first consider the case with either the boundary condition (4.1.3) or (4.1.4), and let $X = B^{2+\alpha}(TM)$ or $X = B_0^{2+\alpha}(TM)$.

LEMMA 4.1.3. *Let $X = B^{2+\alpha}(TM)$ or $X = B_0^{2+\alpha}(TM)$, let $(u,p) \in X \times C^{1+\alpha}(M)$ be a solution of (4.1.6) with corresponding conditions, and let $g \in C^\alpha(TM)$. Then*

$$\|u\|_{C^{2+\alpha}} + \|p\|_{C^{1+\alpha}} \le C\|g\|_{C^\alpha}, \tag{4.1.7}$$

where $C > 0$ is a constant.

PROOF. We proceed by applying the general regularity result for elliptic systems of equations by Agmon–Doglis–Nirenberg [**2**] to the above Stokes problem (4.1.6) with different boundary conditions.

In [**102**], the ellipticity of (4.1.6) is verified, and the Dirichlet boundary conditions (4.1.4) satisfy the Complementary Boundary Conditions. By Theorem 9.3 and Remark 2 on p. 74 of [**2**], it remains to check the Complementary Boundary Conditions required there for the free boundary conditions (4.1.3).

Let $u = (u_1, u_2), u_3 = \frac{1}{\mu}p, v = (u_1, u_2, u_3)$. The principal part of (4.1.6) is

$$\ell(D)v = \begin{pmatrix} (\frac{\partial}{\partial x_1})^2 + (\frac{\partial}{\partial x_1})^2 & 0 & -\frac{\partial}{\partial x_2} \\ 0 & (\frac{\partial}{\partial x_1})^2 + (\frac{\partial}{\partial x_2})^2 & -\frac{\partial}{\partial x_2} \\ \frac{\partial}{\partial x_1} & \frac{\partial}{\partial x_2} & 0 \end{pmatrix} \begin{pmatrix} u_1 \\ u_2 \\ u_3 \end{pmatrix}.$$

The boundary operator is

$$B(D)v = \begin{pmatrix} n_1 & n_2 & 0 \\ \tau_1 n_1 \frac{\partial}{\partial x_1} + \tau_1 n_2 \frac{\partial}{\partial x_2} & \tau_2 n_1 \frac{\partial}{\partial x_1} + \tau_2 n_2 \frac{\partial}{\partial x_2} & 0 \end{pmatrix} \begin{pmatrix} u_1 \\ u_2 \\ u_3 \end{pmatrix}.$$

For a vector $\xi = (\xi_1, \xi_2)$, the corresponding matrices of the above differential operators are given by

$$L(\xi) = \begin{pmatrix} \xi_1^2 + \xi_2^2 & 0 & \xi_1 \\ 0 & \xi_1^2 + \xi_2^2 & \xi_2 \\ -\xi_1 & -\xi_2 & 0 \end{pmatrix},$$

$$B(\xi) = \begin{pmatrix} n_1 & n_2 & 0 \\ \tau_1 n_1 \xi_1 + \tau_1 n_2 \xi_2 & \tau_2 n_1 \xi_1 + \tau_2 n_2 \xi_2 & 0 \end{pmatrix}.$$

For the normal and tangent vectors n, τ on the boundary and a parameter t, we have

$$L(\tau + tn) = \begin{pmatrix} (1+t^2) & 0 & \tau_1 + tn_1 \\ 0 & (1+t^2) & \tau_2 + tn_2 \\ -(\tau_1 + tn_1) & -(\tau_2 + tn_2) & 0 \end{pmatrix},$$

$$B(\tau + tn) = \begin{pmatrix} n_1 & n_2 & 0 \\ t\tau_1 & t\tau_2 & 0 \end{pmatrix}.$$

Hence, we have

$$B(\tau + tn) \times L(\tau + tn) = \begin{pmatrix} n_1(1+t^2) & n_2(1+t^2) & t \\ t\tau_1(1+t^2) & t\tau_2(1+t^2) & t \end{pmatrix}.$$

It is easy to see that the algebraic equation

$$\det\ \ell(\tau + tn) = 0$$

has exactly two roots with positive imaginary part and these roots are all equal to $t^+ = i$. Therefore,

$$M^+(\tau + tn) = (t - t^+)^2 = (t - i)^2.$$

Obviously the algebraic equation system

$$\begin{cases} n_1(1+t^2)c_1 + n_2(1+t^2)c_2 = 0, \\ t\tau_1(1+t^2)c_1 + t\tau_2(1+t^2)c_2 = 0 \quad (\text{mod } M^+ = (t-i)^2), \\ tc_1 + tc_2 = 0, \end{cases}$$

has only a zero solution $c_1 = c_2 = 0$, and the Complementary Boundary Conditions hold. The proof is complete.

□

The L_p estimates for the solutions of the Stokes equations with either the free boundary conditions (4.1.6) or the homogeneous Dirichlet boundary conditions can also be obtained in the same fashion using Theorem 10.5 on p. 78 of [**2**] (see also [**9, 26, 97, 25, 102**]).

LEMMA 4.1.4. *Let $(u, p) \in W^{2,p}(TM) \times W^{1,p}(M) (p \geq 2)$ be a solution of (4.1.6) with boundary conditions (4.1.3) or (4.1.4), and let $g \in L^p(TM)$. Then*

$$\|u\|_{W^{2,p}} + \|p\|_{W^{1,p}} \leq c\|g\|_{L^p}, \tag{4.1.8}$$

where $c > 0$ is a constant.

Now we return to derive the $C^{2+\alpha}$–estimates for the steady state solutions of the Navier–Stokes equations.

LEMMA 4.1.5. *Let $X = B^{2+\alpha}(TM)$ or $X = B_0^{2+\alpha}(TM)$, let $(u, p) \in X \times C^{1+\alpha}(M)$ be a solution of (4.1.1) and (4.1.2) with corresponding conditions, and let $f \in C^\alpha(TM)$. Then*

$$\|u\|_{C^{2+\alpha}} + \|p\|_{C^{1+\alpha}} \leq C\left[\|f\|_{C^\alpha} + \|f\|_{L^p}^{p+2}\right], \qquad p > \frac{2}{1-\alpha}. \tag{4.1.9}$$

PROOF. Let $g = f - (u \cdot \nabla)u$; then $g \in C^\alpha(TM)$. Lemma 4.1.3 yields that

$$\|u\|_{C^{2+\alpha}} + \|p\|_{C^{1+\alpha}} \leq C\|g\|_{C^\alpha} \leq C\left[\|f\|_{C^\alpha} + \|u\|_{C^{1+\alpha}}^2\right]. \tag{4.1.10}$$

By the Sobolev embedding theorem,

$$\|u\|_{C^{1+\alpha}} \leq C\|u\|_{W^{2,p}}, \qquad p > \frac{2}{1-\alpha}. \tag{4.1.11}$$

By Lemma 4.1.4,

$$\|u\|_{W^{2,p}} \leq C\|g\|_{L^p} \leq C\Big[\|f\|_{L^p} + \|u \cdot Du\|_{L^p}\Big]. \tag{4.1.12}$$

On the other hand,

$$\|u \cdot Du\|_{L^p} \leq \|u\|_{L^{2p}} \cdot \|Du\|_{L^{2p}}, \tag{4.1.13}$$

$$\|Du\|_{L^{2p}} \leq C\ \varepsilon\|D^2u\|_{L^p} + \varepsilon^{-\beta}\|Du\|_{L^2}, \tag{4.1.14}$$

where $\varepsilon > 0$ is an arbitrary constant, and $\beta = \frac{p}{2} - 1$. By the Poincaré inequalities and the $W^{1,2}$-estimates for the stationary solutions of the Navier-Stokes equations, we deduce that

$$\|u\|_{L^{2p}} \leq C\|Du\|_{L^2} \leq C\|f\|_{L^2}. \tag{4.1.15}$$

We infer then from (4.1.12–4.1.15) that

$$\|u\|_{W^{2,p}} \leq C\|f\|_{L^p} + C\varepsilon\|f\|_{L^2} \cdot \|u\|_{W^{2,p}} + C\varepsilon^{-\beta}\|f\|^2_{L^2}.$$

We take $C\varepsilon\|f\|_{L^2} = 1/2$; then

$$\|u\|_{W^{2,p}} \leq C\|f\|_{L^p} + C\|f\|^{\beta+2}_{L^2} \leq C\|f\|^{\frac{p}{2}+1}_{L^p}.$$

Therefore (4.1.9) follows. The proof is complete. □

REMARK 4.1.6. Let $(u,p) \in B^{2+\alpha}_{\varphi}(TM) \times C^{1+\alpha}(M)$ be a solution of (4.1.6) and (4.1.5), and let $\varphi \in C^{2+\alpha}(T\partial M), g \in C^{\alpha}(TM)$. Then

$$\|u\|_{C^{2+\alpha}} + \|p\|_{C^{1+\alpha}} \leq C\Big[\|g\|_{C^\alpha} + \|\phi\|_{C^{2+\alpha}}\Big], \tag{4.1.16}$$

$$\|u\|_{W^{2,p}} + \|p\|_{W^{1,p}} \leq C\Big[\|g\|_{L^p} + \|\phi\|_{W^{2,p}}\Big]. \tag{4.1.17}$$

□

By (4.1.16), (4.1.17) and the $W^{1,2}$-estimates for the stationary solutions of the Navier-Stokes equations with non-homogeneous Dirichlet boundary condition, it is easy to obtain

LEMMA 4.1.7. *Let $(u,p) \in B^{2+\alpha}(TM) \times C^{1+\alpha(M)}$ be a solution of (4.1.1), (4.1.2) and (4.1.5), let $\varphi \in C^{2+\alpha}(T\partial M)$ and let $f \in C^\alpha(TM)$. Then*

$$\|u\|_{C^{2+\alpha}} + \|p\|_{C^{1+\alpha}} \leq C\Big[\|f\|_{C^\alpha} + \|f\|^{p+2}_{L^p} + \|\varphi\|_{C^{2+\alpha}}\Big] \tag{4.1.18}$$

where $C > 0$ is a constant, and $p > 2/(1-\alpha)$.

4.1.3. Sard-Smale theorem on Banach spaces. The proof of Theorem 4.1.1 relies on Sard's theorem on Banach spaces due to Smale [**95**].

Let E_1 and E_2 be two Banach spaces. A map $G : E_1 \to E_2$ is called a completely continuous field if $G = L + H$, where $L : E_1 \to E_2$ is a linear isomorphism and $H : E_1 \to E_2$ is a compact operator. We note that a C^1 completely continuous field $G : E_1 \to E_2$ must be a Fredholm map of index zero.

Let $G : E_1 \to E_2$ be a C^1 completely continuous field. A point $u \in E_1$ is called a regular point of G if $G'(u) : E_1 \to E_2$ is an isomorphism, and u is called a singular point if it is not a regular point. The image of a singular point under G

is called a singular value of G, and the complement of all singular points are called regular values of G. Notice that if $f \in E_2$ is not in the image $G(E_1)$, then f is automatically a regular value of G.

THEOREM 4.1.8. (Smale [**95**]). *Let E_1 and E_2 be two Banach spaces and $G : E_1 \to E_2$ be a C^1 completely continuous field. Then the set of all regular values of G is dense in E_2. Moreover, if $f \in E_2$ is a regular value of G, then $G^{-1}(f)$ is discrete.*

4.1.4. Completely continuous fields defined by the Navier-Stokes equations. Let $X = B^{2+\alpha}(TM)$ or $X = B_0^{2+\alpha}(TM)$, and let

$$G = L + H : X \times C^{1+\alpha}(M) \to C^\alpha(TM) \tag{4.1.19}$$

be a map such that for $(u,p) \in X \times C^{1+\alpha}(M)$,

$$\begin{cases} L(u,p) &= -\mu\Delta u + \nabla p, \\ H(u,p) &= (u \cdot \nabla)u. \end{cases} \tag{4.1.20}$$

We notice that $L : X \times C^{1+\alpha}(M) \to C^\alpha(TM)$ is a bounded linear operator corresponding to the Stokes equation (4.1.6), and $H : X \times C^{1+\alpha}(M) \to C^\alpha(TM)$ is a C^∞ nonlinear operator.

LEMMA 4.1.9. *The operator G is a completely continuous field. Moreover, G is a surjective map, i.e., $G(X \times C^{1+\alpha}(M)) = C^\alpha(TM)$.*

PROOF. It is well known that for any $g \in C^\infty(TM)$, the Stokes problem (4.1.6) with (4.1.3) or (4.1.4) has a unique solution $(u,p) \in C^\infty(TM) \times C^\infty(M)$ (for $p \in C^\infty(M)$ up to a constant). Therefore, $L(X \times C^{1+\alpha}(TM))$ is dense in $C^\alpha(TM)$. It follows from Lemma 4.1.3 that $L : X \times C^{1+\alpha}(M) \to C^\alpha(TM)$ is an isomorphism. The compactness of $H : X \times C^{1+\alpha}(M) \to C^\alpha(TM)$ is obvious.

Notice that for any $f \in C^\infty(TM)$ the Navier-Stokes equation (4.1.1) and (4.1.2) with either (4.1.3) or (4.1.4) has a solution $(u,p) \in C^\infty(TM) \times C^\infty(M)$, and by Lemma 4.1.5, $G : X \times C^{1+\alpha}(M) \to C^\alpha(TM)$ is surjective.

The proof is complete. □

Let $\varphi \in D^{2+\alpha}(TM)$ be an extension of the given boundary condition (4.1.5). For the non-homogeneous Navier-Stokes equations (4.1.1), (4.1.2) and (4.1.5), we define a map

$$G_\varphi = L + H_\phi : B_0^{2+\alpha}(TM) \times C^{1+\alpha}(M) \to C^\alpha(TM), \tag{4.1.21}$$

where the linear map L is as that in (4.1.19), and H_ϕ is

$$H_\varphi(u,p) = (u \cdot \nabla)u + (\varphi \cdot \nabla)u + (u \cdot \nabla)\varphi - \mu\Delta\varphi + (\varphi \cdot \nabla)\varphi.$$

Obviously, if $G_\phi(u,p) = f$ for $(u,p) \in B_0^{2+\alpha}(TM) \times C^{1+\alpha}(M)$ and $f \in C^\alpha(TM)$, then (u,p) is a solution of (4.1.1), (4.1.2) and (4.1.5).

With the same argument as that of Lemma 4.1.7, we have

LEMMA 4.1.10. *The operator G_φ defined by (4.1.21) is a completely continuous field, and is surjective.*

THEOREM 4.1.11. *Let X be either $B^{2+\alpha}(TM)$ or $B_0^{2+\alpha}(TM)$ or $B_\varphi^{2+\alpha}(TM)$, and let $A = G$ or $A = G_\varphi$. Then the set of all regular values of A is open and dense in $C^\alpha(TM)$. Furthermore, if $f \in C^\alpha(TM)$ is a regular value of A, then $A^{-1}(f)$ is nonempty and finite.*

PROOF. Let $\mathcal{R} \subset C^{\alpha}(TM)$ be the set of regular values of A. By the Sard-Smale theorem and Lemmas 4.1.9 and 4.1.10, $\mathcal{R}$ is dense in $C^{\alpha}(TM)$. Thanks to A being surjective, for any $f \in \mathcal{R}$, $A^{-1}(f)$ is nonempty and finite.

Let E_1 and E_2 be two Banach spaces, and let $L(E_1, E_2)$ be the space of all bounded linear operators from E_1 to E_2. It is known that the set of all isomorphisms from E_1 to E_2 is open in $L(E_1, E_2)$.

Let $f \in \mathcal{R}$, and let $A^{-1}(f) = \{v_1, \cdots, v_n\}$. By the inverse function theorem, for each $v_i \in A^{-1}(f)(1 \leq i \leq n)$, there is a neighborhood $U_i \subset X \times C^{1+\alpha}(M)$ of v_i and a neighborhood $\mathcal{F}_i \subset C^{\alpha}(TM)$ of f such that $A : U_i \to \mathcal{F}_i$ is a homeomorphism.

Therefore, there is an open set $\mathcal{O} \subset \mathcal{F}_1 \cap \cdots \cap \mathcal{F}_n$ with $f \in \mathcal{O}$ such that $A^{-1}(\mathcal{O}) = V_1 + \cdots + V_n, V_i \cap V_j = \emptyset (i \neq j), v_i \in V_i(1 \leq i \cdot j \leq n)$, and for any $u \in \Sigma_{i=1}^{n} V_i, G'(u)$ is an isomorphism. Hence, $\mathcal{O} \subset \mathcal{R}$ and $\mathcal{R} \subset C^{\alpha}(TM)$ is open.

The proof is complete. □

REMARK 4.1.12. The result of Theorem 4.1.11 in $W^{2,2}$ for the Dirichlet boundary conditions was obtained by Foias and Temam in [**23**].

4.1.5. Completion of the proof of Theorem 4.1.1. Theorem 4.1.11 tells us that there is an open and dense set $\mathcal{R} \subset C^{\alpha}(TM)$ such that

$$A_0 = A|_{G^{-1}(\mathcal{R})} : G^{-1}(\mathcal{R}) \to \mathcal{R}$$

is an open map; i.e., A_0 maps open sets in $A^{-1}(\mathcal{R})$ to open sets in $\mathcal{R}$. Moreover, since $\mathcal{R}$ is open, $A^{-1}(\mathcal{R})$ is open in $B^{2+\alpha}(TM) \times C^{1+\alpha}(TM)$.

By Theorems 2.1.2, 2.2.9 and 2.2.17, the set of $X_0 \subset X = B^{2+\alpha}(TM)$ of all structurally stable vector fields in X is open and dense in X. Let $K = (X_0 \times C^{1+\alpha}(M)) \cap A^{-1}(\mathcal{R}) \subset X_0 \times C^{1+\alpha}(M)$, and let $\mathcal{F} = A(K) \subset \mathcal{R}$. It is easy to see that $\mathcal{F}$ is open and dense in $\mathcal{R}$. Namely, $\mathcal{F}$ is open and dense in $C^{\alpha}(TM)$.

The proof of Theorems 4.1.1 and 4.1.2 is complete. □

4.2. Properties for Structurally Stable Solutions on the Reynolds Numbers

4.2.1. The case with free and Dirichlet boundary conditions. In this subsection, we discuss structural stability of solutions of the Navier-Stokes equations (4.1.1) and (4.1.2) with the free and Dirichlet boundary conditions with the Reynolds number as a parameter.

The main theorem is related to the following Stokes equations:

$$\begin{cases} -\Delta u + \nabla p = f & \forall x \in \overset{\circ}{M}, \\ \text{div } u = 0 & \forall x \in \overset{\circ}{M}. \end{cases} \tag{4.2.1}$$

THEOREM 4.2.1. *Let $X = B^{2+\alpha}(TM)$ or $X = B_0^{2+\alpha}(TM)$ $(0 < \alpha < 1)$, and let $F \subset C^{\alpha}(TM)$ such that for each $f \in F$, the solution $u \in X$ of the Stokes problem (4.2.1) with corresponding boundary conditions is structurally stable. Then the following assertions hold true:*

1. *F is open and dense in $C^{\alpha}(TM)$,*
2. *For any $f \in F$, there exists $\mu_0 > 0$ such that for any $\mu > \mu_0$, the solutions $u(x, \mu) \in X$ of the Navier-Stokes equations (4.1.1) and (4.1.2) with corresponding boundary conditions are structurally stable. Moreover, for any $\mu_1, \mu_2 > \mu_0$, $u(x, \mu_1)$ and $u(x, \mu_2)$ are topologically equivalent.*

Theorem 4.2.1 follows immediately from Lemma 4.2.2 below.

Consider the following equivalent Navier-Stokes equations

$$-\triangle u + R[(u\cdot\nabla)u + \nabla p - f] = 0 \qquad \forall x \in \overset{\circ}{M}, \tag{4.2.2}$$

where $R > 0$ is the Reynolds number.

LEMMA 4.2.2. *Let $u(\cdot, R) \subset X$ be the solution of (4.2.2). Then $u(\cdot, R)$ is C^∞ with respect to R at $R_0 > 0$, provided that R_0 is not an eigenvalue of the linearized equations of (4.2.2) at $u(\cdot, R_0)$. Moreover, near $R = 0$, $u(\cdot, R)$ can be expanded by*

$$\begin{cases} u(x,R) = Ru_1 + R^3u_3 + o(R^3), \\ p(x,R) = p_0 + R^2p_2 + o(R^2), \end{cases} \tag{4.2.3}$$

where (u_1, p_0) satisfies the Stokes equation (4.2.1) with the corresponding boundary conditions.

PROOF. For a non-eigenvalue R_0 of the linearized equations of (4.2.2) at $u(\cdot, R_0)$, we define a map

$$G : X \times C^{1+\alpha}(M) \times \mathbb{R}^+ \to C^\alpha(TM) \tag{4.2.4}$$

by

$$G(u, p; R) = -\triangle u + R[(u\cdot\nabla)u + \nabla p - f].$$

Obviously, G is C^∞, and $G(v_0, \phi_0; R_0) = 0$, where $v_0 = u(\cdot, R_0)$ and $\phi_0 = p(\cdot, R_0)$. The derivative of G with respect to (u, p) is

$$D_{(u,p)}G(v_0,\phi_0;R_0)(u,p) = -\triangle u + R_0[(v_0\cdot\nabla)u + (u\cdot\nabla)v_0 + \nabla p]. \tag{4.2.5}$$

Since R_0 is not an eigenvalue of the linearized equations of (4.2.2) at $u(\cdot, R_0)$,

$$D_{(u,p)}G(v_0,\phi_0;R_0) : X \times C^{1+\alpha}(M)\times \to C^\alpha(TM)$$

is a homeomorphism. By the implicit function theorem, it follows that $u(\cdot, R)$ and $p(\cdot, R)$ are C^∞ at R_0.

It is well known that for any $R > 0$ sufficiently small, R is not an eigenvalue of the linear operator (4.2.5). Therefore, $u(\cdot, R)$ and $p(\cdot, R)$ are C^∞ for $R > 0$ small.

Then we set

$$\begin{cases} u(\cdot,R) = u_0 + Ru_1 + R^2u_2 + R^3u_3 + u_4(\cdot,R), \\ p(\cdot,R) = p_0 + Rp_1 + R^2p_2 + p_3(\cdot,R). \end{cases} \tag{4.2.6}$$

It is easy to see that

(1) $u_0 = 0$, using the zeroth order equation;
(2) (u_1, p_0) solves the Stokes problem (4.2.1), using the first order equation;
(3) $(u_2, p_1) = 0$, using the second order equation; and
(4) (u_3, p_2) solves

$$\begin{cases} -\triangle u_3 + \nabla p_2 = -(u_1\cdot\nabla)u_1 & \text{in } \overset{\circ}{M}, \\ \operatorname{div} u_3 = 0, & \text{in } \overset{\circ}{M}. \end{cases} \tag{4.2.7}$$

The standard Schauder estimates of (4.2.7) (see Lemma 4.1.5) yield that $(u_3, p_2) \in X \times C^{1+\alpha}(M)$, and (4.2.3) follows.

The proof is complete. □

4.2.2. 2D periodic case. Consider the 2D Navier-Stokes equations with periodic boundary conditions:

$$
\begin{cases} -\mu\Delta u + (u\cdot\nabla)u + \nabla p = f & \text{in } M = \mathbb{T}^2, \\ \operatorname{div} u = 0 & \text{in } M, \\ \displaystyle\int_{\mathbb{T}^2} u(x)dx = 0. \end{cases} \tag{4.2.8}
$$

The corresponding Stokes problem is

$$
\begin{cases} -\mu\Delta u + \nabla p = f & \text{in } M = \mathbb{T}^2, \\ \operatorname{div} u = 0 & \text{in } M, \\ \displaystyle\int_{\mathbb{T}^2} u(x)dx = 0. \end{cases} \tag{4.2.9}
$$

We set

$$
C_0^r(TM) = \{v \in C^r(TM) \mid \int_{\mathbb{T}^2} u(x)dx = 0\}.
$$

Then, similar to Theorem 4.2.1, we have

THEOREM 4.2.3. *Let $F \subset C_0^\alpha(TM)$ such that for each $f \in F$, the solution $u \in X$ of the Stokes problem (4.2.9) is Hamiltonian structurally stable. Then the following assertions hold true:*

(1) *F is open and dense in $C_0^\alpha(TM)$;*
(2) *For any $f \in F$, there exists $\mu_0 > 0$ such that for any $\mu > \mu_0$, the solutions $u(x,\mu) \in X$ of the Navier-Stokes equations (4.2.8) are Hamiltonian structurally stable. Moreover, for any $\mu_1, \mu_2 > \mu_0$, $u(x,\mu_1)$ and $u(x,\mu_2)$ are topologically equivalent.*

4.2.3. Nonhomogeneous boundary conditions. The case of nonhomogeneous boundary conditions differs slightly from the previous two cases. Consider the following Stokes problem with nonhomogeneous boundary conditions:

$$
\begin{cases} -\mu\Delta u + \nabla p = 0 & \text{in } \overset{\circ}{M}, \\ \operatorname{div} u = 0 & \text{in } \overset{\circ}{M}, \\ u = \varphi & \text{on } \partial M. \end{cases} \tag{4.2.10}
$$

Then we have

THEOREM 4.2.4. *Let $\varphi \in C^{2+\alpha}(T\partial M)$. If the solution of (4.2.10) is structurally stable in $B_\varphi^{2+\alpha}(TM)$, then for any $f \in C_0^\alpha(TM)$, there is a $\mu_0 \geq 0$ such that all solutions $u(\cdot,\mu) \in B_\varphi^{2+\alpha}(TM)$ of the 2D Navier-Stokes equations (4.1.1) and (4.1.2) with nonhomogeneous boundary condition (4.1.5) are also structurally stable for any $\mu > \mu_0$.*

PROOF. Let $u(\cdot,R) \in B_\varphi^{2+\alpha}(TM)$ be a solution of (4.2.2) with boundary condition (4.1.5). As in the proof of Lemma 4.2.2, it is easy to see that $u(\cdot,R)$ is C^∞ with respect to the Reynolds number $R > 0$ small enough, and the solutions $(u(\cdot,R), p(\cdot,R))$ of (4.2.2) and (4.1.5) can be expressed near $R = 0$ by

$$
\begin{cases} u(\cdot,R) = u_0 + Ru_1 + R^2u_2 + o(R), \\ p(\cdot,R) = R^{-1}p_{-1} + p_0 + O(R), \end{cases} \tag{4.2.11}
$$

where (u_0, p_{-1}) solves (4.2.10). Then the theorem follows from (4.2.11). □

EXAMPLE 4.2.5. Let $M = B^2 = \{x \in \mathbb{R}^2 \mid x_1^2 + x_2^2 \leq 1\}$, and let $\varphi \in C^\infty(T\partial B^2)$ with $|\varphi| = \text{ const } > 0$. Obviously,

$$\begin{cases} u_0(x) = (\sqrt{c}x_2, -\sqrt{c}x_1), \\ p_{-1} = 0, \end{cases} \tag{4.2.12}$$

is a solution of (4.2.10), and u_0 is structurally stable, where $c = |\varphi| > 0$. By Theorem 4.2.3, for any $f \in C^\alpha_\varphi(TM)$, there is a $\mu_0 \geq 0$ such that all solutions $u(\cdot, \mu) \in B^{2+\alpha}_\varphi(TM)$ of the 2D Navier-Stokes equations (4.1.1) and (4.1.2) with nonhomogeneous boundary condition (4.1.5) are also structurally stable for any $\mu > \mu_0$. Moreover, $u(\cdot, \mu)$ are topologically equivalent to u_0 for any $\mu > \mu_0$.

In particular, if $f = \nabla\psi$ is a gradient field, then $\mu_0 = 0$. In other words, for all $\mu > 0$, the solutions of (4.1.1) and (4.1.2) with $|\varphi| = \text{ const } > 0$ are structurally stable.

4.3. Asymptotic Hamiltonian Structural Stability

In this section, we study the large time Hamiltonian structural stability for the solutions of the Navier-Stokes equations defined on a 2-D torus. Two cases will be addressed. The first case concerns the solutions of the Navier-Stokes equations with forcing with decay. The second case deals with the structural stability of Lyapunov stable solutions.

4.3.1. Navier-Stokes equations. We consider the Navier-Stokes equations defined on the flat torus $M = \mathbb{T}^2 = \mathbb{R}^2/(2\pi\mathbb{Z})^2$:

$$\begin{cases} \dfrac{\partial u}{\partial t} + (u \cdot \nabla)u = \mu\Delta u - \nabla p + f(x,t), \\ \operatorname{div} u = 0, \\ u(x_1 + 2k_1\pi,\ x_2 + 2k_2\pi) = u(x_1, x_2), \\ u(x, 0) = \varphi(x), \end{cases} \tag{4.3.1}$$

where $k_1, k_2 \in \mathbb{Z}$ are integers.

For any $u = (u_1, u_2) \in D^r(TM)$, we have the following Fourier expansion:

$$\begin{aligned} (u_1, u_2) &= \sum_{n,m=-\infty}^{\infty} (a_{nm}, b_{nm}) e^{i(nx_1 + mx_2)}, \\ (a_{-n-m}, b_{-n-m}) &= (\overline{a}_{nm}, \overline{b}_{nm}), \\ na_{nm} + mb_{nm} &= 0. \end{aligned}$$

By the Hodge decomposition, we have

$$\begin{aligned} C^r(TM) &= D^r(TM) \oplus G^r(TM), \\ G^r(TM) &= \{\nabla\phi \mid \phi \in C^{r+1}(M)\}, \\ D^r(TM) &= H^r(TM) \oplus \mathcal{H}. \end{aligned}$$

Here $\mathcal{H}$ contains all harmonic fields and has dimension 2, which is the first Betti number of M. It is easy to see that

$$\mathcal{H} = \{u = (a, b) \mid a, b, \in \mathbb{R}\} = \mathbb{R}^2.$$

In view of the Fourier expansion, if $a_{00} = b_{00} = 0$, then u given by (4.3.1) is a Hamiltonian vector field, in which the Hamiltonian function is given by

$$H = -i \sum_{\substack{n,m=-\infty \\ n\neq 0, m\neq 0}}^{\infty} \frac{1}{m} a_{nm} e^{i(nx_1+mx_2)} - i \sum_{n=-\infty}^{\infty} \frac{1}{n} \left[a_{0n}e^{inx_2} + b_{n0}e^{inx_1}\right].$$

The following theorem is useful for the discussion of asymptotic structural stability of solutions of the Navier-Stokes equations.

THEOREM 4.3.1. *If $\varphi \in H^r(TM)$ and $f \in H^{r-2}(TM) \oplus G^{r-2}(TM)$ $(r \geq 2)$, then the solution $u(x,t)$ of (4.3.1) is a one-parameter family of Hamiltonian vector fields, i.e., $u(\cdot,t) \in H^r(TM)$ for any $t \geq 0$.*

4.3.2. Decay of the solutions and Hamiltonian structural stability. We consider problem (4.3.1) with forcing decaying in time t as t goes to infinity. More precisely, let $f \in C^1([0,1], H^r(TM))$ $(r \geq 3)$ such that for some $n > 1/2$,

$$f(x,t) = t^{-n}(f_0(x) + f_1(x,t)), \tag{4.3.2}$$

$$\lim_{t\to\infty} \|f_1(\cdot,t)\|_{C^r} = 0, \qquad \int_0^\infty \left\|\frac{\partial}{\partial t} f_1\right\|_{L^2}^2 dt < \infty. \tag{4.3.3}$$

In order to study the asymptotic structural stability of (4.3.1), it is necessary to consider the Stokes problem

$$\begin{cases} -\mu\Delta u + \nabla p = f_0 & \text{in } M = \mathbb{T}^2, \\ \operatorname{div} u = 0 & \text{in } M. \end{cases} \tag{4.3.4}$$

Since the Laplace operator $\Delta : H^{r+2}(TM) \to H^r(TM)$ $(r \geq 0)$ is an isomorphism, we infer from Theorem 2.3.4 that there exists an open and dense $\mathcal{F} \subset H^r(TM)$,

$$\mathcal{F} = (-\mu\Delta)(H_1^r(TM)), \tag{4.3.5}$$

such that for any $f_0 \in \mathcal{F}$, the solution $u(x,t)$ of (4.3.4) is Hamiltonian structurally stable, where $H_1^r(TM)$ is the set of Hamiltonian structurally stable fields. Obviously, the set $\mathcal{F}$ is independent of the kinematic viscosity $\mu > 0$.

The main result in this section is as follows:

THEOREM 4.3.2. *Let $f \in C^\infty([0,\infty), H^r(TM))$ $(r \geq 3)$ have the expansion (4.3.2–4.3.3) near $t = \infty$, and let $\varphi \in H^r(TM)$. If $f_0 \in \mathcal{F}$, then there is a time $t_0 > 0$ depending on φ and f such that the solution $u(\cdot,t)$ of (4.3.1) is Hamiltonian structurally stable for all $t > t_0$. Moreover, u is topologically equivalent to the unique solution of (4.3.4) for any $t > t_0$.*

PROOF. Let the solution $u(x,t)$ and $p(x,t)$ be expressed near $t = \infty$ by

$$\begin{cases} u(x,t) = t^{-n}(u_1(x) + v(x,t)), \\ p(x,t) = t^{-n}(p_1(x) + q(x,t)), \end{cases}$$

where (u_1, p_1) is the unique solution of the Stokes problem (4.3.4), and (v,q) satisfies

$$\begin{cases} \frac{\partial v}{\partial t} - \mu\Delta v + \nabla q = nt^{-1}(u_1 + v) \\ \qquad\qquad -t^{-n}[(u_1+v)\cdot\nabla](u_1+v) + f_1, \\ \operatorname{div} v = 0. \end{cases} \tag{4.3.6}$$

Here $f_1 \in H^r(TM)$ $(r \geq 3)$ satisfies (4.3.3). For any $T_0 > 0$ sufficiently large, the initial value for v in (4.3.6) is given by

$$v(x, T_0) = T_0^n u(x, T_0) - u_1(x). \tag{4.3.7}$$

To prove the theorem, it suffices then to prove that

$$\lim_{t \to \infty} \|v(\cdot, t)\|_{C^1} = 0. \tag{4.3.8}$$

For this purpose, first it follows from (4.3.6) and (4.3.7) that

$$\begin{aligned} \frac{d}{dt}\int_{\mathbb{T}^2} v^2 dx =& 2\int_{\mathbb{T}^2} \{-\mu|\nabla v|^2 + f_1 \cdot v + nt^{-1}(u_1 + v) \cdot v \\ &+ t^{-1}[((u_1 + v) \cdot \nabla)(u_1 + v) \cdot v]\}dx, \end{aligned} \tag{4.3.9}$$

with initial condition

$$\int_{\mathbb{T}^2} v^2(x, T_0)dx = \int_{\mathbb{T}^2} [T_0^n u(x, T_0) - u_1(x)]^2 dx. \tag{4.3.10}$$

Then we infer from (4.3.9) that for $t > T_0$ sufficiently large,

$$\begin{cases} \dfrac{d}{dt}\|v\|_{L^2}^2 \leq -C\|v\|_{L^2}^2 + g(t), \\ \lim\limits_{t \to \infty} g(t) = 0, \end{cases}$$

for some constant $C > 0$. By the Gronwall inequality and using (4.3.10), we deduce that

$$\lim_{t \to \infty} \|v\|_{L^2} = 0. \tag{4.3.11}$$

By Sobolev's embedding theorems, it suffices for us to prove

$$\|v\|_{W^{3,2}} = \left[\int_{\mathbb{T}^2} \sum_{k=0}^{3} |D^k v|^2 dx\right] \leq C, \quad C > 0 \text{ a constant}, \tag{4.3.12}$$

which implies

$$\lim_{t \to \infty} \|v\|_{W^{2,p}} = 0 \qquad \forall\, 1 \leq p < \infty.$$

Hence, (4.3.8) follows.

To prove (4.3.12), we infer from (4.3.6) that

$$\begin{aligned} &\frac{d}{dt}\|v'\|_{L^2}^2 + \mu\|\nabla v'\|_{L^2}^2 \\ &= \int_{\mathbb{T}^2} \{nt^{-n-1}((u_1 + v) \cdot \nabla)(u_1 + v) \cdot v' \\ &\quad - [nt^{-2}(u_1 + v) + nt^{-1}v' - t^{-n}(v' \cdot \nabla)v]v' \\ &\quad - [t^{-n}(v \cdot \nabla)v' + f_1']v'\}dx. \end{aligned} \tag{4.3.13}$$

Here, $v_0 = v(x, T_0)$, $q_0 = q(x, T_0)$, $f_1^0 = f_1(x, T_0)$, and $v' = \partial v/\partial t$ is given by

$$\begin{aligned} v'(x, T_0) =& \mu\Delta v(x, T_0) - \nabla q_0 + nT_0^{-1}(u_1 + v_0) \\ &- t^{-n}((u_1 + v_0) \cdot \nabla)(u_1 + v_0) + f_1^0. \end{aligned} \tag{4.3.14}$$

For $t \geq T_0 > 0$ large enough, we obtain from (4.3.13–4.3.14) that

$$\frac{d}{dt}\|v'\|_{L^2}^2 + \frac{\mu}{2}\|\nabla v'\|_{L^2}^2 \leq C\|f_1'\|_{L^2}^2 + C\|\nabla v\|_{L^2}^2\|v'\|_{L^2}^2 + Ct^{-1-n}\|\nabla v\|_{L^2}^2\|v\|_{L^2}^2.$$

Then by the Gronwall inequality, we have

$$\frac{d}{dt}\left[\|v'\|_{L^2}\cdot e^{-\int_0^t\|\nabla v\|_{L^2}^2dt}\right]\leq C\|f_1'\|_{L^2}^2+Ct^{-n-1}\|v\|_{L^2}^2\|\nabla v\|_{L^2}^2.$$

Hence, we obtain

$$\|v'\|_{L^2}\leq C, \tag{4.3.15}$$

where $C>0$ is a constant independent of t.

Now (4.3.6) can be rewritten as

$$\begin{cases}-\mu\Delta v+\nabla q=F(x,t),\\ \operatorname{div}v=0,\end{cases} \tag{4.3.16}$$

where

$$F=nt^{-1}(u_1+v)-t^{-n}[(u_1+v)\cdot\nabla](u_1+v)+f_1-\frac{\partial v}{\partial t}.$$

Since $v,v'\in L^\infty([0,\infty),L^2(TM))$, we deduce from (4.3.6) that $v\in L^\infty([0,\infty),W^{1,2}(TM))$. Using the same method as in the proof of the regularity theorem, Theorem 3.3.6 in [**102**], we obtain from (4.3.16) that $v\in L^\infty([0,\infty),W^{2,2}(TM))$.

Moreover, let $z=D^3v$. Then by (4.3.6), z is governed by the equation

$$\begin{cases}\frac{\partial z}{\partial r}-\mu\Delta z+\nabla p=nt^{-1}(D^3u_1+z)\\ \qquad\qquad -t^{-n}D^3[((u_1+v)\cdot\nabla)(r_1+v)]+D^3f_1,\\ \operatorname{div}z=0.\end{cases} \tag{4.3.17}$$

Using $v\in L^\infty([0,\infty),W^{2,2}(TM))$ and (4.3.17), it is easy to show that $z\in L^\infty([0,\infty),L^2(TM))$, and the estimate (4.3.12) holds true. The proof is complete. □

4.3.3. Structural stability vs. Lyapunov stability. In this section, we are mainly concerned with the relationship between the Lyapunov stability and the asymptotic Hamiltonian structural stability.

Let $f\in C^\infty([0,\infty),H^r(TM))$ $(r\geq 3)$ such that

$$\lim_{t\to\infty}f(x,t)=\lambda f_0(x)\quad\text{in } H^r(TM), \tag{4.3.18}$$

$$\int_0^\infty\int_{\mathbb{T}^2}\left|\frac{\partial f}{\partial t}\right|^2dxdt<\infty, \tag{4.3.19}$$

where $\lambda>0$ is a parameter.

The Lyapunov Stability Theorem amounts to saying that for given f_0, if $\lambda>0$ is small enough, then the solution $u(x,t)$ of (4.3.1) converges to the steady state of (4.3.1), namely

$$\begin{cases}\lim_{t\to\infty}\|u(\cdot,t)-u_0\|_{C^1}=0,\\ \lim_{t\to\infty}\|p(\cdot,t)-p_0\|_{C^0}=0,\end{cases} \tag{4.3.20}$$

where (u_0,p_0) is a solution of the stationary equations of (4.3.1).

For convenience, we restate the Lyapunov Stability Theorem as follows.

THEOREM 4.3.3 (Lyapunov Stability Theorem). *Let $f\in C^\infty([0,\infty),H^r(TM))$ $(r\geq 3)$ satisfying (4.3.18) and (4.3.19). If $\lambda=\mu^2$ with μ being the kinematic viscosity, then there exists a constant $\alpha>0$ independent of μ such that as $\|f_0\|_{H^r}<\alpha$, for any $\varphi\in H^r(TM)$, the solution $u(x,t)$ of (4.3.1) satisfies (4.3.20). Moreover,* *the stationary solution u_0 of (4.3.1) has the form $u_0 = \mu v_0(x)$, where $v_0(x)$ solves the equation*

$$\begin{cases} -\Delta v + (v \cdot \nabla)v + \nabla p = f_0(x), \\ div\, u = 0. \end{cases} \tag{4.3.21}$$

PROOF. Let $f(x,t) = \mu^2 f_0(x) + h(x,t)$. By (4.3.18) and (4.3.19), we have

$$\lim_{t\to\infty} \|h(\cdot,t)\|_{C^r} = 0 \qquad (r \geq 3), \tag{4.3.22}$$

$$\int_0^\infty \int_T \left|\frac{\partial h}{\partial t}\right|^2 dxdt < \infty. \tag{4.3.23}$$

Let $u_0 = \mu v_0$ and $v(x,t) = u(x,t) - \mu v_0(x)$. It is easy to see that v_0 satisfies equation (4.3.21), and $v(x,t)$ is governed by the following equation:

$$\begin{cases} \dfrac{\partial v}{\partial t} + (v \cdot \nabla)v = \mu[\Delta v - (v_0 \cdot \nabla)v - (v \cdot \nabla)v_0] \\ \qquad\qquad\qquad - \nabla p + h(x,t), \\ \operatorname{div} u = 0, \\ v(x,0) = \varphi(x) - \mu v_0(x). \end{cases} \tag{4.3.24}$$

Then the following estimate can be obtained in the same fashion as in Lemma 4.1.5:

$$\|v_0\|_{C^{2+r}} \leq C\left[\|f_0\|_{C^r} + \|f_0\|_{C^r}^p\right], \qquad \text{for some } p > 8 \tag{4.3.25}$$

where $C > 0$ is independent of μ.

Hence, there exists a constant $\alpha > 0$ independent of μ such that when $\|f_0\|_{C^r} < \alpha$, we have

$$\begin{aligned} &\int_{\mathbb{T}^2} [-\Delta v + (v_0 \cdot \nabla)v + (v \cdot \nabla)v_0] \cdot v dx \\ = &\int_{\mathbb{T}^2} \left[|\nabla v|^2 + (v \cdot \nabla)v_0 \cdot v\right] dx \\ \geq &\ \rho \int_{\mathbb{T}^2} |\nabla v|^2 dx, \quad \text{for some } 0 < \rho < 1, \end{aligned}$$

which yields from (4.3.24) that

$$\lim_{t\to\infty} \|v(\cdot,t)\|_{L^2} = 0.$$

Moreover, by (4.3.22)–(4.3.24), it is easy to show that

$$\|v(\cdot,t)\|_{W^{3,2}} \leq C, \quad C > 0 \text{ a constant},$$

which implies that $\lim_{t\to\infty} \|v(\cdot,t)\|_{C^1} = 0$. The proof is complete. □

We are now in position to state the main result on the relationship between the Lyapunov stability and the asymptotic Hamiltonian structural stability for solutions of the Navier-Stokes equations (4.3.1).

THEOREM 4.3.4. *Let $f \in C^\infty([0,\infty), H^r(TM))$ $(r \geq 3)$ satisfying (4.3.18) and (4.3.19). Let $\mathcal{F} \subset H^r(TM)$ be the structurally stable solution set of the Stokes equation (4.3.4) as given by (4.3.5). Let $\lambda = \mu^k$ $(k > 2)$ and $\|f_0\|_{C^r} = 1$. Then for $\mu > 0$ sufficiently small, there exists a time $t_0 > 0$ such that, for all $t > t_0$,* *the solution $u(x,t)$ of (4.3.1) is Hamiltonian structurally stable, and is topologically equivalent to the solution $w(x)$ of the Stokes problem*

$$\begin{cases} -\Delta w + \nabla p = f_0(x), \\ \operatorname{div} w = 0. \end{cases}$$

PROOF. By Theorem 4.3.3, $u(x,t) \to u_0(x)$ $(t \to \infty)$ in the C^1-norm, where u_0 satisfies the equation

$$\begin{cases} -\mu\Delta u_0 + (u_0 \cdot \nabla)u_0 + \nabla p = \mu^k f_0(x) \quad (k > 2) \\ \operatorname{div} u_0 = 0. \end{cases} \tag{4.3.26}$$

If u_0 is Hamiltonian structurally stable, then it is clear that $u(x,t)$ are topologically equivalent to u_0 for all $t > 0$ sufficiently large.

Let u_1 be a solution of the Stokes problem

$$\begin{cases} -\mu\Delta u_1 + \nabla p = \mu^k f_0, \\ \operatorname{div} u_1 = 0. \end{cases}$$

Let $u_1 = \mu^{k-1} w$; then, the above Stokes problem is equivalent to

$$\begin{cases} -\Delta w + \nabla p = f_0, \\ \operatorname{div} w = 0. \end{cases} \tag{4.3.27}$$

Let the solution $u_0(x)$ of (4.3.26) have the form

$$u_0 = \mu u_1 + \mu v(x,\mu) = \mu^k w + \mu v(x,\mu).$$

Then $v(x,\mu)$ solves the equation

$$\begin{cases} -\Delta v + (v \cdot \nabla)v + \mu^{k-1}[(w \cdot \nabla)v + (v \cdot \nabla)w] \\ \qquad\qquad + \nabla p = \mu^{2k-2}(w \cdot \nabla)w, \\ \operatorname{div} u = 0. \end{cases} \tag{4.3.28}$$

From (4.3.27) we get

$$\|w\|_{C^{r+2}} \leq C\|f_0\|_{C^r}, \tag{4.3.29}$$

where $C > 0$ is a constant independent of μ. Equation (4.3.28) can be written as

$$\begin{cases} -\Delta v + \mu^{k-1}[(w \cdot \nabla)v + (v \cdot \nabla)w] + \nabla p = F(x), \\ \operatorname{div} v = 0, \end{cases} \tag{4.3.30}$$

where $F(x) = \mu^{2k-2}(w \cdot \nabla)w - (v \cdot \nabla)v$. For equation (4.3.30), the L^p-estimates are valid (see [**102**]):

$$\begin{aligned} \|v\|_{W^{2,p}} &\leq C\left[\|F\|_{L^p} + \|v\|_{L^p}\right] \\ &\leq C\left[\|v\|_{L^p} + \|\nabla v\|_{L^p} \cdot \|v\|_{L^p} + \mu^{2k-2}\|w\|^2_{C^1}\right] \end{aligned} \tag{4.3.31}$$

where $p^{-1} + q^{-1} = 1$. From (4.3.31) we have

$$\|v\|_{W^{2,p}} \leq C\left[\|v\|_{L^p} + \mu^{2k-2}\|f_0\|^2_{C^0}\right]. \tag{4.3.32}$$

On the other hand, we derive from (4.3.30) that

$$\|\nabla v\|^2_{L^2} = \int_{\mathbb{T}^2} |\nabla v|^2 dx \leq \int_{\mathbb{T}^2} \left[\mu^{2k-2}(w \cdot \nabla)w + \mu^{k-1}(v \cdot \nabla)w\right] v dx.$$

Then for $\mu > 0$ sufficiently small, by (4.3.29) we have

$$\left(\frac{1}{2} - \mu^{-1}C\right) \|\nabla v\|_{L^2}^2 \leq C\mu^{2k-2}\|w\|_{C^1}^4. \tag{4.3.33}$$

It follows from (4.3.32) and (4.3.33) that

$$\|v\|_{W^{2,p}} \leq C\mu^{2k-2}\left[\|f_0\|_{C^0}^2 + \|f_0\|_{C_0}^4\right] \qquad \forall\, p > 1.$$

Thus, for the solution u_0 of (4.3.26) we get

$$u_0 = \mu^k[w(x) + O(\mu^{k-2})] \qquad \text{in } H^1(TM)\ (k > 2).$$

Thus, for all $\mu > 0$ sufficiently small, u_0 is topologically equivalent to the solution $w(x)$ of (4.3.27). By assumption, $w(x)$ is Hamiltonian structurally stable, hence the proof of the theorem is complete. □

REMARK 4.3.5. The same assertions as in Theorems 4.3.2–4.3.4 are valid for the Navier-Stokes equations with the homogeneous Dirichlet boundary condition and the free-boundary condition.

4.3.4. Some Examples. In the above stability theorems, it is important to know the structurally stable solution set $\mathcal{F}$ of the Stokes problem defined by (4.3.5). For this purpose, we give some examples of the $f \in \mathcal{F}$.

First, when $f = (\sin x_2, \frac{1}{2}\cos x_1)$, the solution of (4.3.27) is as follows:

$$u(x_1, x_2) = (\sin x_2, \cos x_1).$$

Obviously, the solution $u(x)$ has exactly four singular points:

$$\left(\frac{\pi}{2}, 0\right), \left(\frac{2}{3}\pi, 0\right), \left(\frac{\pi}{2}, \pi\right), \left(\frac{2}{3}\pi, \pi\right)$$

which are nondegenerate. By Theorem 2.4.3, the block structure of u must be as $\mathbb{T}^2 = T \cup A_1 \cup A_2$, i.e., u has exactly two S-blocks, and the saddle points are self-connected. Hence, $u(x)$ is Hamiltonian structurally stable.

In general, let

$$f(x_1, x_2) = (\alpha_1 \sin x_2, \alpha_2 \cos kx_1), \quad k = 1, 2, \ldots,$$

where $\alpha_1, \alpha_2 \neq 0$ and $\alpha_1 \neq \alpha_2$; then the solution of (4.3.27) is Hamiltonian structurally stable, and is given by

$$u(x_1, x_2) = (\alpha_1 \sin x_2,\ \alpha_2 k^{-2} \cos kx_1). \tag{4.3.34}$$

For the case where $\alpha_1 = \alpha_2 k^{-2} = 1$, the structure of (4.3.34) is schematically shown by Figure 4.3.1.

Similarly, for the functions given by $f(x) = \{\alpha_1 \sin(x_2+\theta_1),\ \alpha_2 \sin(kx_1+\theta_2)\}$ or $f(x) = \{\alpha_1 \sin(kx_2+\theta_1),\ \alpha_2 \sin(x_1+\theta_2)\}$, the solutions of (4.3.27) are Hamiltonian structurally stable.

4.4. Asymptotic Block Stability

4.4.1. Main Theorems. We know that when the harmonic vector fields appear in the data (φ, f), the solution u of (4.3.1) is not Hamiltonian. Thanks to the block stability theorem, Theorem 3.2.3, we shall consider the asymptotic block stability instead of the Hamiltonian structural stability.

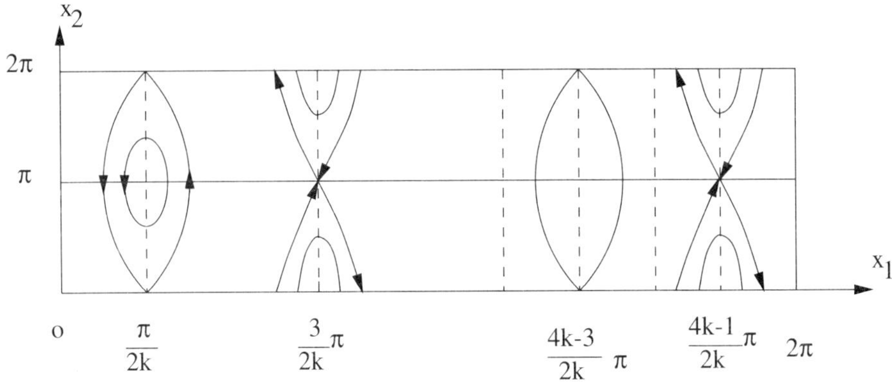

FIGURE 4.3.1. Flow structure of the vector field given by (4.3.34) for the case where $\alpha_1 = \alpha_2 k^{-2} = 1$, $\alpha_1, \alpha_2 \neq 0$ and $\alpha_1 \neq \alpha_2$.

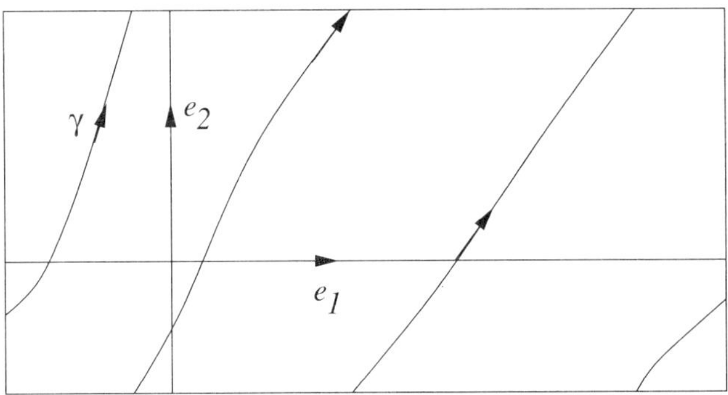

FIGURE 4.4.1. Two elements e_1 and e_2 in $H_1(\mathbb{T}^2, \mathbb{Z})$.

Assume that

$$\begin{cases} \varphi = \varphi_H + a, \\ f = f_H(x,t) + b(t), \\ \displaystyle\int_0^\infty |b(t)|dt < \infty, \end{cases} \tag{4.4.1}$$

where φ_H and f_H are Hamiltonian vector fields, $a = (a_1, a_2)$ is a constant vector, and $b = \{b_1(t), b_2(t)\}$ is a continuous harmonic field on t. Denote by

$$k(t) = a + \int_0^t b(\tau)d\tau, \qquad (k(t) = (k_1(t), k_2(t))).$$

We recall that a T-block of a stable Hamiltonian vector field $u \in H^r(TM)$ consists of closed orbits of u in its interior. All closed orbits in the T-block as the elements in homology $H_1(\mathbb{T}^2, \mathbb{Z})$ are equal to each other, which are nonzero. This will characterize the stable Hamiltonian vector fields as follows.

We know that a torus is homeomorphic to a rectangle with its two pairs of edges identified as shown in Figure 4.4.1. We take two closed curves e_1 and e_2 with orientations (see Figure 4.4.1) in M as two basic elements in $H_1(\mathbb{T}^2, \mathbb{Z})$.

Then, for any $\gamma \in H_1(\mathbb{T}^2, \mathbb{Z})$, we have

$$\gamma = ne_1 + me_2, \quad n, m \in \mathbb{Z}.$$

If $\gamma \subset M$ is a closed curve with orientation, then

$$\begin{aligned} n &= \text{the intersection number of } \gamma \text{ with } e_1, \\ m &= \text{the intersection number of } \gamma \text{ with } e_2. \end{aligned}$$

Let $u \in H^r(TM)$ be a stable Hamiltonian vector field, and let $\Omega \subset M = \mathbb{T}^2$ be the T-block of u. We say that u is of (n, m) type if, for any closed orbit $\gamma \subset \overset{\circ}{\Omega}$ of u, we have $\gamma = \pm(ne_1 + me_2)$.

Now we return to consider the asymptotic block stability of the Navier-Stokes equations (4.3.1). The main theorems in this section are given by the following, which are the counterparts of Theorem 4.3.2 and Theorem 4.3.4, respectively.

THEOREM 4.4.1. *Let (f_H, φ_H) satisfy the hypotheses of Theorem 4.3.2, and let $|k(t)| \neq 0 \ \forall\ t > 0$. Then the following assertions hold true:*

(1) *If $\lim_{t\to\infty} k(t) = C = (C_1, C_2) \neq 0$, then there is a $t_0 > 0$ such that the solution $u(x,t)$ of (4.3.1) is block stable for any $t > t_0$, whose block structure only consists of a T-block $\Omega(t)$; i.e., $\mathbb{T}^2 = \Omega(t)$, and $u(x,t)$ are ergodic on $\mathbb{T}^2$ for almost all $t > t_0$ provided $b(t) \not\equiv 0$, or C_1/C_2 is irrational.*
(2) *Let the solution u_0 of the Stokes equation (4.3.4) be of (n, m) type ($|n| + |m| \neq 0$). If $k(t) = o(t^{-n})$ $(t \to \infty)$ and $k_1(t)/k_2(t) \neq \left|\frac{n}{m}\right|$ for all $t > 0$ sufficiently large, then there is a $t_0 > 0$ such that for all $t > t_0$, the solutions of (4.3.1) are block stable. Their block structures are isomorphic to the unique stable block structure of the solution of (4.3.4) with a harmonic perturbation.*

THEOREM 4.4.2. *Let (f_H, φ_H) satisfy the hypotheses of Theorem 4.3.4, and let $|k(t)| \neq 0 \ \forall\ t > 0$. Then we have*

(1) *If $\lim_{t\to\infty} k(t) = C \gg \mu$, then assertion 1 in Theorem 4.4.1 holds.*
(2) *Assume the solution of (4.3.4) is of (n, m) type; if $k(t) \to 0$ $(t \to \infty)$ and $k_1(t)/k_2(t) \neq \left|\frac{n}{m}\right|$ for all $t > 0$ sufficiently large, then assertion 2 in Theorem 4.4.1 holds.*

REMARK 4.4.3. In Assertion 2 of Theorems 4.4.1–4.4.2, the phrase "the unique stable block structure of the solution of (4.3.4) with a harmonic perturbation" means that for a stable Hamiltonian solution $u_0(x)$ of (4.3.4) and almost all harmonic vector fields $a = (a_1, a_2)$ sufficiently small, the vector fields $u_0 + a$ are block stable, and the block structure in the sense of isomorphism is unique. For example, if $u_0 = (\sin x_2, \cos kx_1)$ $(k \geq 1)$, then $u_0 + a$ have the block structure as shown in Figure 4.4.2 for all $a = (a_1, a_2)$ with $a_1, a_2 \neq 0$ sufficiently small.

To prove Theorems 4.4.1–4.4.2, we need a lemma that can be directly derived from the connection lemma, Lemma 2.3.1.

LEMMA 4.4.4. *Let $M = \mathbb{T}^2$, let $u, v \in D^r(TM)$ with v sufficiently small, and let $L \subset M$ be a (n, m) type closed orbit line. If the equality (2.3.1) is not valid on L, then for any $p \in L$, the orbit line of $u + v$ passing through p, if it is closed, is not of the type (n, m).*

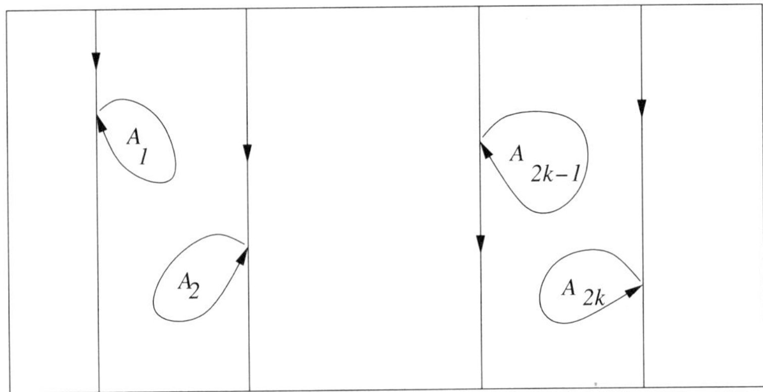

FIGURE 4.4.2. Block stable pattern given by $(\sin x_2, \cos kx_1) + a$ with $a = (a_1, a_2)$ and $a_1, a_2 \neq 0$ sufficiently small.

4.4.2. Proof of Theorems 4.4.1–4.4.2. We prove only Theorem 4.4.1, while Theorem 4.4.2 can be proved in the same fashion.

Let $u = u_H + k$, where $u_H \in C^1([0,\infty), H^r(TM))$ and k is the harmonic part of u. By the Hodge decomposition, problem (4.3.1) is equivalent to

$$\begin{cases} \dfrac{\partial u_H}{\partial t} + (k \cdot \nabla)u_H + (u_H \cdot \nabla)u_H = \mu\Delta u_H - \nabla p + f_H, \\ u_H(x,0) = \varphi_H, \end{cases} \tag{4.4.2}$$

$$\begin{cases} \dfrac{dk}{dt} = b(t), \\ k(0) = a. \end{cases} \tag{4.4.3}$$

It is easy to see that

$$k(t) = a + \int_0^t b(t)$$

is the solution of (4.4.3).

As in the proof of Theorem 4.3.2, we can show from (4.4.2) that

$$\lim_{t\to\infty} \|u_H(\cdot,t)\|_{C^1} = 0.$$

Hence the solution $u = k(t) + u_H(x,t)$ has no singular points on $\mathbb{T}^2$ provided $k(t) \to C \neq 0$ as $t \to \infty$. This proves Assertion 1 in the theorem.

If $k(t) = o(t^{-n})$, then we can infer from (4.4.2) that

$$u_H(x,t) = t^{-n}u_1 + t^{-n}v(x,t) \tag{4.4.4}$$

where u_1 satisfies (4.3.4), and v satisfies

$$\begin{aligned} \frac{\partial v}{\partial t} - \mu\Delta v + \nabla p =& nt^{-1}(u_1 + v) - t^{-n}[(u_1 + v) \cdot \nabla](u_1 + v) \\ & - t^{-n}(k(t) \cdot \nabla)(u_1 + v) + f_1(x,t). \end{aligned} \tag{4.4.5}$$

As in the proof of Theorem 4.3.2, we can deduce from (4.4.5) that

$$\lim_{t\to\infty} \|v(\cdot,t)\|_{C^1} = 0. \tag{4.4.6}$$

By assumption, u_1 is Hamiltonian structurally stable, which is of the (n, m) type. Hence, from (4.4.4) and (4.4.6), we get that $u_H(x, t)$ are Hamiltonian structurally stable with the (n, m) type for all $t > 0$ sufficiently large.

Finally, we shall show that the solution $u = u_H(x,t) + k(t)$ has the stable block structure, which is isomorphic to the structure that the S-blocks of u_H are broken and transformed into the D-blocks.

By Theorem 2.4.3, for $t > 0$ sufficiently large, $u_H(x, t)$ have the block decomposition as

$$\mathbb{T}^2 = \Omega(t) + \cup_{i=0}^{I} D_i(t) + \cup_{k=1}^{K} A_k(t).$$

Since the D-blocks $D_i(t)$ of u_H are structurally stable, under the small perturbation $k(t)$, the topological structures of $D_i(t)$ are invariant.

We know that ∂A_k consists of two saddle self-connections Γ_i^k $(i = 1, 2)$, and by assumption

$$\Gamma_1^k = \pm(ne_1 + me_2), \qquad \Gamma_2^k = \pm(ne_1 + me_2).$$

Moreover, each closed orbit γ in $\overset{\circ}{\Omega}$ near Γ_i^k is homological to Γ_i^k; namely, $\gamma = \Gamma_i^k$ in $H_1(\mathbb{T}^2, \mathbb{Z})$. Notice that

$$\int_\gamma k(t) \times d\ell = 0$$

holds true if and only if

$$\frac{k_1(t)}{k_2(t)} = \frac{|n|}{|m|}\ .$$

Hence, by Lemma 4.4.4, there exists a $t_0 > 0$ sufficiently large such that for any $t > t_0$, the saddle points q_k of $u_H(x, t) + k(t)$ near that of u on ∂A_k do not have the (n, m) type of saddle self-connections. On the other hand, we see that the closed orbits ℓ in A_k are of $(0, 0)$ type and satisfy

$$\int_\ell k(t) \times d\ell = 0.$$

Hence, the saddle points q_k of $u_H + k$ must have a $(0, 0)$ type of saddle self-connection. Thus, the S-blocks A_k of u_H are broken under the perturbation k, and are transformed into the D-blocks. Therefore, the block structure of $u_H + k$ is stable for any $t > 0$ sufficiently large. The proof is complete. □

4.5. Periodic Structure of Solutions of the Navier-Stokes Equations

4.5.1. Flow Invariant Eigenspaces of the Navier-Stokes Equations. The Navier-Stokes equations defined on the torus $\mathbb{T}^2 = \mathbb{R}^2/(2\pi\mathbb{Z})^2$ are as follows:

$$\begin{cases} \dfrac{\partial u}{\partial t} + (u \cdot \nabla)u = \mu\Delta u - \nabla p + f(x,t), \\ \operatorname{div} u = 0, \\ u(x_1 + 2k_1\pi,\ x_2 + 2k_2\pi) = u(x_1, x_2), \\ u(x, 0) = \varphi(x), \end{cases} \tag{4.5.1}$$

where $k_1, k_2 \in \mathbb{Z}$ are integers.

We know that if $f \in C^\infty([0, \infty), \mathcal{H}^\infty(TM))$ and $\varphi \in \mathcal{H}^2(TM)$, problem (4.5.1) has a unique solution $u \in C^\infty((0, \infty), \mathcal{H}^\infty(TM))$. In addition, any $u = (u_1, u_2) \in$

$W^{k,2}_H(TM)$ can be Fourier expanded as follows:

$$\begin{aligned}
&(u_1,u_2)=\sum_{n,m=-\infty}^{\infty}(a_{nm},b_{nm})e^{i(nx_1+mx_2)},\\
&(a_{-n-m},b_{-n-m})=(\bar{a}_{nm},\bar{b}_{nm}),\\
&(a_{00},b_{00})=(0,0),\\
&na_{nm}+mb_{nm}=0.
\end{aligned}$$

The space $W^{l,2}_H(TM)$ can be equipped with the following norm:

$$\|u\|_{W^{l,2}}=\left[\sum_{n,m=-\infty}^{\infty}(n^2+m^2)^{l/2}\left(|a_{nm}|^2+|b_{nm}|^2\right)\right]^{1/2}.$$

The Hamiltonian function of u reads

$$H(x)=-i\sum_{\substack{n,m=-\infty\\ n\neq 0,m\neq 0}}^{\infty}\frac{1}{m}\,a_{nm}e^{i(nx_1+mx_2)}-i\sum_{n=-\infty}^{\infty}\frac{1}{n}\left[a_{0n}e^{inx_2}+b_{n0}e^{inx_1}\right].$$

Let $\{\rho_k|0<\rho_1\leq\rho_2\leq\cdots\leq\rho_k\leq\cdots\}$ and $\{e_k|k=1,2,\cdots\}\subset\mathcal{H}^\infty(TM)$ be the eigenvalues (counting multiplicities) and eigenvectors of the following Stokes problem:

$$\begin{cases}-\Delta e_k=\rho_k e_k,\\ e_k=\operatorname{curl} h_k.\end{cases}\tag{4.5.2}$$

For simplicity, we assume that the corresponding h_k ($1\leq k\leq\infty$) form an orthonomal basis for $L^2(M)$. Namely,

$$\int_{\mathbb{T}^2}h_ih_j dx=\delta_{ij},\quad i,j=1,2,\cdots,$$

and for any $h\in W^{l,2}(M)$,

$$h=\sum_{i=1}^{\infty}\alpha_i h_i,\qquad ||h||_{W^{l,2}}=\left(\sum_{i=1}^{\infty}\rho_i^l|\alpha_i|^2\right)^{1/2}.$$

Let $0<\Lambda_1<\Lambda_2<\cdots<\Lambda_k<\cdots$ be the distinct eigenvalues of the above Stokes problem, and let $E_k\subset\mathcal{H}^\infty(TM)$ be the eigenspace corresponding to Λ_k ($k\geq 1$). For simplicity, let

$$\begin{aligned}
E_k&=\operatorname{span}\left\{e_{\alpha_{k-1}+1},\ldots,e_{\alpha_{k-1}+m_k}\right\},\\
\widetilde{E}_k&=\operatorname{span}\left\{h_{\alpha_{k-1}+1},\ldots,h_{\alpha_{k-1}+m_k}\right\},
\end{aligned}$$

where m_k is the multiplicity of the eigenvalue Λ_k, and $\alpha_{k-1}+m_k=\alpha_k$.

A special property of the 2-D Navier-Stokes equations on a torus is the invariance of the eigenspaces of (4.5.2). We consider the equations

$$\begin{cases}\dfrac{\partial u}{\partial t}=\mu\Delta u-(u\cdot\nabla)u-\nabla p+f_k,\\ \operatorname{div} u=0,\\ u|_{t=0}=\varphi,\end{cases}\tag{4.5.3}$$

where f_k is an eigenvector of (4.5.2) corresponding to Λ_k.

THEOREM 4.5.1. *For problem (4.5.3), the following assertions hold true:*

(1) $v_k = \mu^{-1}\Lambda_k^{-1} f_k$ *is a stationary solution of (4.5.3), and the eigenspace* $E_k \subset \mathcal{H}^\infty(TM)$ *corresponding to* Λ_k *is flow invariant. Namely, if the initial value* $\varphi \in E_k$, *then the solution of (4.5.3) satisfies* $u(\cdot, t, \varphi) \in E_k$ *for any* $t \geq 0$.

(2) *The stationary solution* v_k *is stable in* E_k, *i.e., the solution* $u(\cdot, t, \varphi) \to v_k$ *as* $t \to \infty$ *provided* $\varphi \in E_k$.

PROOF. By the Hodge decomposition, we have

$$W^{l,2}(TM) = W_H^{l,2}(TM) \oplus G^l(TM) \oplus H,$$

where $H = \{(a_1, a_2) \mid a_1, a_2 \in \mathbb{R}^1\}$ is the space consisting of the harmonic vector fields on $\mathbb{T}^2$, and

$$G^l(TM) = \left\{\nabla p \mid p \in W^{l+1,2}(M)\right\}.$$

We need to prove that

$$P[(u \cdot \nabla)u] = 0, \quad \forall\ u \in E_k, \tag{4.5.4}$$

where $P : W^{l,2}(TM) \to W_H^{l,2}(TM)$ is the projection. In fact, for any $v = \operatorname{curl} \psi \in W_H^{l,2}(TM)$, we have

$$\int_{\mathbb{T}^2} (u \cdot \nabla)u \cdot v dx = \int_{\mathbb{T}^2} \psi \left[\frac{\partial}{\partial x_2}(u \cdot \nabla)u_1 - \frac{\partial}{\partial x_1}(u \cdot \nabla)u_2\right] dx. \tag{4.5.5}$$

Since $u \in E_k$ and $u = \operatorname{curl} h$, we have

$$-\Delta h = \Lambda_k h. \tag{4.5.6}$$

It follows from (4.5.5) and (4.5.6) that,for any $v = \operatorname{curl} \psi \in W_H^{l,2}(TM)$,

$$\int_{\mathbb{T}^2} (u \cdot \nabla)u \cdot v dx = -\int_{\mathbb{T}^2} \psi \left[\frac{\partial h}{\partial x_2}\frac{\partial}{\partial x_1}\Delta h - \frac{\partial h}{\partial x_1}\frac{\partial}{\partial x_2}\Delta h\right] dx = 0.$$

Hence, Assertion 1 follows.

By (4.5.4), equation (4.5.3) restricted on E_k is equivalent to the following equation:

$$\begin{cases} \dfrac{\partial u}{\partial t} = \mu \Delta u + f_k, \\ u(x, 0) = \varphi, \quad \varphi \in E_k. \end{cases} \tag{4.5.7}$$

Since $-\mu\Delta u = \Lambda_k u$, the solution of (4.5.7) reads

$$u(t) = e^{-\mu\Lambda_k t}\varphi + \mu^{-1}\Lambda_k^{-1}(1 - e^{-\mu\Lambda_k t}) f_k, \tag{4.5.8}$$

and Assertion 2 follows. The proof is complete. □

We end this section with a global Lyapunov stability theorem, due to Marchioro [**70**], for any forcing $f_1 \in E_1$, where E_1 is the eigenspace corresponding to the first eigenvalue $\Lambda_1 = 1$.

$$\begin{cases} \dfrac{\partial u}{\partial t} = \mu\Delta u - (u \cdot \nabla)u - \nabla p = f_1, \\ \operatorname{div} u = 0, \\ u(x, 0) = \varphi, \end{cases} \tag{4.5.9}$$

where $\varphi \in \mathcal{H}^2(TM)$. It is easy to see that the eigenspace E_1 of (4.5.2) has dimension 4, and is generated by

$$E_1 = \text{span}\left\{(\sin x_2, 0), (\cos x_2, 0), (0, \sin x_1), (0, \cos x_1)\right\}.$$

By Theorem 4.5.1, the function $v_0 = \mu^{-1} f_1$ is the stationary solution of (4.5.9). We have the following asymptotic Lyapunov stability theorem, which is due to Marchioro [**70**]. For convenience, we provide here a simplified proof, and some of the technicalities will be used later.

THEOREM 4.5.2. *For any $\varphi \in \mathcal{H}^\infty(TM)$, the stationary solution v_0 of (4.5.9) is asymptotically stable in the $W^{l,2}$-norm for any $l \geq 0$; i.e., for the solution $u(\cdot, t, \varphi)$ of (4.5.9), we have*

$$\lim_{t\to\infty} \|u(\cdot, t, \varphi) - v_0\|_{W^{l,2}} = 0. \tag{4.5.10}$$

PROOF. Let $u = \text{curl}\,\psi$. Without loss of generality, we assume that $f_1 = \lambda e_1 = \lambda$ curl h_1. Then the function ψ satisfies

$$\begin{cases} \dfrac{\partial \Delta\psi}{\partial t} = \mu\Delta^2\psi + J[\psi, \Delta\psi] - \lambda h_1, \\ \psi(x, 0) = \psi_0, \end{cases} \tag{4.5.11}$$

where

$$J[f, g] = \frac{\partial f}{\partial x_1}\frac{\partial g}{\partial x_2} - \frac{\partial g}{\partial x_1}\frac{\partial f}{\partial x_2}.$$

Let $\Psi = \psi - \mu^{-1}\lambda h_1$. Then, from (4.5.11), we derive

$$\begin{cases} \dfrac{\partial \Delta\Psi}{\partial t} = \mu\Delta^2\Psi + \lambda\mu^{-1} J[h_1, \Delta\Psi + \Psi] + J[\Psi, \Delta\Psi], \\ \Psi(x, 0) = \Psi_0 = \psi_0 - \lambda\mu^{-1} h_1. \end{cases} \tag{4.5.12}$$

Taking the L^2 inner product between (4.5.12) and $\Delta\Psi + \Psi$, we deduce that

$$\frac{d}{dt}\int_{\mathbb{T}^2}\left[|\Delta\Psi|^2 + \Delta\Psi\cdot\Psi\right] = 2\mu\int_{\mathbb{T}^2}\left[\Delta^2\Psi\cdot\Delta\Psi + |\Delta\Psi|^2\right]. \tag{4.5.13}$$

Let $\Psi = \sum_{j=1}^\infty \Psi_j h_j$. Since the multiplicity $m_1 = 4$ for the first eigenvalue $\Lambda_1 = 1$, we infer from (4.5.13) that

$$\begin{aligned} \frac{d}{dt}\sum_{j=5}^\infty \rho_j(\rho_j - 1)\Psi_j^2 &= -2\mu\sum_{j=5}^\infty \rho_j^2(\rho_j - 1)\Psi_j^2 \\ &\leq -2\rho_5\mu\sum_{j=5}^\infty \rho_j(\rho_j - 1)\Psi_j^2, \end{aligned} \tag{4.5.14}$$

where $\rho_5 = 2$. By the Gronwall inequality, we infer from (4.5.14) that

$$\sum_{j=5}^\infty \rho_j(\rho_j - 1)\Psi_j^2 \leq Ce^{-2\rho_5\mu t},$$

where $C = \|\Delta^2\Psi_0 + \Delta\Psi_0\|_{L^2}$ is a constant. Thus we have

$$\|\Psi - P_1\Psi\|_{W^{2,2}} \leq Ce^{-\mu\rho_5 t}, \tag{4.5.15}$$

where $C > 0$ is a constant independent of Ψ, and

$$P_1 : W_H^{l,2}(M) \longrightarrow \widetilde{E}_1$$

is the orthogonal projection. Let $\widetilde{\Psi} = \Psi - P_1\Psi$. Thanks to $\Delta P_1\Psi = -P_1\Psi$, we derive from (4.5.12) that

$$\frac{d}{dt}P_1\Psi = -\mu P_1\Psi - P_1J[P_1\Psi, \Delta\widetilde{\Psi} + \widetilde{\Psi}] - P_1J[\widetilde{\Psi}, \Delta\widetilde{\Psi}] - \lambda\mu^{-1}P_1J[h_1, \Delta\widetilde{\Psi} + \widetilde{\Psi}],$$

which yields

$$P_1\Psi = e^{-\mu t}\left\{P_1\Psi_0 - \int_0^t e^{\mu t}[P_1J(P_1\Psi - \lambda\mu^{-1}h_1, \Delta\widetilde{\Psi} + \widetilde{\Psi}) - P_1J(\widetilde{\Psi}, \Delta\widetilde{\Psi})]dt\right\}. \tag{4.5.16}$$

We now estimate each term on the right-hand side of (4.5.16). By (4.5.15), we have

$$\begin{aligned}
||P_1J[\widetilde{\Psi}, \Delta\widetilde{\Psi}]||_{L^2} &= \left(\sum_{i=1}^4 |\int_{\mathbb{T}^2} J(\widetilde{\Psi}, \Delta\widetilde{\Psi})h_i|^2\right)^{1/2} \\
&= \left(\sum_{i=1}^4 |\int_{\mathbb{T}^2} J(\widetilde{\Psi}, h_i)\Delta\widetilde{\Psi}|^2\right)^{1/2} \\
&\leq C\|\widetilde{\Psi}\|^2_{W^{2,2}} \leq Ce^{-2\mu\rho_5 t}, \\
||P_1J(h_1, \Delta\widetilde{\Psi} + \widetilde{\Psi})||_{L^2} &\leq Ce^{-\mu\rho_5 t}, \\
||P_1J(P_1\Psi, \Delta\widetilde{\Psi} + \widetilde{\Psi})||_{L^2} &\leq Ce^{-\mu\rho_5 t}\|P_1\Psi\|_{L^2}.
\end{aligned}$$

Hence, it follows from (4.5.16) that

$$\|P_1\Psi\|_{L^2} \leq Ce^{-\mu t} - Ce^{-\mu t}\int_0^t e^{-\mu(\rho_5 - 1)\tau}\|P_1\Psi\|_{L^2}d\tau$$

where $\rho_5 = 2$, which yields

$$\|P_1\Psi\|_{L^2} \leq Ce^{-\mu t}. \tag{4.5.17}$$

Hence we obtain from (4.5.15) and (4.5.17) that

$$\lim_{t\to\infty} \|\psi - \mu^{-1}\lambda h_1\|_{W^{2,2}} = \lim_{t\to\infty} \|\Psi\|_{W^{2,2}} = 0,$$

which implies that (4.5.10) holds true for $l = 1$. The proof for general l can be achieved using standard energy estimates and the result for the case with $l = 1$; we omit the details. □

4.5.2. Taylor Vortex Structure of Eigenvectors. It is easy to see that the Taylor fields defined as in (3.3.1) are the eigenvectors of (4.5.2). Actually, many of the eigenvectors of (4.5.2) are of the Taylor vortex type of structure, and we explore this connection in this section.

For any $n, m \geq 1$, let

$$\begin{aligned}
V_{nm} &= \text{span}\left\{e^1_{nm}, e^2_{nm}, e^3_{nm}, e^4_{nm}\right\}, \\
e^1_{nm} &= (m\sin nx_1\cos mx_2,\ -n\cos nx_1\sin mx_2), \\
e^2_{nm} &= (m\cos nx_1\sin mx_2,\ -n\sin nx_1\cos mx_2), \\
e^3_{nm} &= (m\cos nx_1\cos mx_2,\ n\sin nx_1\sin mx_2), \\
e^4_{nm} &= (m\sin nx_1\sin mx_2,\ n\cos nx_1\cos mx_2).
\end{aligned}$$

For any $n \geq 1$, let

$$\begin{aligned} V_{n0} &= \text{span}\left\{e^1_{n0}, e^2_{n0}, e^3_{n0}, e^4_{n0}\right\}, \\ e^1_{n0} &= \{\sin nx_2, 0\}; \quad e^2_{n0} = \{\cos nx_2, 0\}; \\ e^3_{n0} &= \{0, \sin nx_1\}; \quad e^4_{n0} = \{0, \cos nx_1\}. \end{aligned}$$

Then it is easy to check that these vector fields enjoy the following three basic properties:

(1) The eigenspace E_k of (4.5.2) corresponding to the eigenvalue Λ_k is given by

$$E_k = \bigcup_{\substack{n^2+m^2=\Lambda_k \\ n\geq 1, m\geq 0}} V_{nm}.$$

Namely, the above defined vector fields generate all eigenvectors of the Stokes problem.

(2) For any $n, m \geq 1$, any $u = (u_1, u_2) = \sum_{i=1}^4 a_i e^i_{nm} \in V_{nm}$ can be equivalently expressed as

$$\left\{\begin{aligned} u_1 &= m\lambda_1 \sin(nx_1 + mx_2 + \theta_1) + m\lambda_2 \sin(nx_1 - mx_2 + \theta_2), \\ u_2 &= -n\lambda_1 \sin(nx_1 + mx_2 + \theta_1) + n\lambda_2 \sin(nx_1 - mx_2 + \theta_2), \end{aligned}\right. \tag{4.5.18}$$

where

$$\left\{\begin{aligned} \lambda_1 &= \frac{1}{2}\left[(a_1 + a_2)^2 + (a_3 - a_4)^2\right]^{1/2}, \\ \lambda_2 &= \frac{1}{2}\left[(a_1 - a_2)^2 + (a_3 + a_4)^2\right]^{1/2}, \\ \theta_1 &= \sin^{-1}\frac{a_3 - a_4}{2\lambda_1}, \\ \theta_2 &= \sin^{-1}\frac{a_3 + a_4}{2\lambda_2}. \end{aligned}\right. \tag{4.5.19}$$

(3) For any $n \geq 1$, any $u = \sum_{i=1}^4 a_i e^i_{n0} \in V_{n0}$ can be rewritten as

$$u = (\lambda_1 \sin(nx_2 + \theta_1),\ \lambda_2 \sin(nx_1 + \theta_2)), \tag{4.5.20}$$

where

$$\begin{aligned} \lambda_1 &= \sqrt{a_1^2 + a_2^2}, \qquad & \lambda_2 &= \sqrt{a_3^2 + a_4^2}, \\ \theta_1 &= \sin^{-1}\frac{a_2}{\lambda_1}, \qquad & \theta_2 &= \sin^{-1}\frac{a_4}{\lambda_1}. \end{aligned} \tag{4.5.21}$$

The following theorem illustrates the Taylor vortex type of structure of the eigenvectors of (4.5.2) in V_{nm} and E_k, which is useful in our discussion of the structural evolution of solutions of the Navier-Stokes equations on the torus.

THEOREM 4.5.3. *Let $u = \sum_{i=1}^4 a_i e^i_{nm} \in V_{nm}$ $(n, m \geq 1)$ (resp. $u = \sum_{i=1}^4 a_i e^i_{n0} \in V_{n0}$), and let $\lambda_1, \lambda_2 \neq 0$. Then u is regular, and the following assertions hold true.*

1. If $\lambda_1 = \lambda_2$, then the set $S \subset \mathbb{T}^2$ of all saddle-connections of u consists of only one connected component, and u is topologically equivalent to the Taylor field v_{nm} defined by (3.3.1) (resp. to the vector field $v = (\sin nx_2, \sin nx_1)$).

2. If $\lambda_1 \neq \lambda_2$, then

(a) *the set S consists of exactly $2R$ connected components Γ_i $(i = 1, \cdots, 2R)$, where R is the greatest common factor of n and m (resp. $R = n$);*

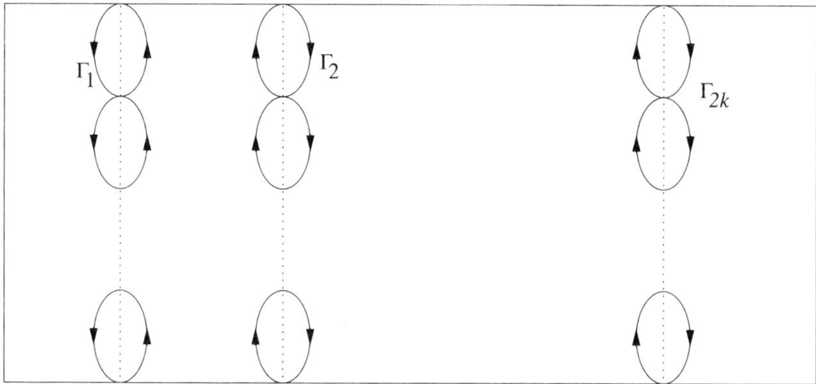

FIGURE 4.5.1. Schematic illustrating the topological structure of fields in V_{nm}.

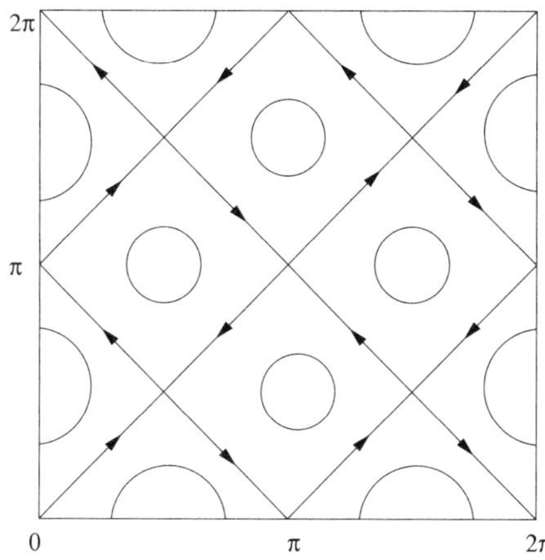

FIGURE 4.5.2. Phase diagram for $v = (\sin 2x_2, \sin 2x_1)$.

(b) *all vector fields* $u \in V_{nm}$ $(n \geq 1,\ m \geq 0)$ *with* $\lambda_1 \neq \lambda_2$ *are topologically equivalent; their topological structures can be characterized by the* $2R$ *connected components* $\Gamma_i \subset S$ $(1 \leq i \leq 2R)$ *as shown in Figure 4.5.1;*

(c) *each* $u \in V_{nm}$ $(n \geq 1,\ m \geq 0)$ *with* $\lambda_1 \neq \lambda_2$ *has exactly* $4nm$ *centers and* $4nm$ *saddles;*

(d) *each connected component* Γ_i $(1 \leq i \leq 2R)$ *in the set* S *of saddle-connections of* $u \in V_{nm}$ *(resp.* $u \in V_{n0}$*) contains exactly* $2Rm_1n_1$ *(resp.* n*) saddles, where* m_1 *and* n_1 *are coprime and are defined by* $(m, n) = R(m_1, n_1)$*; and*

(e) *each connected component* Γ_i $(1 \leq i \leq 2R)$ *encloses* $2Rm_1n_1$ *centers (resp.* n *centers).*

REMARK 4.5.4. By the above theorem, the saddle connection set S of $v = (\sin nx_2, \sin nx_1)$ contains exactly one component; see Figure 4.5.2.

REMARK 4.5.5. Obviously, all vector fields in V_{nm} (except in V_{10}) are not Hamiltonian structurally stable, because they are not saddle self-connected. However, Theorem 4.5.3 tells us that the regular vectors in V_{nm} have exactly two types of topological structure, called the Taylor vortex type of structure.

PROOF OF THEOREM 4.5.3. *Step 1.* It is easy to check that as $\lambda_1, \lambda_2 \neq 0$, the vector field (4.5.18) has $4nm$ saddle points and $4nm$ centers, and the vector field (4.5.20) has $2n^2$ saddle points and $2n^2$ centers. It suffices to prove the case where $u \in V_{nm}$ $(n, m \geq 1)$, as the case where $u \in V_{n0}$ can be proved in the same fashion.

It is easy to see that the topological structures of vector fields (4.5.18) are independent of the phrase shift by (θ_1, θ_2). Indeed, under the transformation

$$z_1 = x_1 + \frac{\theta_1 + \theta_2}{2n},$$
$$z_2 = x_2 + \frac{\theta_1 - \theta_2}{2m},$$

the vector fields (4.5.18) are transformed into the form with $(\theta_1, \theta_2) = 0$. It suffices then to consider the case where $u = (u_1, u_2)$ is expressed as

$$\begin{cases} u_1 = m\lambda_1 \sin(nx_1 + mx_2) + m\lambda_2 \sin(nx_1 - mx_2), \\ u_2 = -n\lambda_1 \sin(nx_1 + mx_2) + n\lambda_2 \sin(nx_1 - mx_2). \end{cases} \tag{4.5.22}$$

It is easy to calculate that the $4nm$ saddle points of (4.5.22) are

$$\left(\frac{k_1}{n}\pi, \frac{k_2}{m}\pi\right), \quad 0 \leq k_1 \leq 2n-1,\ 0 \leq k_2 \leq 2m-1, \tag{4.5.23}$$

and the $4nm$ centers of (4.5.22) are

$$\left(\frac{2k_1+1}{2n}\pi, \frac{2k_2+1}{2m}\pi\right), \quad 0 \leq k_1 \leq 2n-1,\ 0 \leq k_2 \leq 2m-1. \tag{4.5.24}$$

The Hamiltonian function of (4.5.22) is given by

$$H(x_1, x_2) = -\lambda_1 \cos(nx_1 + mx_2) + \lambda_2 \cos(nx_1 - mx_2). \tag{4.5.25}$$

Obviously, if $\lambda_1 = \lambda_2 = \lambda$, the vector field (4.5.22) is as follows:

$$u = (2m\lambda \sin nx_1 \cos mx_2, -2n\lambda \cos nx_1 \sin mx_2),$$

which is topologically equivalent to the Taylor field v_{nm} defined by (3.3.1). We have then proved Assertion 1.

Step 2. To prove Assertion 2, we assume, without loss of generality, that $\lambda_2 > \lambda_1 > 0$. Let

$$A_1 = \left\{\left(\frac{k_1}{n}\pi, \frac{k_2}{m}\pi\right) \middle| k_1 \pm k_2 = \text{even}\right\},$$
$$A_2 = \left\{\left(\frac{k_1}{n}\pi, \frac{k_2}{m}\pi\right) \middle| k_1 \pm k_2 = \text{odd}\right\}.$$

Then, if $\lambda_1 \neq \lambda_2$, the Hamiltonian function (4.5.25) satisfies

$$\begin{cases} H(x) = H(y), & \forall\, x, y \in A_1, \text{ or } x, y \in A_2, \\ H(x) \neq H(y), & \forall\, x \in A_1, \text{ and } y \in A_2. \end{cases} \tag{4.5.26}$$

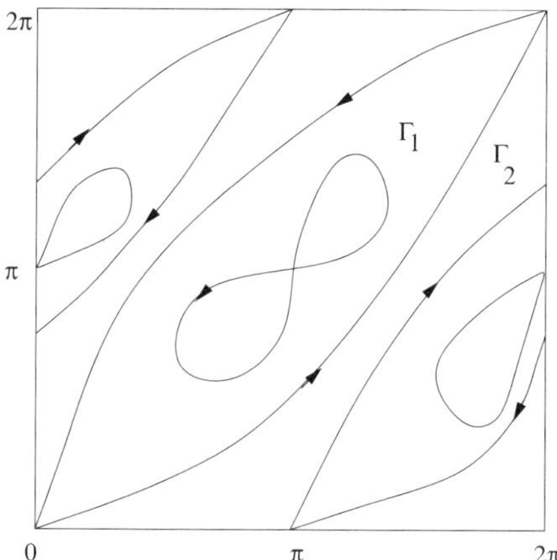

FIGURE 4.5.3. Schematic illustrating Step 2 of the proof of Theorem 4.5.3.

It follows from (4.5.26) that each saddle point $(\frac{k_1}{n}\pi, \frac{k_2}{m}\pi)$ of (4.5.22) cannot be connected to the four adjacent saddle points

$$\left(\frac{k_1 \pm 1}{n}\pi, \frac{k_2}{m}\pi\right) \quad \text{and} \quad \left(\frac{k_1}{n}\pi, \frac{k_2 \pm 1}{m}\pi\right).$$

Let $R = \frac{n}{n_1} = \frac{m}{m_1}$ be the greatest common factor of n and m. Then it is easy to see that for each $r = 1, \ldots, 2R$, the following set of saddle points are possibly on the same connected component of the saddle connection set S:

$$\left\{\left(\frac{i+r-1}{n}\pi, \frac{i+r}{m}\pi\right) \Big| 0 \leq i < 2m_1 n\right\}. \tag{4.5.27}$$

We shall prove that for each $r = 1, \ldots, 2R$, the saddle points given in (4.5.27) must be on one connected component of the saddle-connections. For simplicity, we proceed with the case where $(n, m) = (1, 1)$; the proof for the general case is similar.

If the conclusion is not true, then the phase diagram of (4.5.22) with $(n, m) = (1, 1)$ can only be as shown in Figure 4.5.3.

By (4.5.26), the Hamiltonian function has the same value at $(0, 0)$ and (π, π), i.e., $H(0, 0) = H(\pi, \pi)$. Hence, on the saddle-connections Γ_1 and Γ_2 in Figure 4.5.3, $H(x)$ has the same value:

$$H(x) = H(y), \quad \forall\, x \in \Gamma_2, \forall y \in \Gamma_1. \tag{4.5.28}$$

Since the domain D enclosed by Γ_1 and Γ_2 is on the right and left sides of Γ_1 and Γ_2, respectively, traveling their orientations, we have

$$H(x) \geq H(z) \geq H(y), \quad \text{for } x \in \Gamma_2,\ z \in D,\ y \in \Gamma_1. \tag{4.5.29}$$

This yields that H is constant in D, a contradiction. Thus, we deduce that the saddle points of u will be connected in the manner as shown in Figure 4.5.4. Assertion 2(a) is proven.

Assertion 2(b) follows immediately from Assertion 2(a), since the structure in Figure 4.5.4 is obviously topologically equivalent to the structure as shown in Figure 4.5.5. The remainder of the proof is obvious. The proof of Theorem 4.5.3 is complete. □

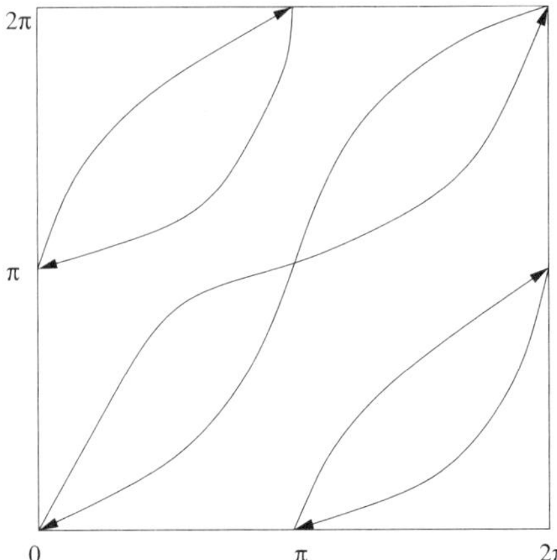

FIGURE 4.5.4. A flow pattern derived in Step 2 of the proof of Theorem 4.5.3.

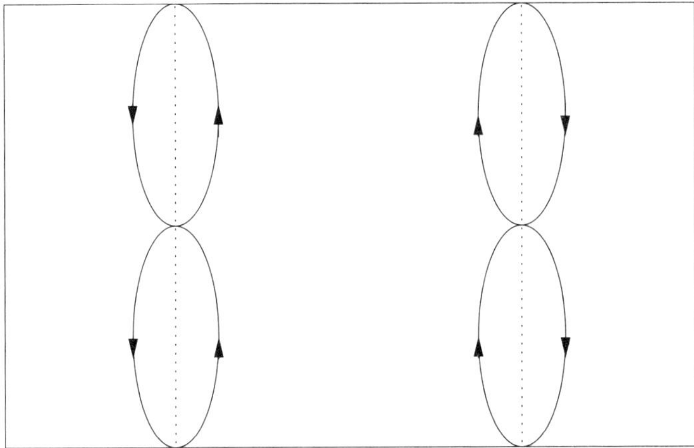

FIGURE 4.5.5. A flow pattern, toplogically equivalent to that shown in Figure 4.5.4.

4.5.3. Navier-Stokes equations with data in eigenspaces. In this section, we shall discuss the asymptotic structure of solutions of the equations (4.5.3) with the data in the eigenspace E_k.

We start with the case where both the forcing f_k and the initial velocity φ are in V_{nm}. The following theorem follows directly from Theorems 4.5.1 and 4.5.3.

THEOREM 4.5.6. *Let $n^2 + m^2 = \Lambda_k$, and let the data $f_k, \phi \in V_{nm}$ be given in the form (4.5.18) or (4.5.20) with corresponding (λ_1, λ_2) denoted by $(\lambda_1^f, \lambda_2^f)$ and $(\lambda_1^\phi, \lambda_2^\phi)$, respectively. Then there exists a $t_0 \geq 0$ such that for any $t > t_0$, the solution $u(\cdot, t)$ of (4.5.3) is topologically equivalent to the structure of the following vector fields:*

(1)

$$v = \begin{pmatrix} m\sin(nx_1+mx_2)+\frac{1}{2}m\sin(nx_1-mx_2) \\ -n\sin(nx_1+mx_2)+\frac{1}{2}n\sin(nx_1-mx_2) \end{pmatrix}^t \quad \text{for } n,m\geq 1, \tag{4.5.30}$$

$$v = \left(\sin nx_2, \frac{1}{2}\sin nx_1\right) \quad \text{for } n\geq 1, m=0, \tag{4.5.31}$$

if $|\lambda_1^f-\lambda_2^f|+|\lambda_1^\phi-\lambda_2^\phi|\neq 0$,

(2) *the Taylor field* v_{nm} *(resp.* $v=(\sin nx_2,\sin nx_1)$ *for* $m=0$*) if* $\lambda_1^f=\lambda_2^f$ *and* $\lambda_1^\phi=\lambda_2^\phi$.

Now we consider the case where the forcing $f_k\in V_{nm}$ and $\varphi\in E_k$, where $n^2+m^2=\Lambda_k$. Without loss of generality, we assume that $f_k\in V_{nm}$ is given by either (4.5.18) or (4.5.20) with $\theta_1=\theta_2=0$. The general case differs only by a phase shift. In this case, all saddle points of $f_k\in V_{nm}$ (resp. $f_k\in V_{n0}$) are given by

$$\left\{\left(\frac{k_1}{n}\pi, \frac{k_2}{m}\pi\right) \;\middle|\; 0\leq k_1\leq 2n-1,\ 0\leq k_2\leq 2m-1\right\}$$

$$\left(\text{resp. } \left\{\left(\frac{k_1}{n}\pi, \frac{k_2}{n}\pi\right) \;\middle|\; k_1+k_2=\text{even},\ 0\leq k_1,\ k_2\leq 2n-1\right\}\right).$$

THEOREM 4.5.7. *Let* $f_k\in V_{nm}$, $n^2+m^2=\Lambda_k$ *(resp.* $f_k\in V_{n0}$, $n^2=\Lambda_k$*) such that* $\lambda_1\neq\lambda_2$ *and* $\theta_1=\theta_2=0$. *Let* $\varphi=\text{curl}\,\psi\in E_k$ *such that*

$$\psi\left(\frac{k_1}{n}\pi, \frac{k_2}{m}\pi\right) \neq \psi\left(\frac{j_1}{n}\pi, \frac{j_2}{m}\pi\right), \quad \forall\,(k_1,k_2)\neq(j_1,j_2) \tag{4.5.32}$$

$$\left(\textit{resp. } \psi\left(\frac{k_1}{n}\pi, \frac{k_2}{n}\pi\right) \neq \psi\left(\frac{j_1}{n}\pi, \frac{j_2}{n}\pi\right), \quad \forall\,(k_1,k_2)\neq(j_1,j_2)\right).$$

Then there is a $t_0\geq 0$ *such that, for all* $t>t_0$, *the solutions* $u(\cdot,t)$ *of (4.5.3) are structurally stable, and topologically equivalent to the vector field having the block structure*

$$\mathbb{T}^2=\Omega+\cup_{r=1}^{2R}A_r \quad (\textit{resp. } \mathbb{T}^2=\Omega+\cup_{r=1}^{2n}A_r), \tag{4.5.33}$$

where R *is the greatest common factor of* n *and* m, Ω *is the* T*-block,* A_r *are the* S*-blocks, and each* A_j *contains exactly* $2mn/R$ *circle cells (resp.* n *circle cells) having the same orientation as* A_j. *For instance, as* $(n,m)=(1,1)$, *the block structure (4.5.33) is shown schematically in Figure 3.3.12.*

PROOF. By the proof of Theorem 4.5.1, we see that the solution of (4.5.3) can be expressed as

$$u=\Lambda_k^{-1}\mu^{-1}f_k+e^{-\mu\Lambda_k t}\varphi-\mu^{-1}\Lambda_k^{-1}f_ke^{-\mu\Lambda_k t}. \tag{4.5.34}$$

Obviously, the vector fields $u(\cdot,t)$ are regular for all $t>0$ sufficiently large. By Theorem 4.5.3, if the saddle points of $u(\cdot,t)$ are self-connected, then the topological structure of $u(\cdot,t)$ must be equivalent to the form (4.5.33). Hence, the theorem follows from the connection lemma, Lemma 2.3.3, (4.5.32) and (4.5.34). The proof is complete. □

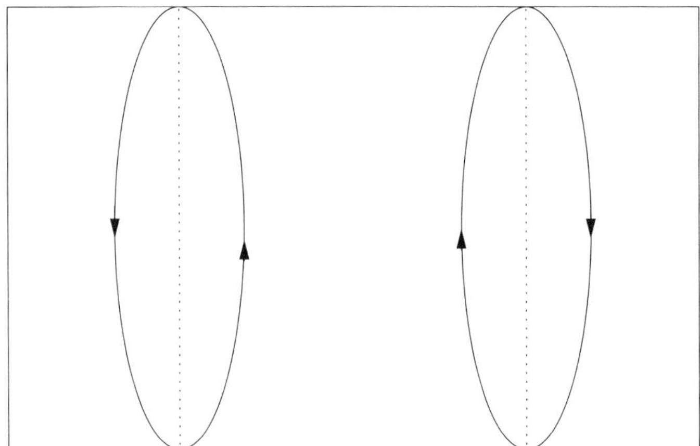

FIGURE 4.5.6. Flow structure of f_1 given by (4.5.35) with $\lambda_1 \neq \lambda_2$.

REMARK 4.5.8. Similar results are also true when $\lambda_1 = \lambda_2$.

REMARK 4.5.9. Let n and m be prime to each other, and let $V_{nm} + V_{mn} \subset E_k$. Let $f_k \in V_{nm}$ with $\lambda_1 \neq \lambda_2$, and let $\varphi = \varphi_1 + \varphi_2$, $\varphi_1 \in V_{nm}$, $\varphi_2 \in V_{mn}$. If $\varphi_2 = \sum_{i=1}^4 b_i e_{mn}^i$ and satisfies $b_1 b_2 \neq b_3 b_4$, then the Hamiltonian function ψ of φ satisfies condition (4.5.32). Likewise, when $e_k \in V_{nm}$ with $\lambda_1 = \lambda_2$, and $\varphi = \sum_{i=1}^4 \alpha_i e_{nm}^i + \sum_{i=1}^4 \beta_i e_{mn}^i$ with $\alpha_1\alpha_2 \neq \alpha_3\alpha_4$ and $\beta_1\beta_2 \neq \beta_3\beta_4$, then the Hamiltonian function of φ satisfies condition (4.5.32).

REMARK 4.5.10. Although the Taylor field v_{nm} with a perturbation, by Theorem 3.3.2, may have in general a few different structures, the solution of (4.5.3), by Theorems 4.5.6–4.5.7, has a unique block structure (4.5.33) for $t > 0$ sufficiently large.

4.5.4. Navier-Stokes equations with forcing in E_1. In this section, we shall consider the asymptotic structure of solutions of the equations (4.5.9) forced by eigenvectors corresponding to the first eigenvalue of the Stokes problem.

By (4.5.20), we see that for any $f_1 \in E_1 = V_{10}$, f_1 can be written as $f_1 = (\lambda_1 \sin(x_2 + \theta_1), \lambda_2 \sin(x_1 + \theta_2))$. Without loss of generality, for simplicity, we take f_1 in the form as follows

$$f_1 = (\lambda_1 \sin x_2,\ \lambda_2 \sin x_1). \tag{4.5.35}$$

THEOREM 4.5.11. *Let $f_1 \in E_1$ be given by (4.5.35).*

(1) *If $\lambda_1 \neq \lambda_2$ and $\lambda_1, \lambda_2 \neq 0$, then for any $\varphi \in \mathcal{H}^3(TM)$ there is a $t_0 > 0$ such that for all $t > t_0$ the solution $u(\cdot, t)$ of (4.5.9) is Hamiltonian structurally stable, and topologically equivalent to $v = (\sin x_2, \frac{1}{2}\sin x_1)$; the topological structure is shown in Figure 4.5.6.*

(2) *If $\lambda_1 = \lambda_2 \neq 0$, or $\lambda_1 \cdot \lambda_2 = 0$ and $(\lambda_1, \lambda_2) \neq 0$, then there exists a manifold $M_0 \subset \mathcal{H}^3(TM)$ with* codim *$M_0 = 2$ such that for any $\varphi \in \mathcal{H}^3(TM)/M_0$, the conclusions above hold true.*

REMARK 4.5.12. The manifold M_0 in Assertion 2 of Theorem 4.5.11 is the spectral manifold introduced by Foias and Saut [**21, 22**]. In fact, the spectral manifolds are analytic.

PROOF OF THEOREM 4.5.11. By Theorem 4.5.2, the solution of (4.5.9) is

$$\begin{cases} u = \mu^{-1} f_1 + v(x,t), \\ \lim_{t\to\infty} \|v(\cdot,t)\|_{C^1} = 0. \end{cases} \tag{4.5.36}$$

As $\lambda_1 \neq \lambda_2$, we can see that f_1 is a stable Hamiltonian vector field, which has the topological structure as shown in Figure 4.5.6. Hence, Assertion 1 follows from (4.5.36).

To prove Assertion 2, we prove only the case where $\lambda_1 = \lambda_2$; the other case can be proved in the same fashion. Let $v = \text{curl}\,\Psi$. Then Ψ satisfies equation (4.5.12). By (4.5.15) and (4.5.16),

$$\Psi = \Phi e^{-\mu t} + O(e^{-2\mu t}), \tag{4.5.37}$$

in $W^{4,2}(M)$, where $\Phi \in \widetilde{E}_1 = \text{span}\{\sin x_1, \cos x_1, \sin x_2, \cos x_2\}$ satisfies

$$\Phi = P_1\Psi_0 - \int_0^\infty e^{\mu t}(P_1 J(\Psi, \Delta\Psi) + \mu^{-1} P_1 J(h_1, \Delta\Psi + \Psi))d\tau, \tag{4.5.38}$$

where $\Psi_0 = \Psi(x,0)$, $h_1 = -\lambda\cos x_2 + \lambda\cos x_1$ $(\lambda = \lambda_1 = \lambda_2)$ is the Hamiltonian function of f_1, and $P_1 : W^{4,2}(M) \to \widetilde{E}_1$ is the projection.

Let

$$A = \{\psi \in \widetilde{E}_1 \mid \psi = \alpha(\sin x_2 - \sin x_1) + \beta(\cos x_2 - \cos x_1),\ \alpha, \beta \in \mathbb{R}^1\}.$$

If $\Phi \in \widetilde{E}_1/A$, then $\mu^{-1}h_1 + e^{-\mu t}\Phi \in \widetilde{E}_1/A$, for any $t \geq 0$. Thus the vector field

$$\begin{aligned} u_1 &= \text{curl}(\mu^{-1}h_1 + e^{-\mu t}\Phi) \\ &= \mu^{-1} f_1 + e^{-\mu t}\text{curl}\,\Phi \\ &= (\gamma_1 \sin(x_2 + \theta_1),\ \gamma_2 \sin(x_1 + \theta_2)) \end{aligned}$$

satisfies $\gamma_1 \neq \gamma_2$ and $\gamma_1 \cdot \gamma_2 \neq 0$ for all $t > 0$ sufficiently large. Hence, it follows from (4.5.36) and (4.5.37) that the solution $u = u_1 + O(e^{-2\mu t})$ of (4.5.9) is Hamiltonian structurally stable.

From (4.5.38) we see that Ψ is a function of the initial value $\Psi_0 \in W^{4,2}(M)$. We only need to prove that there exists a manifold $\widetilde{M} \in W^{4,2}(M)$ such that for any $\Psi_0 \in W^{4,2}(M)/\widetilde{M}$, we have $\Psi \in \widetilde{E}_1/A$.

To this end, we set the mapping $\Psi : W^{4,2}(M) \to \widetilde{E}_1$ by

$$\Psi(\Psi_0) = P\Psi_0 - \int_0^\infty e^{\mu t}(P[S(\tau)\Psi_0, S(\tau)\Delta\Psi_0] + \mu^{-1}P[h_1, (\Delta + I)S(\tau)\Psi_0])d\tau,$$

where $S(t) : W^{4,2}(M) \to W^{4,2}(M)$ is the semi-group of nonlinear operators generated by (4.5.12). This mapping is well defined, since the solution $\Psi = S(t)\Psi_0$ of (4.5.12) satisfies (4.5.15) and (4.5.17).

Let $\widetilde{M} = \Psi^{-1}(A)$. It is clear that $\Psi \in \widetilde{E}_1/A$ if and only if $\Psi_0 \in W^{4,2}(M)/\widetilde{M}$. On the other hand, we have

$$\begin{aligned} \Psi'(\Psi_0) &= P - \int_0^\infty e^{\mu t}\Big(P\left[S(\tau)\Psi_0, S(\tau)\Delta\cdot\right] + P\left[S(\tau)\cdot, S(\tau)\Delta\Psi_0\right] \\ &\quad + \mu^{-1}P\left[h_1, S(\tau)(\Delta + I)\cdot\right]\Big)d\tau. \end{aligned}$$

Because the space A is invariant under $S(t)$, and $(\Delta + I)\widetilde{E}_1 = 0$, for any $\Psi_0 \in A$, the operator

$$\Psi'(\Psi_0)|_{\widetilde{E}_1} = id : \widetilde{E}_1 \longrightarrow \widetilde{E}_1.$$

It implies that, for any $\Psi_0 \in A$, the mapping $\Psi'(\Psi_0) : W^{4,2}(M) \to \widetilde{E}_1$ is surjective.

In fact, in the same manner as used in Foias and Saut [**21**], one can derive that for any $\Psi_0 \in W^{4,2}(M)$, the mapping

$$\Psi'(\Psi_0) : W^{4,2}(M) \longrightarrow \widetilde{E}_1$$

is onto; here we omit the details. Thus, we derive that the set $\widetilde{M} = \Psi^{-1}(A) \subset W^{4,2}(M)$ is a manifold with codim $\widetilde{M} = 2$. The proof is complete. □

4.5.5. Navier-Stokes equations with potential forcing. Finally, we consider the asymptotic structure of solutions of the Navier-Stokes equations

$$\begin{cases} \dfrac{\partial u}{\partial t} + (u \cdot \nabla)u = \mu\Delta u - \nabla p, \quad x \in \mathbb{T}^2, \\ u = \text{curl } \psi \\ u(x,0) = \varphi(x). \end{cases} \tag{4.5.39}$$

Let $0 < \Lambda_1 < \Lambda_2 < \cdots$, be the sequence of distinct eigenvalues of (4.5.2), Λ_k having multiplicity m_k. The results in this section are based on the work by Foias and Saut [**21, 22**], which is briefly summarized in the following theorem.

THEOREM 4.5.13. (Foias and Saut [**21, 22**]) *There exists connected unbounded analytic manifolds M_k $(k = 1, 2, \cdots)$ in $W^{1,2}_H(TM)$ such that*

(1) $W^{1,2}_H(TM) = M_0 \supset M_1 \supset \cdots$, *and M_k has codimension $m_1 + \cdots + m_k$ in M_0.*
(2) *M_k is invariant by the semigroup $S(t)$ generated by (4.5.39).*
(3) *$\varphi \in M_{k-1}/M_k$ if and only if the solution u of (4.5.39) has the expansion*

$$\begin{cases} u = \Psi_k e^{-\mu\Lambda_k a t} + \Psi_{k+1} e^{-\mu\alpha_{k+1} t} + \cdots + \Psi_N e^{-\mu\alpha_n t} + v_N(x,t) \\ \|v\|_{W^{m,2}} = O(e^{-\mu(\alpha_n+\varepsilon)t}) \quad \text{for some } \varepsilon > 0 \\ \Psi_k \in E_k \qquad \text{the eigenspace corresponding to } \Lambda_k, \end{cases} \tag{4.5.40}$$

where $\alpha_j = \sum_{k\le i\le j} a_i\Lambda_k$, $a_i \ge 0$ are integers, and the expansion (4.5.40) satisfies the following properties:
(a) *If $\alpha_j \le \alpha_N$ is an eigenvalue with $\alpha_j \ne a_k\Lambda_k + \cdots + a_{j-1}\Lambda_{j-1}$ for some $a_k, \ldots, a_{j-1} \in N$, then $\Psi_j \in E_j$.*
(b) *Otherwise, Ψ_j satisfies*

$$\frac{d\Psi_j}{dt} + \mu(\Delta - \alpha_j)\Psi_j + \sum_{\alpha_\ell + \alpha_i = \alpha_j} (\Psi_{\alpha_\ell} \cdot \nabla)\Psi_{\alpha_i} + \nabla p = 0. \tag{4.5.41}$$

Now we state and prove the result that gives the asymptotic structure of solutions of the Navier-Stokes equations (4.5.39).

THEOREM 4.5.14. *For each pair of integers $n, m \ge 0$ with $n^2 + m^2 = \Lambda_k$ $(k \ge 1)$, there exists an analytic manifold $\widetilde{M} \subset M_{k-1}$ having codimension $m_k - 4$ in M_{k-1}, and there is an open and dense set $\mathcal{O} \subset \widetilde{M}$ such that for any initial value $\varphi \in \mathcal{O}$, the solution $u(\cdot, t)$ of (4.5.39) is Hamiltonian structurally stable for all $t > 0$ sufficiently large, with block decomposition*

$$\mathbb{T}^2 = \Omega + \bigcup_{j=1}^{2R} A_j, \tag{4.5.42}$$

where $R > 0$ is the maximum common factor of n and m, and each A_j contains exactly $N = 2nm/R$ circle cells having the same orientation as A_j (if $(n,m) = (n,0)$, then $R = n$ and A_j contains n circle cells), where Ω is the T-block, and A_j are the S-blocks.

PROOF. In the expansion (4.5.40), the first coefficient $\Psi_k \in E_k$ depends continuously on $\varphi \in M_{k-1}/M_k$, and Ψ satisfies

$$\Psi_k(\varphi) = P_k\varphi - \int_0^\infty e^{\mu\Lambda_k t} P_k\left((S(t)\varphi \cdot \nabla)S(t)\varphi\right) dt, \tag{4.5.43}$$

where $P_k : W_H^{1,2}(TM) \to E_k$ is the projection. Moreover, the mapping $\Psi_k : M_{k-1} \to E_k$ defined by (4.5.43) is analytic, and for any $\varphi \in M_{k-1}$, the derivative operator $\Psi_k'(\varphi) : M_{k-1} \to E_k$ is surjective; see [**21**]. Hence, $\widetilde{M} = \Psi_k^{-1}(V_{nm}) \subset M_{k-1}$ is an analytic manifold having codimension $m_k - 4$, because V_{nm} has codimension $m_k - 4$ in E_k. It is easy to see that

$$\Psi_k(\varphi) \in V_{nm} \Longleftrightarrow \varphi \in \widetilde{M} \subset M_{k-1}. \tag{4.5.44}$$

By (4.5.4), we infer that if $\alpha_j < \Lambda_{k+1}$, then

$$\sum_{\alpha_\ell + \alpha_i = \alpha_j} (\Psi_{\alpha_\ell} \cdot \nabla)\Psi_{\alpha_i} + \nabla p + 0,$$

which yields from (4.5.41) that $\Psi_j = 0$; see [**22**]. Thus, we deduce that in the special periodic boundary condition of $n = 2$, the solution u of (4.5.39) admits the expansion

$$\begin{cases} u = \Psi_k e^{-\mu\Lambda_k t} + \Psi_{k+1} e^{-\mu\Lambda_{k+1} t} + v(x,t), \\ \|v\|_{W^{m,2}} = O\left(e^{-\mu(\Lambda_{k+1}+\varepsilon)t}\right) \\ \Psi_k \in E_k, \quad \Psi_{k+1} \in E_{k+1}, \end{cases} \tag{4.5.45}$$

and Ψ_{k+1} satisfies

$$\begin{aligned} \Psi_{k+1}(\varphi) &= P_{k+1}\varphi - \int_0^\infty e^{\mu\Lambda_{k+1} t} P_{k+1}((v \cdot \nabla)v \\ &\quad + e^{-\mu\Lambda_k t}(\Psi_k \cdot \nabla)v + e^{-\mu\Lambda_k t}(v \cdot \nabla)\Psi_k) dt, \end{aligned} \tag{4.5.46}$$

where $v = s(t)\varphi$.

Obviously, by Theorem 4.5.3 and (4.5.44), if $\varphi \in \widetilde{M}$ such that the solution (4.5.45) is Hamiltonian structurally stable, then the topological structure of (4.5.45) must be of the form (4.5.42). And the set $O \subset \widetilde{M}$ where $\varphi \in O$ yields the stable solution (4.5.45) for all $t > 0$ sufficiently large is open in $\widetilde{M}$. It remains to prove that O is dense in $\widetilde{M}$.

For each given $\Psi \in V_{nm}$, we denote by $M_k(\Psi) = \Psi_k^{-1}(\Psi) \subset \widetilde{M}$, which has codimension m_k in M_{k-1}. Then, the solution (4.5.45) of (4.5.39) is of the following form:

$$u = \Psi e^{-\mu\Lambda_k t} + \Psi_{k+1} e^{-\mu\Lambda_{k+1} t} + v(x,t) \tag{4.5.47}$$

if and only if the initial value $\varphi \in M_k(\Psi)$, where Ψ_{k+1} satisfies (4.5.46) with Ψ replacing Ψ_k.

We define the mapping $\Psi_{k+1} : M_k(\Psi) \to E_{k+1}$ by

$$(4.5.48) \qquad \begin{cases} \Psi_{k+1}(\varphi) = P_{k+1}\varphi - \displaystyle\int_0^\infty e^{\mu\Lambda_{k+1}t} P_{k+1}[(u_\varphi \cdot \nabla)u_\varphi]dt \\ u_\varphi = e^{-\mu\Lambda_k t}\Psi + S(t)\varphi. \end{cases}$$

By (4.5.4) it is easy to see that the coefficient Ψ_{k+1} in (4.5.47) is given by the mapping (4.5.48).

From Proposition 3 in [**21**], one deduces that the mapping $\Psi_{k+1} : M_k(\Psi) \to E_{k+1}$ is a submersion; i.e., the derivative operator $\Psi'_{k+1}(\varphi) : M_k(\Psi) \to E_{k+1}$ is onto for any $\varphi \in M_k(\Psi)$.

On the other hand, as we take $\Psi = \sum_{i=1}^4 a_i e^i_{nm} \in V_{nm}$ with $\lambda_1 \cdot \lambda_2 \neq 0$ (λ_1 and λ_2 are given by (4.5.19) or (4.5.21)), then there is an open and dense set $\widetilde{O}_\Psi$ in E_{k+1} such that for any $\Psi_{k+1} \in \widetilde{O}_\Psi$, the Hamiltonian function H_{k+1} of Ψ_{k+1} satisfies

$$H_{k+1}(p_i) \neq H_{k+1}(p_j), \quad \forall\, i \neq j,$$

where $\{p_i \mid 1 \leq i \leq 4mn\}$ are the set of all saddle points of Ψ. By Lemma 2.3.10, we infer that the vector fields (4.5.47) are Hamiltonian structurally stable for all $t > 0$ sufficiently large.

Let $O = U_{\Psi \in V'_{nm}} \Psi^{-1}_{k+1}(\widetilde{O}_\Psi)$, where $V'_{nm} = \{\Psi \in V_{nm} \mid \lambda_1 \cdot \lambda_2 \neq 0\}$. Then, $O \subset \widetilde{M}$ is the set desired. The proof is complete. □

REMARK 4.5.15. If $(n,m) = (1,0)$, then $n^2 + m^2 = \Lambda_1 = 1$, and $\widetilde{M} = W^{1,2}_H(TM)$. The set $\mathcal{O} \subset \widetilde{M}$ is given by

$$\begin{aligned} &\mathcal{O} = W^{1,2}_H(TM)/(M_1 + M'_1 + M'_2 + E), \\ &M'_i = \Psi_1^{-1}(E^i_{10}), \quad i = 1,2, \\ &E^1_{10} = \text{span}\{(\sin x_2, 0), (\cos x_2, 0)\}, \\ &E^2_{10} = \text{span}\{(0, \sin x_1), (0, \cos x_1)\}, \\ &E = \{(\lambda \sin(x_2 + \theta_1), \lambda \sin(x_1 + \theta_2)) \mid \lambda, \theta_1, \theta_2 \in \mathbb{R}^1\}. \end{aligned}$$

In the case, for $\varphi \in \mathcal{O}$, the solution $u(\cdot, t)$ of (4.5.39) for all $t > 0$ sufficiently large has the topological structure as shown in Figure 4.5.6.

EXAMPLE 4.5.16. As $(n,m) = (1,1)$, $n^2 + m^2 = \Lambda_2 = 2$, the manifold $\widetilde{M} = M_1$ has codimension 4 in $M_0 = W^{1,2}_H(TM)$. For $\varphi \in O \subset \widetilde{M}$, the solution $u(\cdot, t)$ of (4.5.39) has the topological structure as shown in Figure 3.3.12.

EXAMPLE 4.5.17. For $(n,m) = (2,1)$, $n^2 + m^2 = \Lambda_4 = 5$, the manifold $\widetilde{M} \subset M_3$ has codimension 4 in M_3, and codimension 16 in $M_0 = W^{1,2}_H(TM)$. The common factor $R = 1$ of n and m. The topological structure of the solution $u(\cdot, t)$ of (4.5.39) with $\varphi \in O \subset \widetilde{M}$ for all $t > 0$ sufficiently large is given by $\mathbb{T}^2 = \Omega + A_1 + A_2$, and each of the S-blocks A_1 and A_2 contains 4 circle cells (4 centers). For example, the structure may be given by Figure 4.5.7 or Figure 4.5.8.

4.6. Structure of Solutions of the Rayleigh-Bénard Convection

4.6.1. Boussinesq equations. The Bénard experiment can be modeled by the Boussinesq equations; see, among others, Rayleigh [**87**], Drazin and Reid [**15**]

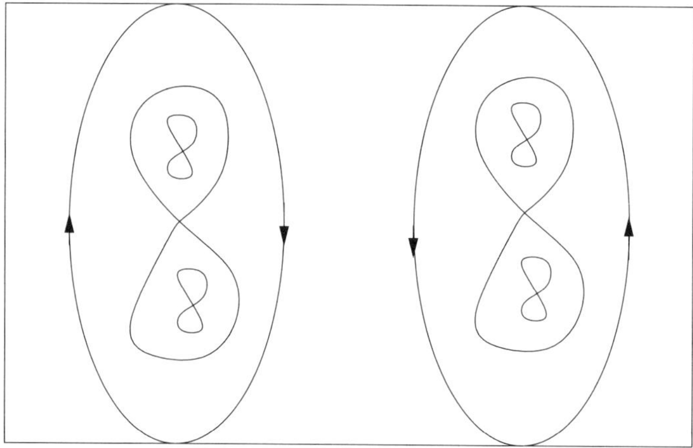

FIGURE 4.5.7. One example of two S-blocks with each containing four circle cells.

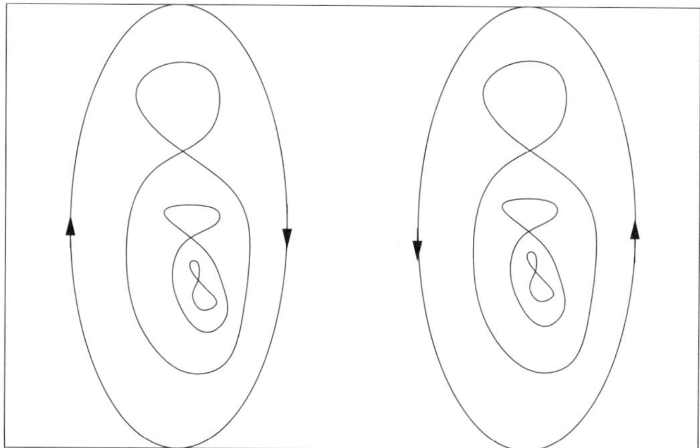

FIGURE 4.5.8. Another example of two S-blocks with each containing four circle cells.

and Chandrasekhar [**10**]. They read

$$\frac{\partial u}{\partial t} + (u \cdot \nabla)u - \nu\Delta u + \rho_0^{-1}\nabla p = -gk[1 - \alpha(T - \bar{T}_0)], \tag{4.6.1}$$

$$\frac{\partial T}{\partial t} + (u \cdot \nabla)T - \kappa\Delta T = 0, \tag{4.6.2}$$

$$\text{div } u = 0, \tag{4.6.3}$$

where g is the acceleration due to gravity, α is the coefficient of thermal expansion of the fluid, $\beta = |dT/dz| = (\bar{T}_0 - \bar{T}_1)/h$ is the vertical temperature gradient with $\bar{T}_0$ the temperature on the lower surface and $\bar{T}_1$ the temperature on the upper surface, h is the depth of the layer of the fluid, $k = (0, 0, 1)$ is the unit vector in x_3-direction, κ is the thermal diffusivity and ν is the kinematic viscosity. The unknown functions are the velocity field $u = (u_1, u_2, u_3)$, the pressure function p, and the temperature function T; see Figure 4.6.1.

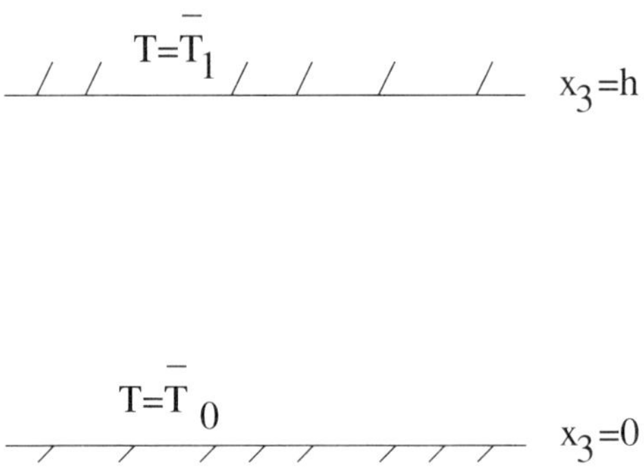

FIGURE 4.6.1. Flow between two plates heated from the bottom: $\bar{T}_0 > \bar{T}_1$.

To make the equations non-dimensional, let

$$\begin{aligned} x &= hx', \\ t &= h^2 t'/\kappa, \\ u &= \kappa u'/h, \\ T &= \beta h(T'/\sqrt{R}) + \bar{T}_0 - \beta h x_3', \\ p &= \rho_0 \kappa^2 p'/h^2 + p_0 - g\rho_0(hx_3' + \alpha\beta h^2 (x_3')^2/2), \\ P_r &= \nu/\kappa, \\ R &= \frac{g\alpha\beta}{\kappa\nu} h^4. \end{aligned}$$

Here R is the Rayleigh number, and P_r is the Prandtl number.

Omitting the primes, equations (4.6.1)–(4.6.1) can be rewritten as follows

$$\frac{1}{P_r}\left[\frac{\partial u}{\partial t} + (u\cdot\nabla)u + \nabla p\right] - \Delta u - \sqrt{R}Tk = 0, \tag{4.6.4}$$

$$\frac{\partial T}{\partial t} + (u\cdot\nabla)T - \sqrt{R}u_3 - \Delta T = 0, \tag{4.6.5}$$

$$\operatorname{div} u = 0. \tag{4.6.6}$$

The non-dimensional domain is $\Omega = D \times (0,1) \subset \mathbb{R}^3$, where $D \subset \mathbb{R}^2$ is an open set. The coordinate system is given by $x = (x_1, x_2, x_3) \in \mathbb{R}^3$.

The Boussinesq equations (4.6.4)–(4.6.6) are basic equations to study the Rayleigh-Bénard problem. They are supplemented with the following initial value conditions

$$(u, T) = (u_0, T_0) \qquad \text{at } t = 0. \tag{4.6.7}$$

Boundary conditions are needed at the top and bottom and at the lateral boundary $\partial D \times (0,1)$. At the top and bottom boundary ($x_3 = 0, 1$), either the so-called rigid or free boundary conditions are given by

$$T = 0, \quad u = 0 \quad \text{(rigid boundary)}, \tag{4.6.8}$$

$$T = 0, \quad u_3 = 0, \quad \frac{\partial(u_1, u_2)}{\partial x_3} = 0 \quad \text{(free boundary)}. \tag{4.6.9}$$

Different combinations of top and bottom boundary conditions are normally used in different physical settings such as *rigid-rigid*, *rigid-free*, *free-rigid*, and *free-free*.

On the lateral boundary $\partial D \times [0,1]$, one of the following boundary conditions is usually used:

(1) Periodic condition:

$$(u,T)(x_1 + k_1L_1, x_2 + k_2L_2, x_3) = (u,T)(x_1,x_2,x_3), \tag{4.6.10}$$

for any $k_1, k_2 \in \mathbb{Z}$.

(2) Dirichlet boundary condition:

$$u = 0, \quad T = 0 \quad (\text{or } \frac{\partial T}{\partial n} = 0); \tag{4.6.11}$$

(3) Free boundary condition:

$$T = 0, \quad u_n = 0, \quad \frac{\partial u_\tau}{\partial n} = 0, \tag{4.6.12}$$

where n and τ are the unit normal and tangent vectors on $\partial D \times [0,1]$ respectively, and $u_n = u \cdot n$, $u_\tau = u \cdot \tau$.

For simplicity, we proceed with the following set of boundary conditions; all results hold true as well for other combinations of boundary conditions.

$$\begin{cases} T = 0, \quad u = 0 \qquad \text{at } x_3 = 0,1, \\ (u,T)(x_1 + k_1L_1, x_2 + k_2L_2, x_3, t) = (u,T)(x,t), \end{cases} \tag{4.6.13}$$

for any $k_1, k_2 \in \mathbb{Z}$.

We recall here the functional setting of equations (4.6.4)–(4.6.6) with initial and boundary conditions (4.6.7) and (4.6.13) and refer the interested readers to Foias, Manley and Temam [**20**] for details. To this end, let

$$\begin{aligned} H = \{(u,T) \in L^2(\Omega)^3 \times L^2(\Omega) \quad | \quad & \text{div} u = 0, u_3|_{x_3=0,1} = 0, \\ & u_i \text{ is periodic in the } x_i \text{ direction } (i = 1,2)\}, \end{aligned} \tag{4.6.14}$$

$$\begin{aligned} V = \{(u,T) \in H_0^1(\Omega)^4 \quad | \quad & \text{div} u = 0, \\ & u_i \text{ is periodic in the } x_i \text{ direction } (i = 1,2)\}, \end{aligned} \tag{4.6.15}$$

where $H_0^1(\Omega)$ is the space of functions in $H^1(\Omega)$, which vanish at $x_3 = 0,1$ and are periodic in the x_i-directions $(i = 1,2)$. Here, $H^1(\Omega)$ is the usual Sobolev space.

Then the results concerning the existence of a solution for (4.6.4)–(4.6.6) with initial and boundary conditions (4.6.7) and (4.6.13) are classical. For every $(\phi_0, T_0) \in H$, (4.6.4)–(4.6.6) with (4.6.7) and (4.6.13) possesses a weak solution

$$(u,T) \in L^\infty([0,\tau]; H) \cap L^2(0,\tau; V) \quad \forall \tau > 0. \tag{4.6.16}$$

If $(u_0, T_0) \in V$, (4.6.4)–(4.6.6) with (4.6.7) and (4.6.13) possesses a unique solution on some interval $[0, \tau_1]$,

$$(u,T) \in C([0,\tau_1]; V) \cap L^2(0,\tau_1; H^2(\Omega)^4 \cap V), \tag{4.6.17}$$

where $\tau_1 = \tau_1(M)$ depends on a bound of the V norm of (ϕ_0, T_0):

$$||(u_0, T_0)|| \leq M.$$

In addition, for any $||(\phi_0, T_0)|| \leq \delta$ small, (4.6.4)–(4.6.6) with (4.6.7) and (4.6.13) possesses a unique global (in time) solution

$$(u,T) \in C([0,\tau]; V) \cap L^2(0,\tau; H^2(\Omega)^4 \cap V), \quad \forall \tau > 0. \tag{4.6.18}$$

Thanks to these existence results, we can define a semi-group

$$S(t) : (u_0, T_0) \to (u(t), T(t)),$$

which enjoys the semi-group properties.

4.6.2. Attractor bifurcation. Let H and H_1 be two Hilbert spaces, and let $H_1 \hookrightarrow H$ be a dense and compact inclusion. We consider the following nonlinear evolution equations

$$\frac{du}{dt} = L_\lambda u + G(u, \lambda), \tag{4.6.19}$$

$$u(0) = u_0, \tag{4.6.20}$$

where $u : [0, \infty) \to H$ is the unknown function, $\lambda \in \mathbb{R}$ is the system parameter, and $L_\lambda : H_1 \to H$ are parameterized linear completely continuous fields continuously depending on $\lambda \in \mathbb{R}^1$, which satisfy

$$\begin{cases} L_\lambda = -A + B_\lambda, & \text{a sectorial operator,} \\ A : H_1 \to H, & \text{a linear homeomorphism,} \\ B_\lambda : H_1 \to H, & \text{the parameterized linear compact operators.} \end{cases} \tag{4.6.21}$$

It is easy to see [**35, 79**] that L_λ generates an analytic semi-group $\{e^{-tL_\lambda}\}_{t \geq 0}$. Then we can define fractional power operators L_λ^α for any $0 \leq \alpha \leq 1$ with domain $H_\alpha = D(L_\lambda^\alpha)$ such that $H_{\alpha_1} \subset H_{\alpha_2}$ if $\alpha_1 > \alpha_2$, and $H_0 = H$.

Furthermore, we assume that the nonlinear terms $G(\cdot, \lambda) : H_\alpha \to H$ for some $1 > \alpha \geq 0$ are a family of parameterized C^r bounded operators ($r \geq 1$), continuously depending on the parameter $\lambda \in \mathbb{R}^1$, such that

$$G(u, \lambda) = o(\|u\|_{H_\alpha}), \quad \forall\, \lambda \in \mathbb{R}^1. \tag{4.6.22}$$

In the applications, we are interested in the sectorial operator $L_\lambda = -A + B_\lambda$ such that there exist a real eigenvalue sequence $\{\rho_k\} \subset \mathbb{R}^1$ and an eigenvector sequence $\{e_k\} \subset H_1$ of A:

$$\begin{cases} Ae_k = \rho_k e_k, \\ 0 < \rho_1 \leq \rho_2 \leq \cdots, \\ \rho_k \to \infty \ (k \to \infty) \end{cases} \tag{4.6.23}$$

such that $\{e_k\}$ is an orthogonal basis of H.

For the compact operator $B_\lambda : H_1 \to H$, we also assume that there is a constant $0 < \theta < 1$ such that

$$B_\lambda : H_\theta \longrightarrow H \text{ bounded}, \ \forall\, \lambda \in \mathbb{R}^1. \tag{4.6.24}$$

Let $\{S_\lambda(t)\}_{t \geq 0}$ be an operator semi-group generated by equation (4.6.19) that enjoys the following properties:

(i) For any $t \geq 0$, $S_\lambda(t) : H \to H$ is a linear continuous operator,
(ii) $S_\lambda(0) = I : H \to H$ is the identity on H, and
(iii) For any $t, s \geq 0$, $S_\lambda(t + s) = S_\lambda(t) \cdot S_\lambda(s)$.

Then the solution of (4.6.19) and (4.6.20) can be expressed as

$$u(t) = S_\lambda(t) u_0, \qquad t \geq 0.$$

DEFINITION 4.6.1. *A set $\Sigma \subset H$ is called an invariant set of (4.6.19) if $S(t)\Sigma = \Sigma$ for any $t \geq 0$. An invariant set $\Sigma \subset H$ of (4.6.19) is said to be an attractor if Σ is compact, and there exists a neighborhood $U \subset H$ of Σ such that, for any $\varphi \in U$, we have*

$$\lim_{t\to\infty} dist_H(u(t,\varphi),\Sigma) = 0. \tag{4.6.25}$$

The largest open set U satisfying (4.6.25) is called the basin of attraction of Σ.

DEFINITION 4.6.2. *Consider equation (4.6.19).*

(1) *We say that equation (4.6.19) bifurcates from $(u,\lambda) = (0,\lambda_0)$ to an invariant set Ω_λ, if there exists a sequence of invariant sets $\{\Omega_{\lambda_n}\}$ of (4.6.19), with $0 \notin \Omega_{\lambda_n}$, such that*

$$\lim_{n\to\infty} \lambda_n = \lambda_0,$$
$$\lim_{n\to\infty} \max_{x\in\Omega_{\lambda_n}} |x| = 0.$$

(2) *If the invariant sets Ω_λ are attractors of (4.6.19), then the bifurcation is called attractor bifurcation.*
(3) *If Ω_λ are attractors and are homotopy equivalent to an m-dimensional sphere S^m, then the bifurcation is called S^m-attractor bifurcation.*

Now let the eigenvalues (counting the multiplicity) of L_λ be given by

$$\beta_1(\lambda), \beta_2(\lambda), \cdots, \beta_k(\lambda) \in \mathbb{C},$$

where $\mathbb{C}$ is the complex plane. Suppose that

$$Re\beta_i(\lambda) \begin{cases} < 0 & \text{if } \lambda < \lambda_0 \\ = 0 & \text{if } \lambda = \lambda_0 \\ > 0 & \text{if } \lambda > \lambda_0 \end{cases} \qquad 1 \leq i \leq m, \tag{4.6.26}$$

$$Re\beta_j(\lambda_0) < 0 \qquad \forall\, m+1 \leq j. \tag{4.6.27}$$

Let the eigenspace of L_λ at λ_0 be

$$E_0 = \bigcup_{1\leq i\leq m} \left\{u \in H_1 \mid (L_{\lambda_0} - \beta_i(\lambda_0))^k u = 0,\ k = 1, 2, \cdots \right\}.$$

It is known that $\dim E_0 = m$.

The following dynamic bifurcation theorem for (4.6.19) was proved in [**65**].

THEOREM 4.6.3 (Attractor Bifurcation, [**65**]). *Assume that conditions (4.6.21), (4.6.22), (4.6.26) and (4.6.27) hold true, and $u = 0$ is a locally asymptotically stable equilibrium point of (4.6.19) at $\lambda = \lambda_0$. Then the following assertions hold true:*

(1) *(4.6.19) bifurcates from $(u,\lambda) = (0,\lambda_0)$ an attractor $\mathcal{A}_\lambda$ for $\lambda > \lambda_0$, with $m - 1 \leq \dim \mathcal{A}_\lambda \leq m$, which is connected if $m > 1$;*
(2) *the attractor $\mathcal{A}_\lambda$ is a limit of a sequence of m-dimensional annulus M_k with $M_{k+1} \subset M_k$; especially if $\mathcal{A}_\lambda$ is a finite simplicial complex, then $\mathcal{A}_\lambda$ has the homotopy type of S^{m-1};*
(3) *For any $u_\lambda \in \mathcal{A}_\lambda$, u_λ can be expressed as*

$$u_\lambda = v_\lambda + o(\|v_\lambda\|_{H_1}), \quad v_\lambda \in E_0;$$

(4) *If $G: H_1 \to H$ is compact, and the equilibrium points of (4.6.19) in $\mathcal{A}_\lambda$ are finite, then we have the index formula*

$$\sum_{u_i \in \mathcal{A}_\lambda} ind[-(L_\lambda + G), u_i] = \begin{cases} 2 & \text{if } m = \text{even}, \\ 0 & \text{if } m = \text{odd}. \end{cases}$$

(5) *If $u = 0$ is globally stable for (4.6.19) at $\lambda = \lambda_0$, then for any bounded open set $U \subset H$ with $0 \in U$, there is an $\varepsilon > 0$ such that as $\lambda_0 < \lambda < \lambda_0 + \varepsilon$, the attractor $\mathcal{A}_\lambda$ bifurcated from $(0, \lambda_0)$ attracts U/Γ in H, where Γ is the stable manifold of $u = 0$ with co-dimension m. In particular, if (4.6.19) has a global attractor for all λ near λ_0, then the ε here can be chosen independently of U.*

To apply the above dynamic bifurcation theorems, it is crucial to verify the asymptotic stability of the critical states. We now establish a theorem to verify the needed asymptotic stability for equations with symmetric linear parts. Let the linear operator L_λ in (4.6.19) be symmetric, i.e.,

$$\langle L_\lambda u, v \rangle_H = \langle u, L_\lambda v \rangle_H, \qquad \forall\, u, v \in H_1.$$

Then all eigenvalues of L_λ are real numbers. Let the eigenvalues $\{\beta_k\}$ of L_λ at $\lambda = \lambda_0$ satisfy

$$\begin{cases} \beta_i = 0 & 1 \le i \le m \quad (m \ge 1), \\ \beta_j < 0 & m + 1 \le j < \infty. \end{cases} \tag{4.6.28}$$

Set

$$\begin{aligned} E_0 &= \{u \in H_1 \mid L_{\lambda_0} u = 0\}, \\ E_1 &= E_0^\perp = \{u \in H_1 \mid \langle u, v \rangle_H = 0 \quad \forall\, v \in E_0\}, \\ P_1 &: H \longrightarrow E_1 \text{ the projection.} \end{aligned}$$

By (4.6.28), $\dim E_0 = m$.

THEOREM 4.6.4 (Asymptotic Stability Theorem). *Let L_λ in (4.6.21) be symmetric with the spectrum condition given by (4.6.28) holds true, and let $G_{\lambda_0}: H_1 \to H$ satisfy the following orthogonal condition:*

$$\langle G_{\lambda_0} u, u \rangle_H = 0, \qquad \forall\, u \in H_1. \tag{4.6.29}$$

Then exactly one and only one of the following two assertions holds true:

(1) *There exists a sequence of invariant sets $\{\Gamma_n\} \subset E_0$ of (4.6.19) at $\lambda = \lambda_0$ such that*

$$0 \notin \Gamma_n, \qquad \lim_{n \to \infty} \mathrm{dist}(\Gamma_n, 0) = 0;$$

(2) *the trivial steady state solution $u = 0$ for (4.6.19) at $\lambda = \lambda_0$ is locally asymptotically stable under the H-norm.*

Furthermore, if (4.6.19) has no invariant sets in E_0 except the trivial one $\{0\}$, then $u = 0$ is globally asymptotically stable.

4.6.3. Attractor bifurcation of the Rayleigh-Bénard problem. The linearized equations of (4.6.4)–(4.6.6) are given by

$$\begin{cases} -\Delta u + \nabla p - \sqrt{R}Tk = 0, \\ -\Delta T - \sqrt{R}u_3 = 0, \\ \operatorname{div} u = 0, \end{cases} \tag{4.6.30}$$

where R is the Rayleigh number. These equations are supplemented with the same boundary conditions (4.6.13) as the nonlinear Boussinesq system. This eigenvalue problem for the Rayleigh number R is symmetric. Hence, we know that all eigenvalues R_k with multiplicities m_k of (4.6.30) with (4.6.13) are real numbers, and

$$0 < R_1 < \cdots < R_k < R_{k+1} < \cdots . \tag{4.6.31}$$

The first eigenvalue R_1, also denoted by $R_c = R_1$, is called the critical Rayleigh number. Let the multiplicity of R_c be $m_1 = m$ $(m \geq 1)$, and the first eigenvectors $\Psi_1 = (e_1(x), T_1), \cdots, \Psi_m = (e_m, T_m)$ of (4.6.30) be orthonormal:

$$\langle \Psi_i, \Psi_j \rangle_H = \int_\Omega [e_i \cdot e_j + T_i T_j] dx = \delta_{ij}.$$

For simplicity, let E_0 be the first eigenspace of (4.6.30) with (4.6.13):

$$E_0 = \left\{ \sum_{k=1}^{m} \alpha_k \Psi_k \mid \alpha_k \in \mathbb{R},\ 1 \leq k \leq m \right\}. \tag{4.6.32}$$

The main results in this section are the following theorems.

THEOREM 4.6.5. *For the Bénard problem (4.6.4)–(4.6.6) with (4.6.13), the following assertions hold true:*

(1) *When the Rayleigh number is less than or equal to the critical Rayleigh number: $R \leq R_c$, the steady state $(u, T) = 0$ is globally asymptotically stable.*
(2) *The equations bifurcate from $((u, T), R) = (0, R_c)$ an attractor $\mathcal{A}_R$ for $R > R_c$, with $m - 1 \leq \dim \mathcal{A}_R \leq m$, which is connected when $m > 0$.*
(3) *For any $(u, T) \in \mathcal{A}_R$, the velocity field u can be expressed as*

$$u = \sum_{k=1}^{m} \alpha_k e_k + o\left(\sum_{k=1}^{m} \alpha_k e_k \right), \tag{4.6.33}$$

where e_k are the velocity fields of the first eigenvectors in E_0.
(4) *The attractor $\mathcal{A}_R$ has the homotopy type of an m-dimensional sphere S^m provided $\mathcal{A}_R$ is a finite simplicial complex.*
(5) *There are an open neighborhood $U \subset H$ of $(u, T) = 0$ and an $\varepsilon > 0$ such that as $R_c < R < R_c + \varepsilon$, the attractor $\mathcal{A}_R$ attracts U/Γ in H, where Γ is the stable manifold of $(u, T) = 0$ with co-dimension m.*

THEOREM 4.6.6. *If the first eigenvalue of L_{λ_0} is simple, i.e., $\dim E_0 = 1$, then the bifurcated attractor $\mathcal{A}_R$ of the Bénard problem (4.6.4)–(4.6.6) with (4.6.13) consists of exactly two points, $\bar{\phi}_1, \bar{\phi}_2 \in H_1 = V \cap H^2(\Omega)^4$, given by*

$$\bar{\phi}_1 = \alpha \Psi_1 + o(|\alpha|), \quad \bar{\phi}_2 = -\alpha \Psi_1 + o(|\alpha|),$$

for some $\alpha \neq 0$, where Ψ_1 is the first eigenvector generating E_0 in (4.6.32). Moreover, for any bounded open set $U \in H$ with $0 \in U$, there is an $\varepsilon > 0$ such that when $R_c < R < R_c + \varepsilon$, U can be decomposed into two open sets U_1 and U_2 such that

(1) $\bar{U} = \bar{U}_1 + \bar{U}_2$, $U_1 \cap U_2 = \emptyset$ *and* $0 \in \partial U_1 \cap \partial U_2$,

(2) $\bar{\phi}_i \subset U_i$ $(i = 1, 2)$, *and*

(3) *for any* $\phi_0 \in U_i$ $(i = 1, 2)$, $\lim_{t\to\infty} S_\lambda(t)\phi_0 = \bar{\phi}_i$, *where* $S_\lambda(t)\phi_0$ *is the solution of the Bénard problem (4.6.4)–(4.6.6) with (4.6.13) with initial data* $\phi_0 = (u_0, T_0)$.

A few remarks are now in order.

REMARK 4.6.7. As we shall see in the next section, (4.6.33) in Theorem 4.6.5 is crucial for studying the topological structure of the Rayleigh-Bénard convection.

REMARK 4.6.8. Theorem 4.6.6 corresponds to the classical pitchfork bifurcation. The main advantage of this theorem is that we know the stability of these bifurcated steady states.

REMARK 4.6.9. Both theorems hold true for Boussinesq equations (4.6.4)–(4.6.6) with different combinations of boundary conditions as described in Section 4.6.1.

PROOF OF THEOREM 4.6.5. We proceed in the following steps.

Step 1. First of all, without loss of generality, we assume the Prandtl number

$$P_r = 1; \tag{4.6.34}$$

otherwise, we only have to consider the following form of (4.6.4)–(4.6.6), and the proof is the same:

$$\begin{cases} \dfrac{\partial u}{\partial t} + (u \cdot \nabla)u + \nabla p - P_r \Delta u - \sqrt{R}\sqrt{P_r}\theta k = 0, \\ \dfrac{\partial \theta}{\partial t} + (u \cdot \nabla)\theta - \sqrt{R}\sqrt{P_r} u_3 - \Delta\theta = 0, \\ \operatorname{div} u = 0, \end{cases} \tag{4.6.35}$$

where $\theta = \sqrt{P_r}T$.

Now let H be the function space defined by (4.6.14) and let H_1 be the intersection of H with the H^2 Sobolev space, i.e.,

$$H_1 = H \cap (H^2(\Omega))^4.$$

Then let $G : H_1 \to H$, and let $L_\lambda = -A + B_\lambda : H_1 \to H$ be defined by

$$\begin{cases} G(\phi) = (-P[(u \cdot \nabla)u], \ -(u \cdot \nabla)T), \\ A\phi = (-P(\Delta u), -\Delta T), \\ B_\lambda \phi = \lambda(P(Tk), u_3), \end{cases} \tag{4.6.36}$$

for any $\phi \in H$. Here $\lambda = \sqrt{R}$, and $P : L^2(\Omega)^3 \to H$ is the Leray projection. Then it is easy to see that these operators enjoy the following properties:

(1) the linear operators A, B_λ and L_λ are all symmetric operators,

(2) the nonlinear operator G is orthogonal, i.e.,

$$\langle G(\phi), \phi \rangle_H = 0. \tag{4.6.37}$$

(3) conditions (4.6.21)–(4.6.24) hold true for these operators defined in (4.6.36).

Then the Boussinesq equations (4.6.4)–(4.6.6) can be rewritten in the following operator form

$$\frac{d\phi}{dt} = L_\lambda \phi + G(\phi), \qquad \phi = (u, T). \tag{4.6.38}$$

Step 2. Now, we need to check conditions (4.6.26) and (4.6.27). Consider the eigenvalue problem

$$L_\lambda \phi = \beta(\lambda)\phi, \qquad \phi = (u, T) \in H_1. \tag{4.6.39}$$

This eigenvalue problem is equivalent to

$$\begin{cases} -\Delta u + \nabla p - \lambda T k + \beta(\lambda) u = 0, \\ -\Delta T - \lambda u_3 + \beta(\lambda) T = 0, \\ \operatorname{div} u = 0. \end{cases} \tag{4.6.40}$$

It is known that the eigenvalues β_k $(k = 1, 2, \cdots)$ of (4.6.40) are real numbers satisfying

$$\begin{cases} \beta_1(\lambda) \geq \beta_2(\lambda) \geq \cdots \geq \beta_k(\lambda) \geq \cdots, \\ \lim_{k \to \infty} \beta_k(\lambda) = -\infty, \end{cases} \tag{4.6.41}$$

and the first eigenvalue $\beta_1(\lambda)$ of (4.6.40) and the first eigenvalue $\lambda_1 = \sqrt{R_c}$ of (4.6.30) have the relation:

$$\beta_1(\lambda) \begin{cases} < 0 & \text{if } 0 \leq \lambda < \lambda_1, \\ = 0 & \text{if } \lambda = \lambda_1. \end{cases} \tag{4.6.42}$$

Step 3. To prove (4.6.26) and (4.6.27), by (4.6.41) and (4.6.42), it suffices to prove that

$$\beta_1(\lambda) > 0 \quad \text{if } \lambda > \lambda_1. \tag{4.6.43}$$

We know that the first eigenvalue $\beta_1(\lambda)$ of (4.6.40) has the minimal property

$$-\beta_1(\lambda) = \min_{(u,T) \in H_1} \frac{\int_\Omega \left[|\nabla u|^2 + |\nabla T|^2 - 2\lambda T u_3 \right] dx}{\int_\Omega [T^2 + u^2] dx}. \tag{4.6.44}$$

It is clear that the first eigenvectors $(e, \varphi) \in H_1$ satisfy

$$\int_\Omega \left[|\nabla e|^2 + |\nabla \varphi|^2 - 2\lambda e_3 \varphi \right] dx \begin{cases} = 0 & \text{if } \lambda = \lambda_1, \\ < 0 & \text{if } \lambda > \lambda_1. \end{cases} \tag{4.6.45}$$

Then (4.6.43) from (4.6.44) and (4.6.45). Thus conditions (4.6.26) and (4.6.27) are achieved.

Step 4. Finally, in order to use Theorem 4.6.3 to prove Theorem 4.6.5, we need to show that $(u, T) = 0$ is a globally asymptotically stable equilibrium point of (4.6.4)–(4.6.6) at the critical Rayleigh number $\lambda_1 = \sqrt{R_c}$. By Theorem 4.6.4, it suffices to prove that equations (4.6.4)–(4.6.6) have no invariant sets except the steady state $(u, T) = 0$ in the first eigenspace E_0 of (4.6.30).

We know that the Boussinesq equations (4.6.4)–(4.6.6) have a bounded absorbing set in H; hence, all invariant sets have the same bound in H as the absorbing set. Assume (4.6.4)–(4.6.6) have an invariant $B \subset E_0$ with $B \neq \{0\}$ at $\lambda_1 = \sqrt{R_c}$.

Then, restricted in B, which contains eigenfunctions of the linear part corresponding to the eigenvalue 0, the Boussinesq equations (4.6.4)–(4.6.6) can be rewritten as

$$\begin{cases} \dfrac{\partial u}{\partial t} + (u \cdot \nabla)u + \nabla p = 0, \\ \dfrac{\partial T}{\partial t} + (u \cdot \nabla)T = 0. \end{cases} \tag{4.6.46}$$

It is easy to see that for the solutions $(u, T) \in B$ of (4.6.46), $(\widetilde{u}, \widetilde{T}) = \alpha(u(\alpha t), T(\alpha t)) \in \alpha B \subset E_0$ are also solutions of (4.6.46). Namely, for any real number $\alpha \in R$, the set $\alpha B \subset E_0$ is an invariant set of (4.6.38). Thus, we infer that (4.6.4)–(4.6.6) have an unbounded invariant set, which is a contradiction to the existence of an absorbing set. Hence, the invariant set B can only consist of $(u, T) = 0$. The proof is complete. □

PROOF OF THEOREM 4.6.6. By Theorem 4.6.5, it suffices to prove that the stationary equations of (4.6.4)–(4.6.6) will bifurcate to exactly two singular points in H_1 as $R > R_c$. We use the Lyapunov-Schmidt method to prove this assertion.

Since the operator $L_\lambda : H_1 \to H$ defined by (4.6.36) is a symmetric completely continuous field, H_1 can be decomposed into

$$\begin{aligned} H_1 &= E_1^\lambda \oplus E_2^\lambda, \\ E_1^\lambda &= \{\alpha \Psi_1(\lambda) \mid \alpha \in \mathbb{R},\ \Psi_1(\lambda) \text{ the first eigenvector of } L_\lambda + G\}, \\ E_2^\lambda &= \{\phi \in H_1 \mid \langle \phi, \Psi_1 \rangle_H = 0\}. \end{aligned}$$

Furthermore, E_1^λ and E_2^λ are invariant subspaces of $L_\lambda + G$.

Let $P_1 : H_1 \to E_1^\lambda$ be the canonical projection, and let

$$\phi = x\Psi_1 + y, \quad x \in \mathbb{R}, \quad y \in E_2^\lambda.$$

Then equations $L_\lambda \phi + G(\phi) = 0$ can be decomposed into

$$\beta(\lambda)x + \langle G(\phi), \Psi_1(\lambda) \rangle_H = 0, \tag{4.6.47}$$

$$L_\lambda y + P_1 G(u) = 0. \tag{4.6.48}$$

By assumption, the eigenvalues $\beta_j(\lambda)$ of $L_\lambda \phi = \beta(\lambda)\phi$ satisfy $\beta_j(\lambda_1) \neq 0$ for $j \geq 2$, and $\lambda_1 = \sqrt{R_c}$. Hence, the restriction

$$L_\lambda \mid_{E_2^\lambda} : E_2^\lambda \longrightarrow E_2^\lambda$$

is invertible. By the implicit function theorem, from (4.6.48), it follows that y is a function of x:

$$y = y(x, \lambda), \tag{4.6.49}$$

which satisfies (4.6.48). Since $G(u) = G(x\Psi_1 + y)$ is an analytic function of u, the function (4.6.49) is also analytic. Hence, the function

$$f(x, \lambda) = \langle G(x\Psi_1 + y(x, \lambda)), \Psi_1 \rangle_H \tag{4.6.50}$$

is analytic. Thus, equation (4.6.47) has the expansion

$$\beta(\lambda)x + f(x, \lambda) = \beta(\lambda)x + \alpha(\lambda)x^k + o(|x|^k) = 0, \tag{4.6.51}$$

for some $\alpha(\lambda) \in \mathbb{R}$ such that $\alpha(\lambda_1) \neq 0$ and $k > 1$, where $\lambda_1 = R_c$ is the critical Rayleigh number. By assumption,

$$\beta(\lambda) \begin{cases} < 0 & \text{if } \lambda < \lambda_1, \\ = 0 & \text{if } \lambda = \lambda_1, \\ > 0 & \text{if } \lambda > \lambda_1. \end{cases}$$

In addition, by Theorem 4.6.5, if $\lambda \leq \lambda_1$ (i.e., $R \leq R_c$) and $\lambda_1 - \lambda$ is small, equations (4.6.47) and (4.6.48) have no non-zero solutions, which implies that $\alpha(\lambda_1) < 0$ and $k =$ odd.

Thus, we derive that equation (4.6.51) has exactly two solutions:

$$x_{\pm} = \pm \left(\frac{\beta(\lambda)}{|\alpha|}\right)^{1/k} + o\left(\left(\frac{\beta(\lambda)}{|\alpha|}\right)^{1/k}\right)$$

for $\lambda > \lambda_1$ with $\lambda - \lambda_1$ sufficiently small. Namely, we have proved that if $\lambda > \lambda_1$, or $R > R_c$, with $\lambda - \lambda_1$ sufficiently small, the stationary equations of (4.6.4)–(4.6.6) bifurcate from $(\phi, \lambda) = (0, \lambda_1)$ to exactly two solutions:

$$\phi_\lambda = x_\pm \Psi_1 + o(|x_\pm|).$$

Thus, this theorem is proved. □

4.6.4. 2-D Rayleigh-Bénard Convection: Asymptotic and Structural Stabilities of the Bifurcated Solution. The main objective of this subsection is to study the dynamic bifurcation and the structural stability of the bifurcated solutions of the 2-D Boussinesq equations related to the Rayleigh-Bénard convection. It is easy to see that both Theorems 4.6.5 and 4.6.6 hold true for the 2D Boussinesq equations with any combination of boundary conditions. Hence, we focus in this section on structural stability in the physical space of the bifurcated solutions, justifying the roll pattern formation in the Rayleigh-Bénard convection.

Technically speaking, when L_2/L_1 is small, the wave number $k_2 = 0$. Hence, the 3-D Bénard problem is reduced to the two dimensional one. Furthermore, due to the symmetry on the xy-plane of the honeycomb structure of the Bénard convection, from the viewpoint of a cross section, the 3-D Bénard convection can be well understood by the two dimensional version.

For consistency, we always assume that the domain $\Omega = [0, L] \times [0, 1]$ with coordinate system $x = (x_1, x_3)$. The 2-D Boussinesq equations for the 2-D Bénard convection take the same form as the 3-D Boussinesq equations (4.6.4)–(4.6.6):

$$\begin{cases} \frac{1}{P_r}\left[\frac{\partial u}{\partial t} + (u \cdot \nabla)u + \nabla p\right] - \Delta u - \sqrt{R}Tk = 0, \\ \frac{\partial T}{\partial t} + (u \cdot \nabla)T - \sqrt{R}u_3 - \Delta T = 0, \\ \text{div } u = 0, \end{cases} \tag{4.6.52}$$

where the velocity field is replaced by $u = (u_1, u_3)$, and the operators are the corresponding 2-D operators in the $x = (x_1, x_3)$ coordinate system. For simplicity,

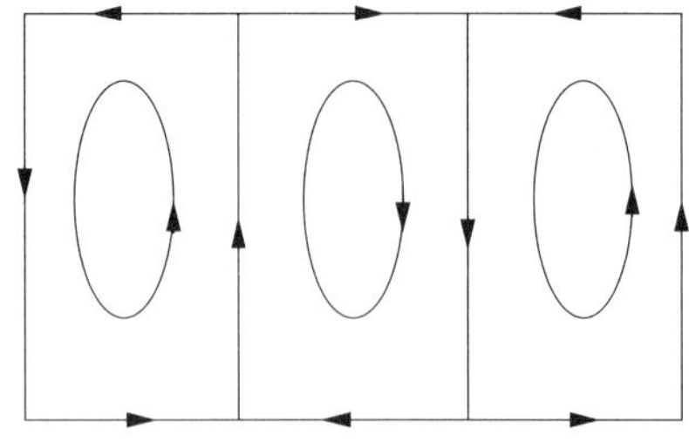
(a)

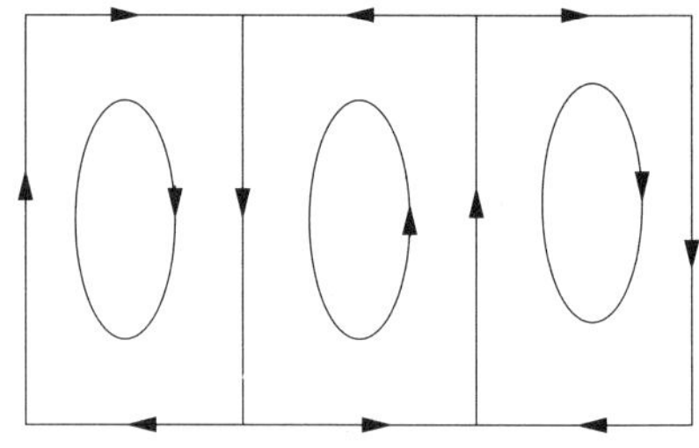
(b)

FIGURE 4.6.2. Rolls with reverse orientations.

we consider here only the free-free boundary conditions as follows:

$$
\begin{cases} u\cdot n=0, & \dfrac{\partial u_\tau}{\partial n}=0, \qquad \text{on } \partial\Omega, \\ T=0 \text{ at } x_3=0,1, & \dfrac{\partial T}{\partial x_1}=0, \text{ at } x_1=0,L. \end{cases} \tag{4.6.53}
$$

In this case, the function space H, defined by (4.6.14), is replaced by

$$
H=\{(u,T)\in L^2(\Omega)^3 \quad | \quad \mathrm{div} u=0, u_3|_{x_3=0,1}=0, \quad u_1|_{x_1=0,L}=0\}.
$$

For equation (4.6.52) with the free boundary condition, the wave number k and the critical Rayleigh number are

$$
\begin{aligned} k &\simeq a_c L/\pi=\frac{L}{\sqrt{2}}, \\ R_c &= \pi^4(k^2+L^2)^3/L^4; \end{aligned}
$$

the first eigenspace E_0 is one-dimensional, and is given by

$$
\begin{cases} E_0=\text{Span }\{\Psi_1=(e_1,T_1)\}, \\ e_1=\left(-\dfrac{L}{k}\sin\dfrac{k\pi x_1}{L}\cos\pi x_3, \cos\dfrac{k\pi x_1}{L}\sin\pi x_3\right), \\ T=\dfrac{1}{k}\sqrt{L^2+k^2}\cos\dfrac{k\pi x_1}{L}\sin\pi x_3. \end{cases} \tag{4.6.54}
$$

The topological structure of e_1 in (4.6.54) consists of k vortices as shown in Figure 4.6.2(a) and (b).

By the structural stability theorem, the first eigenvectors (4.6.54) are structurally stable; therefore, from Theorem 4.6.6, we immediately obtain the following result.

THEOREM 4.6.10. *For any bounded open set $U\subset H$ with $0\in U$, there is an $\varepsilon>0$ such that, as the Rayleigh number $R_c<R<R_c+\varepsilon$, U can be decomposed into two open sets U_1 and U_2 depending on R such that*

(1) $\bar U=\bar U_1+\bar U_2, \quad U_1\cap U_2=\emptyset, \quad 0\in\partial U_1\cap\partial U_2;$

(2) *for any initial value $\phi_0\in U_i$ $(i=1,2)$, there exists a time $t_0>0$ such that the solution $S_R(t)\phi_0$ of (4.6.52) with (4.6.53) is topologically equivalent to*

either the structure as shown in Figure 4.6.2(a) or that as shown in (b) for all $t > t_0$.

Notes for Chapter 4

1. Section 4.1 is based on the paper [**52**]. One of the ideas used in this section comes from the genericity for the steady state solutions of the Navier-Stokes equations, first proved by C. Foias and R. Temam [**23**] using the infinite dimensional Sard's theorem. They proved that there is an open and dense set of the forcing in a proper function $\mathcal{F}$ such that for any $f \in \mathcal{F}$, there are only a finite number of steady state solutions.

2. Sections 4.3 and 4.4 are based on [**62**]. Section 4.5 is based on [**68**], where the structure of the solutions of the Navier-Stokes equations on the spectral manifold, introduced by C. Foias and J. C. Saut [**21, 22**], is classified.

3. Section 4.5 is based on Ma and Wang [**68**].

4. Section 4.6 introduces a notion of dynamic bifurcation and its applications to Rayleigh-Bénard convection. This topic is fully discussed in the book by the authors [**66**]. The main focus here is to give another example of linking the solutions of the governing equations of fluid flows to the structure and its transitions in the physical spaces. This section is based on Ma and Wang [**65, 60, 64**].

CHAPTER 5

Structural Bifurcation for One-Parameter Families of Divergence-Free Vector Fields

The main objective of this chapter is to develop a rigorous theory to address *when*, *where* and *how* the boundary layer separation occurs, and to apply the theory to some fluid problems. From the mathematical point of view, these are still unresolved problems; see A. Chorin and J. Marsden [**11**] and W. Jäger, P. Lax and C. Morawetz [**39**].

The physical and numerical descriptions of these problems go back to the pioneering work of Prandtl [**84**] in 1904. Basically, in the boundary layer, the shear flow can detach/separate from the boundary, generating a slow back flow and leading to more complicated turbulent behavior; see Figure 0.3.6 following [**31**].

Mathematically speaking, the natural tool to study these problems is the kinematic structural bifurcation theory together with its connections to the dynamics of incompressible fluid flows. This structural bifurcation theory is based on the kinematic theory developed in early chapters of this book, and the key ingredient is detailed orbit analysis, providing classifications of the field near the bifurcation point. This chapter addresses studies on both the kinematic structural bifurcation theory and its links to the dynamics, together with applications to boundary layer separation of incompressible viscous fluid flows.

5.1. Necessary Conditions for Structural Bifurcation

We know that solutions of the Navier-Stokes equations or the Euler equations of fluid flows are one-parameter families of divergence-free vector fields. The main objective of this section is to study the structural bifurcation of one-parameter families of divergence-free vector fields on a 2D compact manifold, and applications to 2D Navier-Stokes or Euler equations of incompressible fluid flows.

Let $M \subset \mathbb{R}^m$ $(m = 2, 3)$ be a C^r $(r \geq 1)$ 2D compact manifold with boundary, and let X be either $D^r(TM)$, or $B^r(TM)$, or $B^r_0(TM)$, or $B^r_\varphi(TM)$.

DEFINITION 5.1.1. *Let $u \in C^1([0,T], X)$ be a one-parameter family of vector fields in X. The vector field $u_0 = u(\cdot, t_0)$ $(0 < t_0 < T)$ is called a bifurcation point of u at time t_0 if, for any $t^- < t_0$ and $t_0 < t^+$ with t^- and t^+ sufficiently close to t_0, the vector field $u(\cdot, t^-)$ is not topologically equivalent to $u(\cdot, t^+)$. In this case, we say that $u(x,t)$ has a bifurcation at t_0 in its global structure.*

DEFINITION 5.1.2. *Let $u \in C^1([0,T], X)$. We say that u has a bifurcation in its local structure in a neighborhood $U \subset M$ of $\bar{x}$ at t_0 $(0 < t_0 < T)$ if, for any $t^- < t_0$ and $t_0 < t^+$ with t^- and t^+ sufficiently close to t_0, the vector fields $u(\cdot, t^-)$ and $u(\cdot, t^+)$ are not topologically equivalent locally in $U \subset M$.*

We remark here that bifurcation in the vector field's local structure does not imply bifurcation in its global structure. In fact, one can easily construct examples showing that flow structure changes in some local area $U \subset M$, but not on the whole manifold M.

From the local and global structure stability theorems in Chapter 2, we easily obtain the following necessary conditions for the appearance of structural bifurcation for one-parameter divergence-free vector fields.

THEOREM 5.1.3. *Let $u \in C^1([0,T], D^r(TM))$ $(r \geq 1)$.*

(1) *If $u(\cdot,t)$ has a bifurcation in its local structure in a neighborhood $U \subset M$ of $\bar{x}$ at t_0 $(0 < t_0 < T)$, then $\bar{x}$ must be a degenerate singular point of $u(x,t)$ at t_0.*
(2) *For $M \subset \mathbb{R}^2$, if $u(\cdot,t)$ has a bifurcation in its global structure at t_0 $(0 < t_0 < T)$, then $u(x,t_0)$ does not satisfy at least one of the structural stability conditions (1)–(3) in Theorem 2.1.2.*

THEOREM 5.1.4. *Let $u \in C^1([0,T], B_0^r(TM))$ $(r \geq 2)$.*

(1) *If $u(x,t)$ has a bifurcation in its local structure in an arbitrarily small neighborhood $U \subset M$ of $\bar{x}$ at t_0 $(0 < t_0 < T)$, then $\bar{x} \in \overset{\circ}{M}$ (respectively $\bar{x} \in \partial M$) must be a degenerate singular point (respectively degenerate ∂-singular point) of $u(x,t)$ at t_0.*
(2) *If $u(x,t)$ has a bifurcation in its global structure at t_0 $(0 < t_0 < T)$, then $u(x,t_0)$ does not satisfy at least one of the conditions (1)–(3) in Theorem 2.2.9.*

The above theorem suggests that we study the structure of divergence-free vector fields near degenerate singular points. For this purpose, we recall the definition of indices of singular points of a vector field. Let $p \in M$ be an isolated singular point of $v \in C_n^r(TM)$; then

$$\text{ind}(v,p) = \deg(v,p),$$

where $\deg(v,p)$ is the Brouwer degree of v at p.

Let $p \in \partial M$ be an isolated singular point of v, and let $\widetilde{M} \subset \mathbb{R}^2$ be an extension of M, i.e., $M \subset \widetilde{M}$ such that $p \in \widetilde{M}$ is an interior point. In a neighborhood of p in $\widetilde{M}$, v can be extended by reflection to $\widetilde{v}$ such that p is an interior singular point of $\widetilde{v}$, thanks to the no-normal flow condition $v \cdot n|_{\partial M} = 0$. Then we define the index of v at $p \in \partial M$ by

$$\text{ind}(v,p) = \frac{1}{2}\text{ind}(\widetilde{v},p).$$

Let $p \in M$ be an isolated singular point of $v \in C_n^r(TM)$. An orbit γ of v is said to be a stable orbit (resp. an unstable orbit) connecting to p if the limit set $w(x) = p$ (resp. $\alpha(x) = p$) for any $x \in \gamma$.

We now introduce a singularity classification theorem for incompressible vector fields, which is useful for us to discuss the structure bifurcation.

THEOREM 5.1.5 (Singularity Classification Theorem). *Let $p \in M$ be an isolated singular point of $v \in D^r(TM)$ $(r \geq 1)$. Then p is connected only to a finite number of orbits and the stable and unstable orbits connected to p alternate when tracing a closed curve around p. Furthermore,*

(1) *when* $p \in \overset{\circ}{M}$, *p is connected by* $2n$ $(n \geq 0)$ *orbits, n of which are stable, and the other n unstable, while the index of p is*

$$ind(v, p) = 1 - n; \tag{5.1.1}$$

(2) *when* $p \in \partial M$, *p is connected by* $n+2$ $(n \geq 2)$ *orbits, two of which are on the boundary* ∂M, *and the index of p is*

$$ind(v, p) = -\frac{n}{2}. \tag{5.1.2}$$

PROOF. We proceed in a few steps.

Step 1. Let Γ be the set of all orbits connected to p. Since the flow of v is area-preserving, it is easy to see that Γ has no interior point, i.e., $\overset{\circ}{\Gamma} = \emptyset$. Assuming otherwise, we shall derive a contradiction.

Let $x \in \overset{\circ}{\Gamma}$; then there exists a neighborhood $\mathcal{O} \subset M$ of $\bar{x}$ such that $\mathcal{O} \subset \overset{\circ}{\Gamma}$. Without loss of generality, we assume that

$$\widetilde{\mathcal{O}} = \{x \in \mathcal{O} \mid \omega(x) = p\}$$

has positive measure. Then

$$\lim_{t \to \infty} |\Phi(\widetilde{\mathcal{O}}, t)| = 0,$$

a contradiction to v being incompressible.

Step 2. We shall show that Γ only has a finite number of orbits. Assuming otherwise, there is a sequence of orbits $\gamma_j \subset \Gamma (j = 1, 2, \cdots)$ and an orbit $\gamma_0 \subset \Gamma$ such that $\gamma_j \to \gamma_0$ in a neighborhood U of p. Without loss of generality, we assume that γ_0 is a stable orbit of p. Take $x \in \gamma_0 \cap U$ an arc Σ transversal to γ_0 at x, and N_0 sufficiently large such that Σ is transversal to γ_j for all $j \geq N_0$. If all $\gamma_j (j \geq N_0)$ are stable orbits of p, i.e., there is no unstable orbit of p in a neighborhood of γ_0, then there must exist a neighborhood Σ_0 of x in Σ such that $\omega(z) = p$ for any $z \in \Sigma_0$, which implies that $\overset{\circ}{\Gamma} \neq \emptyset$, a contradiction.

Assume that there is a subsequence γ_{j_k} of $\gamma_j (j_k \geq N_0)$ such that each $\gamma_{j_k} (1 \leq k)$ is an unstable orbit of p. Let $x_{j_k} = \gamma_{j_k} \cap \Sigma$; then, by $\gamma_{j_k} \to \gamma_0 (k \to \infty)$, $x_k \to \bar{x}$. Since the flow leaves p on the orbits γ_{jk}, and approaches p on γ_0, $\bar{x}$ is a singular point. Since $\bar{x} \in \gamma_0 \cap U$ is taken arbitrarily, any point in $\gamma_0 \cap U$ is a singular point, a contradiction to p being an isolated singular point. Therefore, p is connected to only a finite number of orbits.

Step 3. Let γ_1, γ_2 be two adjacent orbits of v connected to p. We take an arc Σ transversal to v and intersecting with γ_1 and γ_2 at x_1 and x_2, respectively, such that there are no singular points of v and orbits of v connected to p in the domain D enclosed by γ_1, γ_2 and the arc Σ; see Figure 5.1.1. Let Q_1, Q_2 be sufficiently small neighborhoods of x_1 and x_2 in D. If both γ_1 and γ_2 are stable orbits of p, then for any $z \in Q_1$ $(z \neq x_1)$, the orbit $\Phi(z, t) \subset D$ for $0 \leq t < t_0$, and $z_0 = \Phi(z, t_0) \in \Sigma$, but $z_0 \notin \bar{Q}_2$, which means that there is an unstable orbit of p in D, a contradiction. Therefore, if γ_1 and γ_2 are adjacent and γ_1 is stable (or unstable), then γ_2 must be unstable (or stable).

Since the stable and unstable orbits of a singular point p are adjacently placed, it is easy to see that when $p \in \overset{\circ}{M}$, p must be connected to $2n (n \geq 0)$ orbits, n of which are stable orbits and the other n of which are unstable orbits. Moreover,

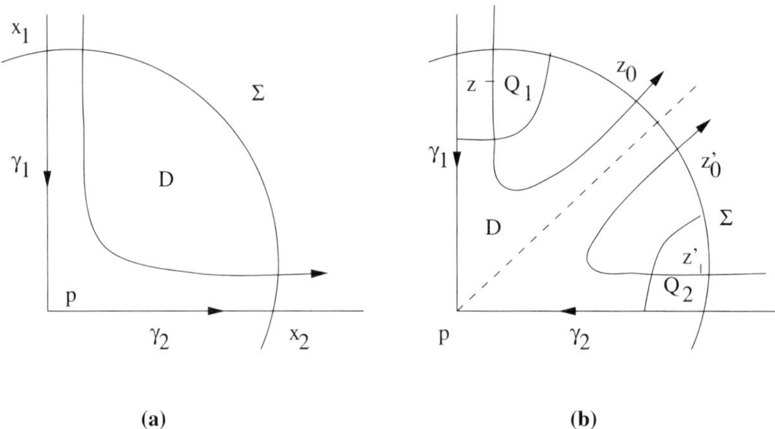

FIGURE 5.1.1. Schematic illustrating the proof of Theorem 5.1.5.

when $p \in \partial M$ is a singular point, p must be connected to two orbits on ∂M and p must be connected to n $(n \geq 0)$ orbits in the interior of M.

Step 4. Finally, by the Brouwer degree theory, it is well known that if $p \in \overset{\circ}{M}$ is an isolated singular point of a vector field $v \in C^r(TM)(r \geq 0)$ connected to $2n(n \geq 0)$ orbits, then the index of p is $1 - n$, i.e., (5.1.1) holds true.

Let $p \in \partial M$ have $n + 2$ $(n \geq 0)$ orbits. Then we can see that, for the reflective extension vector field $\widetilde{v}$ of v on the extension manifold $\widetilde{M}$ of M, $p \in \widetilde{M}$ is an isolated interior singular point of $\widetilde{v}$, which has $2n+2$ orbits. Hence, (5.1.2) follows from the definition and (5.1.1).

The proof is complete. □

5.2. Structural Bifurcation for Flows with No-Normal Flow Boundary Conditions

5.2.1. Main Results. Let $M \subset \mathbb{R}^2$ be a C^r $(r \geq 1)$ 2D compact manifold with boundary, and let X be either $D^r(TM)$ or $B^r(TM)$. Let $u \in C^1([0,T], X)$ be a one-parameter family of divergence-free vector fields.

A natural method for studying structural bifurcations of such a family $u(\cdot, t)$ is to Taylor expand it near t_0, and then to analyze its structure in a neighborhood of t_0 by taking the first-order approximation. Let

$$(5.2.1) \qquad \begin{cases} u(x,t) = u^0(x) + (t - t_0)u^1(x) + o(|t - t_0|), \\ u^0(x) = u(x, t_0), \\ u^1(x) = \dfrac{\partial u(x,t)}{\partial t}|_{t=t_0}. \end{cases}$$

We remark here that in most applications, including in particular applications to fluid problems, u has local analyticity near $\bar{x}$ and t_0, which ensures that u can always be Taylor expanded.

ASSUMPTION 5.2.1. *Let $\bar{x} \in \partial M$ be an isolated degenerate singular point of u^0, $u^0 \in C^{k+1}$ near $\bar{x} \in \partial M$ for some $k \geq 2$. Assume that*

$$u^0(\bar{x}) = 0, \tag{5.2.2}$$

$$\text{ind}(u^0, \bar{x}) \neq -\frac{1}{2}, \tag{5.2.3}$$

$$\frac{\partial^k u^0_\tau(\bar{x})}{\partial^k \tau} \neq 0, \tag{5.2.4}$$

$$u^1(\bar{x}) \neq 0. \tag{5.2.5}$$

A few remarks are now in order.

REMARK 5.2.2. Condition (5.2.5) is a natural condition. If $u^1(\bar{x}) = 0$, then one needs to consider higher-order terms in the Taylor expansion (5.2.1), and similar results will still be true.

REMARK 5.2.3. Condition (5.2.3) amounts to saying that $\bar{x}$ is a degenerate singular point of u^0, thanks to the singularity classification theorem, Theorem 5.1.5.

REMARK 5.2.4. Let k be the smallest integer satisfying condition (5.2.4). It is obvious that $k \geq 2$ since $\bar{x}$ is a degenerate singular point, which implies that $\partial(u^0_\tau(\bar{x}))/\partial\tau = 0$. This condition amounts to saying that u^0_τ can be Taylor expanded near $\bar{x}$ on ∂M. Hence, it avoids rapid oscillation of the vector field $u(x,t)$ near $\bar{x}$.

REMARK 5.2.5. The combination of (5.2.4–5.2.5) yields that there exists a neighborhood $\Gamma \subset \partial M$ of $\bar{x}$ such that for any $z \in \Gamma, z \neq \bar{x}$,

$$\frac{d}{d\tau}\left(\frac{|u^0(z)|}{|u^1(z)|}\right) \neq 0. \tag{5.2.6}$$

The main theorems of this section are as follows.

THEOREM 5.2.6. *Let $u \in C^1([0,T], X)$ be a one-parameter family of divergence-free vector fields satisfying Assumption 5.2.1. Then, in a neighborhood $\Gamma \subset \partial M$ of $\bar{x}$, the singular points of $u(\cdot, t_0 \pm \epsilon)$ are nondegenerate for any $\epsilon > 0$ sufficiently small. Moreover,*

(1) *if ind$(u^0, \bar{x})$ = fraction, then each of $u(\cdot, t_0 \pm \epsilon)$ has only one singular point on $\Gamma \subset \partial M$; and*
(2) *if ind$(u^0, \bar{x})$ = integer, then one of $u(\cdot, t_0 \pm \epsilon)$ has two singular points on Γ, and the other one has no singular points on Γ for any $\epsilon > 0$ sufficiently small.*

THEOREM 5.2.7 (Structural Bifurcation Theorem). *Let $u \in C^1([0,T], X)$ satisfy Assumption 5.2.1. The following assertions hold true:*

(1) *u has a local structural bifurcation at $(\bar{x}, t_0)$;*
(2) *If $\bar{x} \in \partial M$ is a unique degenerate singular point of u^0 on ∂M, then u has a global structure bifurcation at $t = t_0$.*

Before we give the proofs of these theorems, we proceed first with some prototypical examples.

EXAMPLE 5.2.8. When $\bar{x} \in \partial M$ is a singular point of $u^0(x) = u(x, t_0)$ with ind $(u^0, \bar{x}) = 0$, the bifurcation occurs as shown in Figure 5.2.1; see also Figure 0.3.7.

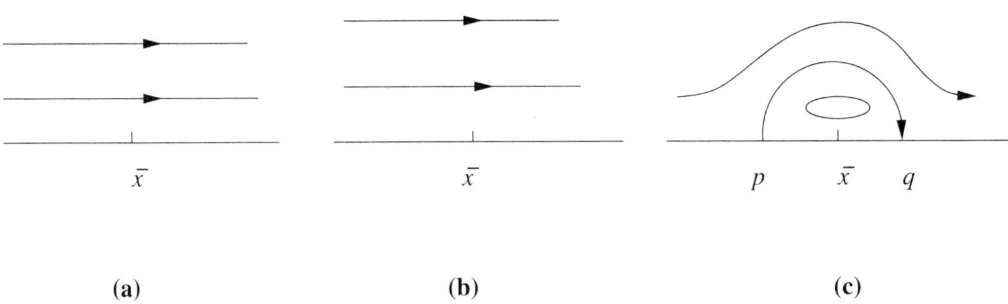

FIGURE 5.2.1. Boundary layer separation of shear flow.

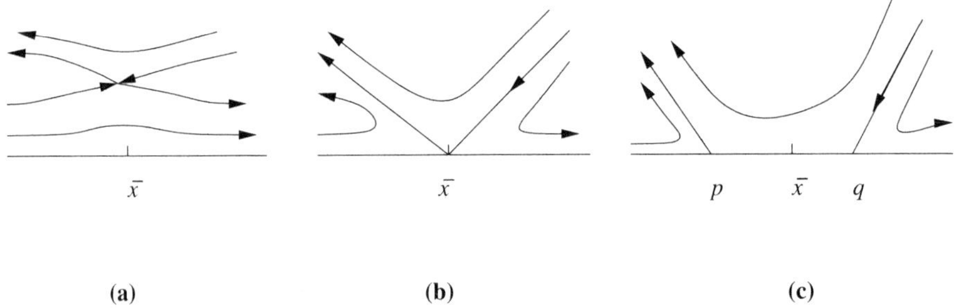

FIGURE 5.2.2. Schematic illustrating the structural bifurcation in the case where ind $(u^0, \bar{x}) = -1$.

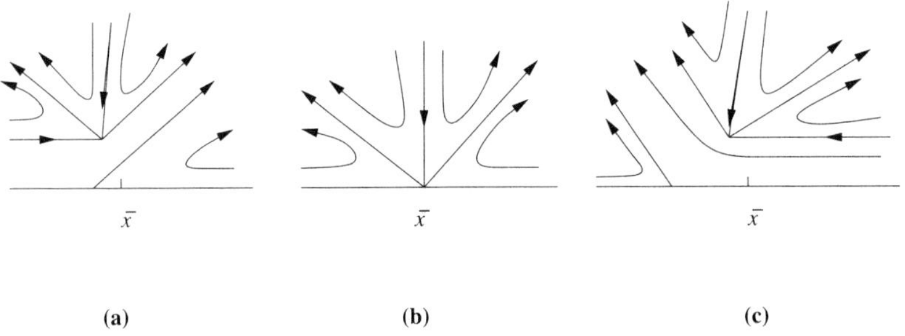

FIGURE 5.2.3. Schematic illustrating the structural bifurcation in the case where ind $(u^0, \bar{x}) = -3/2$.

EXAMPLE 5.2.9. For the case when ind $(u^0, \bar{x}) = -1$, by Theorem 5.2.6, there are two orbits of u^0 connected to $\bar{x} \in \partial M$. The structural evolution of $u(x,t)$ in a neighborhood of $\bar{x}$ is as shown in Figure 5.2.2.

EXAMPLE 5.2.10. When ind $(u^0, \bar{x}) = -3/2$, there are three orbits of u^0 connected to $\bar{x} \in \partial M$. The structural evolution of $u(x,t)$ in a neighborhood of $\bar{x}$ is described by Figure 5.2.3. □

5.2.2. Proof of Theorem 5.2.6. In order to investigate the structural bifurcation of u at t_0, it suffices to consider the topological structure of the vector fields $u^0 \pm \epsilon u^1$ for $\epsilon > 0$ sufficiently small.

The proof of the theorem is achieved in a few lemmas as follows.

LEMMA 5.2.11. *Under Assumption 5.2.1, there exists a small neighborhood $\Gamma \subset \partial M$ of $\bar{x}$ such that all singular points of $u^0 \pm \epsilon u^1$ on Γ are non-degenerate for any $\epsilon > 0$ sufficiently small.*

PROOF. Take a small neighborhood $\Gamma \subset \partial M$ of $\bar{x}$ such that (5.2.6) holds true for any $z \in \Gamma, z \neq \bar{x}$. Let $z_0 \in \Gamma \subset \partial M$ such that $z_0 \neq \bar{x}$ and

$$u^0(z_0) \pm \epsilon u^1(z_0) = 0, \qquad \epsilon > 0. \tag{5.2.7}$$

Let (x_1, x_2) be an orthogonal coordinate system with the origin at z_0, the x_1-axis tangent to ∂M at z_0, and with the x_2-axis pointing inward in the normal direction. Near $\bar{x}$, the vector fields u^i $(i = 0, 1)$ are expressed as $u^i = (u_1^i(x_1, x_2), u_2^i(x_1, x_2))$.

Let $v(x) = (v_1(x), v_2(x))$ be any vector field defined near $\bar{x}$ such that $v \cdot n|_{\partial M} = 0$; consequently, $v_2(0) = 0$ in these local coordinates. We have

$$\begin{aligned}\frac{\partial v_2(0)}{\partial x_1} &= \lim_{\substack{\Delta x = (\Delta x_1, \Delta x_2) \to 0 \\ \Delta x \in \partial M}} \frac{v_2(\Delta x_1, \Delta x_2) - v_2(0)}{\Delta x_1} \\ &= \lim_{\substack{\Delta x = (\Delta x_1, \Delta x_2) \to 0 \\ \Delta x \in \partial M}} \frac{v_\tau(\Delta x_1, \Delta x_2) \sin(\tau, x_1)}{\Delta x_1}.\end{aligned}$$

Here τ is a unit tangent vector at $\Delta x \in \partial M$, $v_\tau = v \cdot \tau$, and

$$\lim_{\substack{\Delta x = (\Delta x_1, \Delta x_2) \to 0 \\ \Delta x \in \partial M}} \frac{\sin(\tau, x_1)}{\Delta x_1} = k(0),$$

$$\lim_{\substack{\Delta x = (\Delta x_1, \Delta x_2) \to 0 \\ \Delta x \in \partial M}} v_\tau(\Delta x) = v_1(0),$$

where $k(0)$ is the curvature of ∂M at $x = 0$. Hence

$$\frac{\partial v_2(0)}{\partial x_1} = v_1(0) k(0). \tag{5.2.8}$$

We infer then from (5.2.7) and (5.2.8) that

$$\frac{\partial (u_2^0(0) \pm \epsilon u_2^1(0))}{\partial x_1} = k(0)(u_1^0(0) \pm \epsilon u_1^1(0)) = 0.$$

Consequently, the Jacobian matrix of $u^0 \pm \epsilon u^1$ at $x = 0$ (i.e., z_0) is

$$D(u^0 \pm \epsilon u^1)(0) = \begin{pmatrix} \frac{\partial u_1^0}{\partial x_1} \pm \epsilon \frac{\partial u_1^1}{\partial x_1} & * \\ 0 & \frac{\partial u_2^0}{\partial x_2} \pm \epsilon \frac{\partial u_2^1}{\partial x_2} \end{pmatrix}_{x=0}.$$

Thanks to the non-divergence of $u(x, t)$, it is easy to see that

$$\frac{\partial u_1^0(0)}{\partial x_1} \pm \epsilon \frac{\partial u_1^1(0)}{\partial x_1} = -\left[\frac{\partial u_2^0(0)}{\partial x_2} \pm \epsilon \frac{\partial u_2^1(0)}{\partial x_2}\right].$$

Therefore, it suffices to prove that

$$\frac{\partial u_1^0(0)}{\partial x_1} \pm \epsilon \frac{\partial u_1^1(0)}{\partial x_1} \neq 0. \tag{5.2.9}$$

Again, since the x_1-axis is tangent to ∂M at $x = 0$,

$$\frac{\partial u_1^0(0)}{\partial x_1} = \frac{d}{d\tau}|u^0(z_0)|, \quad \frac{\partial u_1^1(0)}{\partial x_1} = \frac{d}{d\tau}|u^1(z_0)|, \tag{5.2.10}$$

where $|u^i| = \sqrt{|u_1^i|^2 + |u_2^i|^2}$. By (5.2.7),

$$\epsilon = \mp \frac{|u^0(z_0)|}{|u^1(z_0)|}. \tag{5.2.11}$$

From (5.2.10) and (5.2.11), we obtain:

$$\frac{\partial u_1^0(0)}{\partial x_1} \pm \epsilon \frac{\partial u_1^1(0)}{\partial x_1} = \frac{d}{d\tau}|u^0(z_0)| - \frac{|u^0(z_0)|}{|u^1(z_0)|}\frac{d}{d\tau}|u^1(z_0)|.$$

By $u^1(\bar{x}) \neq 0$, we know that there is a neighborhood $\Gamma \subset \partial M$ of $\bar{x}$ such that for any $z_0 \in \Gamma$, $u^1(z_0) \neq 0$. Hence, (5.2.9) is equivalent to

$$\frac{d}{d\tau}\left(\frac{|u^0|}{|u^1|}\right)|_{z_0 \in \Gamma} \neq 0, \tag{5.2.12}$$

which is true thanks to (5.2.6). Therefore, (5.2.9) holds true, and the lemma is proved. □

LEMMA 5.2.12. *Under Assumption 5.2.1, if ind$(u^0, \bar{x})$ = integer, then, for any $\varepsilon > 0$ small, one of the two vector fields $u^0 \pm \epsilon u^1$ has no singular points on Γ, while the other one has at least two singular points on Γ with one on each side of $\bar{x}$.*

PROOF. By the Singularity Classification Theorem, Theorem 5.1.5, the condition that ind$(u^0, \bar{x})$ = integer is equivalent to n being even. It is easy to see that one of the two boundary orbits connected to $\bar{x}$ is stable, and the other boundary orbit is unstable.

For $\bar{x} \in \partial M$, we take an orthogonal coordinate system (z_1, z_2) with $\bar{x}$ as the origin, its z_1-axis tangent to ∂M at $\bar{x}$, and with the z_2-axis pointing inward. In the z-coordinate system, $u^i = (u_1^i(z), u_2^i(z))$, $i = 0, 1$.

Since one of the two boundary orbits connected to $\bar{x}$ is stable and the other is unstable, we have

$$\text{sign } u_1^0(z^-) = \text{sign } u_1^0(z^+), \tag{5.2.13}$$

where $z^\pm = (z_1^\pm, z_2^\pm) \in \Gamma \subset \partial M$ and $z_1^- < 0 < z_1^+$. Without loss of generality, we assume that

$$u_1^0(z) > 0, \quad u_1^1(z) > 0 \text{ for } z \in \Gamma, \quad \text{and } z \neq 0. \tag{5.2.14}$$

Since $u_1^0(0) = 0$, $u_1^1(0) \neq 0$, we infer from (5.2.14) that for any $\epsilon > 0$ small, $u^0 + \epsilon u^1$ has no singular point on Γ.

On the other hand, by

$$\frac{u_2^0(z)}{u_1^0(z)} = \frac{u_2^1(z)}{u_1^1(z)} \qquad \forall\, z \in \Gamma \text{ and } z \neq 0, \tag{5.2.15}$$

we see that $u^0(z) - \epsilon u^1(z) = 0$ is equivalent to $u_1^0(z) - \epsilon u_1^1(z) = 0$. By (5.2.14) and $u_1^0(0) = 0$, $u_1^1(0) > 0$, we deduce that for any $\epsilon > 0$ sufficiently small, $u^0 - \epsilon u^1$ has at least two singular points $z^+(\epsilon) = (z_1^+, z_2^+)$, $z^-(\epsilon) = (z_1^-, z_2^-) \in \Gamma$ with $z_1^- < 0 < z_1^+$, such that

$$\lim_{\epsilon \to 0} z^+(\epsilon) = 0, \quad \lim_{\epsilon \to 0} z^-(\epsilon) = 0.$$

The proof of the lemma is complete. □

LEMMA 5.2.13. *Under Assumption 5.2.1, if ind$(u^0, \bar{x})$ = non-integer, then for any $\varepsilon > 0$ small, each $u^0 \pm \epsilon u^1$ has at least one singular point on Γ.*

PROOF. When ind $(u^0, \bar{x})$ = fraction, n = odd. Hence, the two boundary orbits connected to $\bar{x}$ are either all stable or all unstable.

Let (z_1, z_2) be the z-coordinate system as in the proof of Lemma 5.2.12. Then

$$\text{sign } u_1^0(z^-) = -\text{sign } u_1^0(z^+), \tag{5.2.16}$$

where $z^\pm = (z_1^\pm, z_2^\pm) \in \Gamma \subset \partial M$ and $z_1^- < 0 < z_1^+$. We assume that

$$u_1^0(z^-) > 0, \quad u_1^1(0) > 0. \tag{5.2.17}$$

Then (5.2.16) and (5.2.17) yield

$$u_1^0(z^+) < 0 \text{ for } z^+ \in \Gamma. \tag{5.2.18}$$

From (5.2.15), we see that $u^0(z) \pm \epsilon u^1(z) = 0$ are equivalent to $u_1^0(z) \pm \epsilon u_1^1(z) = 0$. By (5.2.17), (5.2.18) and $u_1^1(z) > 0 \; \forall \; z \in \Gamma$, we obtain that for any $\epsilon > 0$ sufficiently small, there is at least a point $z_0^+ \in \Gamma$ (resp. a point $z_0^- \in \Gamma$) such that

$$u^0(z_0^\pm) \pm \epsilon u^1(z_0^\pm) = 0. \tag{5.2.19}$$

□

To complete the proof of Theorem 4.1, it suffices to prove the following lemma.

LEMMA 5.2.14. *Let Assumption 5.2.1 hold true.*

(1) *If n = even, one of $u^0 \pm \epsilon u^1$ has no singular points on Γ and the other has exactly two singular points on $\Gamma \subset \partial M$;*
(2) *If n = odd, $u^0 \pm \epsilon u^1$ have only one singular point on $\Gamma \subset \partial M$.*

PROOF. We prove only the case of $u^0 + \epsilon u^1$ for n = odd; the other case can be proved in exactly the same manner.

Without loss of generality, we assume (5.2.17) as in the proof of the previous lemma. Then singularities of $u^0 + \epsilon u^1$ on Γ can only occur for $z = (z_1, z_2) \in \Gamma$ with $z_1 > 0$. Let $\bar{z}, \tilde{z} \in \Gamma$ be two different singular points of $u^0 + \epsilon u^1$; then

$$\epsilon = \frac{u_1^0(\bar{z})}{u_1^1(\bar{z})} = \frac{u_1^0(\tilde{z})}{u_1^1(\tilde{z})}, \tag{5.2.20}$$

which contradicts $u_1^0(z)/u_1^1(z)$ being monotone on each side of $\bar{x}$ on Γ. Here the monotonicity is due to (5.2.6). It follows that $u^0 + \epsilon u^1$ has only one singular point on Γ. Thus both the lemma and, therewith, the theorem are proved. □

5.2.3. Proof of Theorem 5.2.7.

We divide the proof into two steps.

Step 1. The case where ind $(u^0, \bar{x})$ = integer.

By Theorem 5.2.6, the number of singular boundary points of $u^0 + \varepsilon u^1$ in a small neighborhood $\Gamma \subset \partial M$ of $\bar{x}$ is not the same as that for $u^0 - \varepsilon u^1$. Therefore, $u^0 + \varepsilon u^1$ and $u^0 - \varepsilon u^1$ are not topologically equivalent at $\bar{x} \in \partial M$. Moreover, if $\bar{x}$ is the unique degenerate singular point of u^0 on the boundary ∂M, then $u^0 + \varepsilon u^1$ and $u^0 - \varepsilon u^1$ are not topologically equivalent in a small neighborhood of ∂M. In other words, t_0 is a bifurcation point of the vector field u's global structure.

Step 2. The case where ind $(u^0, \bar{x})$ = fraction.

Let $x^+ = (z_1^+, z_2^+) \in \Gamma \subset \partial M$ (resp. $x^- = (z_1^-, z_2^-) \in \Gamma$) be the singular point of $u^0 + \varepsilon u^1$ (resp. $u^0 - \varepsilon u^1$) and $z_1^- < 0 < z_1^+$ as in Theorem 5.2.6. Then we have

$$\text{ind}(u^0 \pm \epsilon u^1, x^\pm) = -\frac{1}{2}.$$

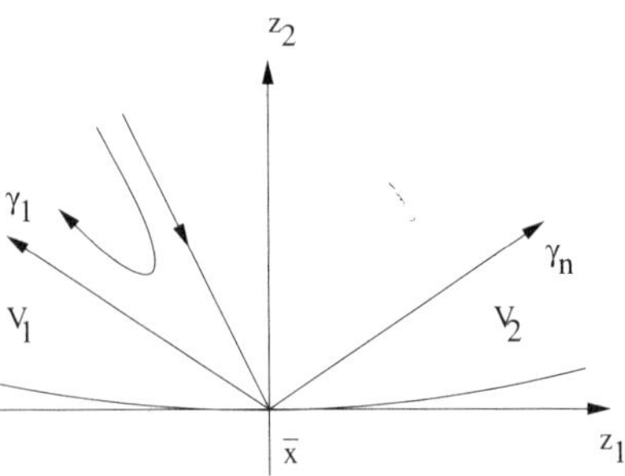

FIGURE 5.2.4. Auxiliary sketch for Step 2 in the proof of Theorem 5.2.7; the curved line tangent to the z_1-axis is the portion of ∂M near $\bar{x}$.

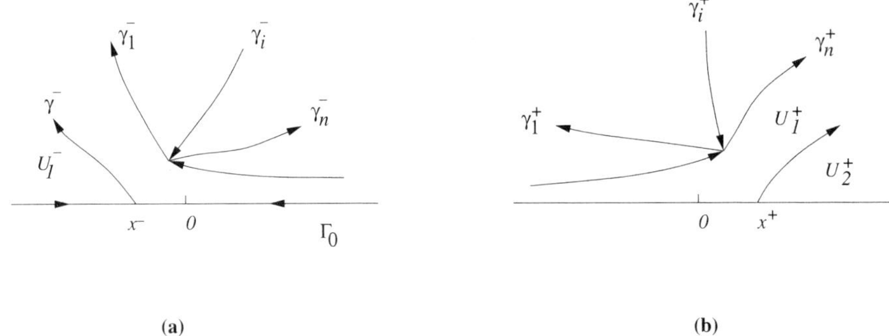

FIGURE 5.2.5. Auxiliary sketch for the final step in the proof of Theorem 5.2.7.

By the Singularity Classification Theorem, Theorem 5.1.5, there is only one orbit $\gamma^+(\epsilon)$ of $u^0 + \epsilon u^1$ in $\overset{\circ}{M}$ connected to x^+ (resp. only one orbit $\gamma^-(\epsilon)$ of $u^0 - \epsilon u^1$ in $\overset{\circ}{M}$ connected to x^-). By the same Theorem 5.1.5, there are n ($n \geq 3$ and odd) orbits γ_i ($1 \leq i \leq n$) of u^0 in $\overset{\circ}{M}$ connected to $\bar{x} \in \partial M$ ($z = 0$); see Figure 5.2.4.

Without loss of generality, we assume that the two orbits near the boundary are stable, and the z_1-component $u_1^1(z)$ of $u^1(z) = \partial u(z, t_0)/\partial t$ satisfies $u_1^1(0) > 0$. Let $V_1 \subset \overset{\circ}{M}$ (resp. $V_2 \subset \overset{\circ}{M}$) be the domain near $\bar{x} \in \partial M$ enclosed by ∂M and γ_1 (resp. by ∂M and γ_n); see Figure 5.2.4.

Since $u_1^1(0) > 0$, the flow of $u^0 + \epsilon u^1$ crosses γ_n transversally and enters into V_2 (resp. the flow of $u^0 - \epsilon u^1$ crosses γ_1 transversally and enters V_1). Therefore, we have

$$\gamma^+(\epsilon) \subset V_2, \qquad \gamma^-(\epsilon) \subset V_1; \tag{5.2.21}$$

see Figure 5.2.5.

Obviously, we have

$$\lim_{\epsilon \to 0} \gamma^-(\epsilon) = \gamma_1, \qquad \lim_{\epsilon \to 0} \gamma^+(\epsilon) = \gamma_n.$$

Moreover, there are n orbits $\Gamma_i^+ = \Gamma_i^+(\epsilon)$ of $u^0 + \epsilon u^1$ in $V - V_1$ (resp. n orbits $\Gamma_i^- = \Gamma_i^-(\epsilon)$ of $u^0 - \epsilon u^1$ in $V - V_2$), which are connected to singular points of $u^0 + \epsilon u^1$ (resp. $u^0 - \epsilon u^1$) such that

$$\lim_{\epsilon \to 0} \gamma_i^+(\epsilon) = \gamma_i \quad (\text{resp. } \lim_{\epsilon \to 0} \gamma_i^-(\epsilon) = \gamma_i),$$

where $V \subset \overset{\circ}{M}$ is a neighborhood of $\bar{x}$.

Let U_1^+ (resp. U_1^-) be the domain enclosed by ∂M and γ^+ (resp. by ∂M and γ^-), and $U_2^+ = V - U_1^+$ (resp. $U_2^- = V - U_1^-$). We infer from (5.2.21) that

$$\gamma_i^+ \subset U_1^+, \qquad \gamma_i^- \subset U_2^- \quad (1 \leq i \leq n). \tag{5.2.22}$$

We know that under topological equivalence, the orbits of $u^0 + \epsilon u^1$ connected to a singular boundary point are mapped to the orbits of $u^0 - \epsilon u^1$ connected to a boundary singular point, preserving orientations. Hence, if $u^0 + \epsilon u^1$ and $u^0 - \epsilon u^1$ are topologically equivalent locally at $\bar{x} \in \partial M$, then the restriction of $u^0 + \epsilon u^1$ to U_1^+ has to be topologically equivalent to the restriction to $u^0 - \epsilon u^1$ in U_1^-. This is in contradiction to (5.2.22), and Assertion 1 of Theorem 5.2.7 thus holds true.

Assertion 2 can be proved in the same fashion, completing therewith the proof of Theorem 5.2.7. □

5.3. Structural Bifurcation for Flows with Dirichlet Boundary Conditions

5.3.1. Structural Bifurcations near Flat Boundary. In this section, we assume the boundary ∂M contains a flat part $\Gamma \subset \partial M$, and consider structural bifurcation near a ∂-singular point $\bar{x} \in \Gamma$. For simplicity, we take a coordinate system (x_1, x_2) with $\bar{x}$ at the origin and with Γ given by

$$\Gamma = \{(x_1, 0) \,\big|\, |x_1| \leq \delta\}$$

for some $\delta > 0$. Obviously, the tangent and normal vectors on Γ are the unit vectors in the x_1- and x_2-directions, respectively.

Let $u \in C^1([0,T], B_0^r(TM))$ $(r \geq 2)$ be a one-parameter family of divergence-free vector fields with the homogeneous Dirichlet boundary condition. In a neighborhood $U \subset M$ of $\bar{x} \in \Gamma$, $u(x,t)$ can be expressed near $x = 0$ by

$$u(x,t) = x_2 v(x,t). \tag{5.3.1}$$

To proceed, we consider the Taylor expansions of both $u(x,t)$ and $v(x,t)$ at t_0 $(0 < t_0 < T)$:

$$\begin{cases} u(x,t) = u^0(x) + (t - t_0)u^1(x) + o(|t - t_0|^2) \\ u^0(x) = u(x, t_0), \\ u^1(x) = \dfrac{\partial u(x,t_0)}{\partial t}, \end{cases} \tag{5.3.2}$$

$$\begin{cases} v(x,t) = v^0(x) + (t - t_0)v^1(x) = o(|t - t_0|)^2, \\ v^0(x) = v(x, t_0), \\ v^1(x) = \dfrac{\partial v(x,t_0)}{\partial t}. \end{cases} \tag{5.3.3}$$

Let $u^i = (u_1^i, u_2^i)$, $v^i = (v_1^i, v_2^i)$, $i = 0, 1$. We start with the following assumption.

ASSUMPTION 5.3.1. *Let $\bar{x} = 0 \in \Gamma$ be an isolated degenerate ∂-singular point of $u^0(x)$, $u^0 \in C^{k+1}$ near $\bar{x} \in \Gamma$ for some $k \geq 2$. Assume that*

$$\frac{\partial u^0(0)}{\partial n} = 0, \tag{5.3.4}$$

$$\text{ind}(v^0, 0) \neq -\frac{1}{2}, \tag{5.3.5}$$

$$\frac{\partial^{k+1} u_1^0(0)}{\partial^k \tau \partial n} \neq 0, \tag{5.3.6}$$

$$\frac{\partial u^1(0)}{\partial n} \neq 0. \tag{5.3.7}$$

Some remarks are now in order.

REMARK 5.3.2. Condition (5.3.4) says that $\bar{x} = 0 \in \Gamma$ is a ∂-singular point of $u^0(x)$. In a 2-D incompressible flow, governed by either the Euler or the Navier-Stokes equations, this condition is equivalent to the leading-order vorticity vanishing at $\bar{x}$. The latter is the so-called Prandtl condition, which he suggested might identify boundary-layer separation points in incompressible flows [**84**].

REMARK 5.3.3. Condition (5.3.5) amounts to saying that there exists a number $m \neq 1$ of interior orbits of v^0 connected to $\bar{x} \in \Gamma$. Since $u^0 = x_2 v^0$, the number of interior orbits of u^0 connected to $\bar{x} \in \Gamma$ is exactly $m \neq 1$ as well. This shows that $\bar{x} \in \Gamma$ is a degenerate ∂-singular point of $u^0(x)$, which is necessary for structural bifurcation, as explained earlier.

REMARK 5.3.4. Condition (5.3.7) states that the first order term u^1 of the Taylor expansion for the normal derivative of u is different from zero. This is just the simplest necessary condition of such a type; if it does not hold, we need to work on a higher-order Taylor expansion, and the corresponding results presented will be true as well. In fluid-mechanics applications, condition (5.3.7) is equivalent to the vorticity associated with u^1 not vanishing at the boundary. In addition, it is easy to see that (5.3.7) is equivalent to

$$\frac{\partial u_1^1(0)}{\partial x_2} = \frac{\partial u_1^1(0)}{\partial n} \neq 0,$$

which shows that the acceleration of the fluid in the tangential direction near $\bar{x}$ is nonzero.

REMARK 5.3.5. Condition (5.3.6) is a technical condition, and amounts to saying that the tangential component u_1^0 of the leading-order term has a nontrivial Taylor expansion. Furthermore, let k be the smallest integer satisfying condition (5.3.6). It is easy to show that $k \geq 2$. In fact, $u^0(x)$ has the Taylor expansion at $x = 0$,

$$u^0(x) = \begin{cases} cx_2 + 2ax_1x_2 + bx_2^2 + x_2h_1(x), \\ \qquad\qquad - ax_2^2 + x_2h_2(x), \end{cases} \tag{5.3.8}$$

with $h_i(x) = o(|x|)$ $(i = 1, 2)$. Since $\bar{x} \in \Gamma$ is a degenerate ∂-singular point of $u^0(x)$, it follows that $c = 0$, $a = 0$, which implies that $k \geq 2$.

The structural bifurcation of u near a degenerate ∂-singular point on flat boundary is described by the following theorems.

THEOREM 5.3.6. *Let $u \in C^1([0,T], B_0^r(TM))$ $(r \geq 2)$ satisfy Assumption 5.3.1. Then there exist a neighborhood*

$$\Gamma_0 = \{(x_1, 0) \mid |x_1| \leq \delta_0\} \subset \Gamma$$

of $\bar{x} = 0$ and an $\varepsilon_0 > 0$ such that all ∂-singular points of $u(\cdot, t_0 \pm \varepsilon)$ are nondegenerate for any $0 < \varepsilon \leq \varepsilon_0$. Moreover,

(1) *if the index $ind(v^0, 0)$ is an integer, then one of $u(x, t_0 \pm \varepsilon)$ has exactly two ∂-singular points on Γ_0, and the other has no ∂-singular points on Γ_0; and*
(2) *if the index $ind(v^0, 0)$ is not an integer, then each of $u(x, t_0 \pm \varepsilon)$ has exactly one ∂-singular point on Γ_0.*

THEOREM 5.3.7 (Structural Bifurcation Theorem). *Let a one-parameter family of divergence-free vector fields $u \in C^1([0,T], B_0^r(TM))$ $(r \geq 2)$ satisfy Assumption 5.3.1. Then*

(1) *the vector field u has a bifurcation in its local structure at $(\bar{x}, t_0)$.*
(2) *if $\bar{x} \in \partial M$ is a unique ∂-singular point of u with the same index as $ind(v^0, 0)$ on ∂M, then $u(x,t)$ has a bifurcation in its global structure at $t = t_0$.*

The proofs of these two theorems are based on analyzing the orbits of u to provide a complete classification of the local structure of u near $(\bar{x}, t_0)$. The bifurcation results follow immediately from the classification.

5.3.2. Proof of Theorem 5.3.6. The proof of the theorem is a direct consequence of the following three lemmas.

LEMMA 5.3.8. *There exist a neighborhood $\Gamma_0 \subset \Gamma$ of $\bar{x} = 0$ and an $\varepsilon_0 > 0$ such that all ∂-singular points of $u(\cdot, t_0 \pm \varepsilon)$ are nondegenerate for any $0 < \varepsilon \leq \varepsilon_0$.*

PROOF. Let $x^* \in \Gamma$, $x^* = (x_1^*)$ and $0 < |x_1^*| \leq \delta_0$ with $0 < \delta_0$ sufficiently small to be chosen later. Then,

$$\frac{\partial u}{\partial n}(x^*, t_0 \pm \varepsilon) = \frac{\partial u}{\partial x_2}(x^*, t_0 \pm \varepsilon) = 0. \tag{5.3.9}$$

By the Taylor expansion (5.3.2) and Condition (5.3.7), it suffices to consider only the first order approximation of (5.3.2). Hence,

$$\frac{\partial u_i^0(x_1^*, 0)}{\partial x_2} \pm \varepsilon \frac{\partial u_i^1(x_1^*, 0)}{\partial x_2} = 0 \qquad (i = 1, 2). \tag{5.3.10}$$

We need to show that

$$\det \begin{pmatrix} \dfrac{\partial^2 u_1^0}{\partial x_1 \partial x_2} \pm \varepsilon \dfrac{\partial^2 u_1^1}{\partial x_1 \partial x_2} & \dfrac{\partial^2 u_1^0}{\partial x_2^2} \pm \varepsilon \dfrac{\partial^2 u_1^1}{\partial x_2^2} \\ \\ \dfrac{\partial^2 u_2^0}{\partial x_1 \partial x_2} \pm \varepsilon \dfrac{\partial^2 u_2^1}{\partial x_1 \partial x_2} & \dfrac{\partial^2 u_2^0}{\partial x_2^2} \pm \varepsilon \dfrac{\partial^2 u_2^1}{\partial x_2^2} \end{pmatrix}_{x=(x_1^*,0)} \neq 0. \tag{5.3.11}$$

For $u \in C^1([0,T], B_0^r(TM))$, $u(x_1, 0, t) = 0$ for all $(x_1, 0) \in \Gamma$ and $0 \leq t \leq T$. Thus we obtain

$$\frac{\partial u_1}{\partial x_1}(x_1, 0, t) = 0, \qquad \forall \, |x_1| \leq \bar{\delta}.$$

Thanks to the fact that u is divergence-free, we have

$$\frac{\partial u_2}{\partial x_2}(x_1,0,t) = -\frac{\partial u_1}{\partial x_1}(x,0,t) = 0, \qquad \forall\, |x_1| \leq \bar{\delta}.$$

Consequently

$$\frac{\partial u_2^0(x_1,0)}{\partial x_2} \pm \varepsilon \frac{\partial u_2^1(x_1,0)}{\partial x_2} = 0, \qquad \forall\, |x_1| \leq \bar{\delta},$$

which yields

$$\frac{\partial^2 u_2^0(x_1^*,0)}{\partial x_1 \partial x_2} \pm \varepsilon \frac{\partial^2 u_2^1(x_1^*,0)}{\partial x_1 \partial x_2} = 0.$$

To verify (5.3.11), it suffices to prove that

$$\frac{\partial^2 u_1^0(x_1^*,0)}{\partial x_1 \partial x_2} \pm \varepsilon \frac{\partial^2 u_1^1(x_1^*,0)}{\partial x_1 \partial x_2} \neq 0,$$

since the sum of the diagonal terms in (5.3.11) is zero, thanks to the fact that $\partial u/\partial x_2$ is divergence-free.

From Condition (5.3.6) and (5.3.7), we get

$$\begin{cases} \dfrac{\partial u_1^0(x_1,0)}{\partial x_2} = \alpha x_1^k + o(|x_1|^k) \quad \alpha \neq 0, \\ \dfrac{\partial u_1^1(x_1,0)}{\partial x_2} = \beta + o(|x_1|) \quad \beta \neq 0. \end{cases} \tag{5.3.12}$$

Therefore, it follows from (5.3.10) that

$$\varepsilon = \pm\frac{\alpha}{\beta} {x_1^*}^k + o(|x_1^*|^k). \tag{5.3.13}$$

Thus, from (5.3.12) and (5.3.13), we obtain that

$$\frac{\partial^2 u_1^0(x_1^*,0)}{\partial x_1 \partial x_2} \pm \varepsilon \frac{\partial^2 u_1^1(x_1^*,0)}{\partial x_1 \partial x_2} = \alpha k {x_1^*}^{k-1} + o\left(|x_1^*|^{k-1}\right),$$

which is different from zero for $0 < |x_1^*| \leq \delta_0$ small. Hence (5.3.11) follows, and the proof of the lemma is complete. □

LEMMA 5.3.9. *If ind$(v^0,0)$ = integer, then one of $u(x,t_0 \pm \varepsilon)$ has no ∂-singular points on Γ_0, and the other one has exactly two ∂-singular ponts on Γ_0 with one ∂-singular point on each side of $\bar{x}=0$.*

PROOF. From (5.3.1), it is easy to see that the zero points of $v(x,t)$ on Γ are equivalent to the ∂-singular points of $u(x,t)$. Hence, we only have to prove the assertion for the vector field $v(x,t_0 \pm \varepsilon)$.

According to (5.3.8), we have

$$v_2(x_1,0,t) = 0, \quad \forall\, x_1 \in \Gamma,\ t \geq 0\ (v = (v_1,v_2)).$$

By (5.3.6) and (5.3.7), we infer from (5.3.1) that

$$\begin{cases} v_1^0(x_1,0) = \alpha x_1^k + o(|x_1|^k), \\ v_1^1(x_1,0) = \beta + g(x_1), \end{cases} \tag{5.3.14}$$

where $\alpha \neq 0$, $\beta \neq 0$, and $g(0)=0$.

On the other hand, if ind$(v^0,0)$ is an integer, then, by Remark 5.3.3, the number n of interior orbits connected to $\bar{x}=0$ of v^0 is even. Hence, one of the two boundary orbits connected to $\bar{x} = 0$ of v^0 is stable, and the other one is unstable. It follows that the exponent k in (5.3.14) is even, i.e., $k = 2m$ $(m \geq 1)$.

Consider the following two equations:

$$\begin{aligned}
(5.3.15) \qquad 0 &= v_1(x_1, 0, t_0 + \varepsilon) = v_1^0(x_1, 0) + \varepsilon v_1^1(x_1, 0) + o(|\varepsilon|) \\
&= \alpha x_1^{2m} + \varepsilon\beta + \varepsilon g(x_1) + o(|\varepsilon|, |x_1|^{2m}), \\
(5.3.16) \qquad 0 &= v_1(x_1, 0, t_0 - \varepsilon) = v_1^0(x_1, 0) - \varepsilon v_1^1(x_1, 0) + o(|\varepsilon|) \\
&= \alpha x_1^{2m} - \varepsilon\beta - \varepsilon g(x_1) + o(|\varepsilon|, |x_1|^{2m}),
\end{aligned}$$

where $m \geq 1$. Without loss of generality, assume that $\alpha, \beta > 0$. Then there is a $\delta_0 > 0$ such that for any $\varepsilon > 0$ sufficiently small, (5.3.16) has only two solutions $x_1^-(\varepsilon) < 0 < x_1^+(\varepsilon)$ in the interval $(-\delta_0, \delta_0)$, and (5.3.15) has no solutions in $(-\delta_0, \delta_0)$. The claim is verified. □

LEMMA 5.3.10. *If* $\text{ind}(v^0, 0) =$ *non-integer, then for any* $\varepsilon > 0$ *sufficiently small, each* $u(x, t_0 \pm \varepsilon)$ *has exactly one* ∂*-singular point on* Γ.

PROOF. Indeed, when $\text{ind}(v^0, 0) = -n/2$ is a fraction, n is odd. Hence the exponent k in (5.3.14) is odd, i.e., $k = 2m + 1$ $(m \geq 1)$. Therefore each of the following two equations

$$\begin{aligned}
\alpha x_1^{2m+1} + \varepsilon\beta + \varepsilon g(x_1) + o\left(|\varepsilon|, |x_1|^{2m+1}\right) = 0, \\
\alpha x_1^{2m+1} - \varepsilon\beta - \varepsilon g(x_1) + o\left(|\varepsilon|, |x_1|^{2m+1}\right) = 0
\end{aligned}$$

has exactly one solution in $(-\delta_0, \delta_0)$. □

5.3.3. Proof of Theorem 5.3.7. As mentioned earlier, u and v have topologically equivalent streamlines in an interior neighborhood of $x = 0$; hence, bifurcation in the local structure of u at $\bar{x} = 0$ is equivalent to that of v. As a result, we only have to consider the local bifurcation for the vector field v. We divide the proof into two steps.

Step 1. Case where ind$(v^0, 0) =$ *integer.* By Theorem 5.3.6, the number of boundary saddle points of $u^0 + \varepsilon u^1$ in a small neighborhood $\Gamma_0 \subset \Gamma \subset \partial M$ of p $(x = 0)$ is not the same as that for $v^0 - \varepsilon v^1$. Therefore, $u^0 + \varepsilon v^1$ and $v^0 - \varepsilon v^1$ are not local topologically equivalent locally near $\bar{x} \in \Gamma \subset \partial M$.

Step 2. Case where ind$(v^0, 0) =$ *fraction.* Without loss of generality, we assume that the two orbits connected to $\bar{x} = 0$ on $\Gamma_0 \subset \partial M$ are stable (i.e., $\alpha < 0$ in (5.3.14)), and $v_1^1(0) > 0$ (i.e., $\beta > 0$ in (5.3.14)). Let $x^+ = (x_1^+, 0)$ and $x^- = (x_1^-, 0) \in \Gamma_0 \subset \partial M$ be the singular points of $v^0 + \varepsilon v^1$ and $v^0 - \varepsilon v^1$, respectively. Hence, $x_1^- < 0 < x_1^+$ as in Theorem 5.3.6. The nondegeneracy of both x^- and x^+ implies that

$$\text{ind}(v^0 \pm \varepsilon v^1, x^\pm) = -\frac{1}{2},$$

and there is exactly one orbit $\gamma^+(\varepsilon)$ of $v^0 + \varepsilon v^1$ in $\overset{\circ}{M}$ connected to x^+ (resp. exactly one orbit $\gamma^-(\varepsilon)$ of $v^0 - \varepsilon v^1$ in $\overset{\circ}{M}$ connected to x^-).

By (5.3.5) and Remark 5.3.3, there are n $(n \geq 3$ and odd) orbits γ_1 $(1 \leq i \leq n)$ of v^0 in $\overset{\circ}{M}$ connected to $\bar{x} \in \Gamma$; see Figure 5.3.1. Let $V_1 \subset \overset{\circ}{M}$ (resp. $V_2 \subset \overset{\circ}{M}$) be the domain near $\bar{x} = 0 \in \Gamma_0 \subset \partial M$ enclosed by ∂M and γ_1 (resp. by ∂M and γ_n); see Figure 5.3.1.

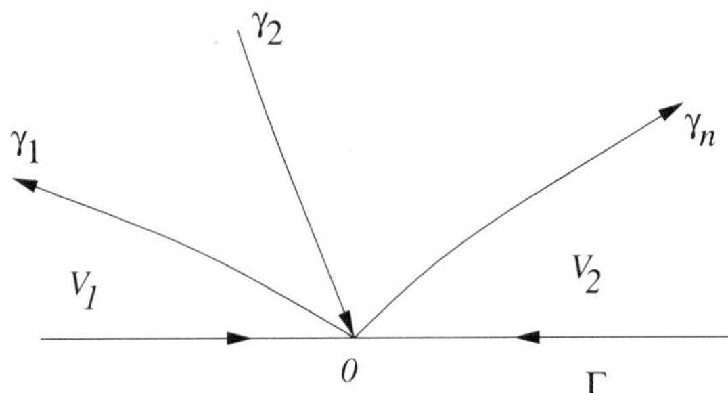

FIGURE 5.3.1. Case of n interior orbits connected to $\bar{x} = 0 \in \Gamma$, where $\Gamma \subset \partial M$ is flat, and $n \geq 3$ is odd.

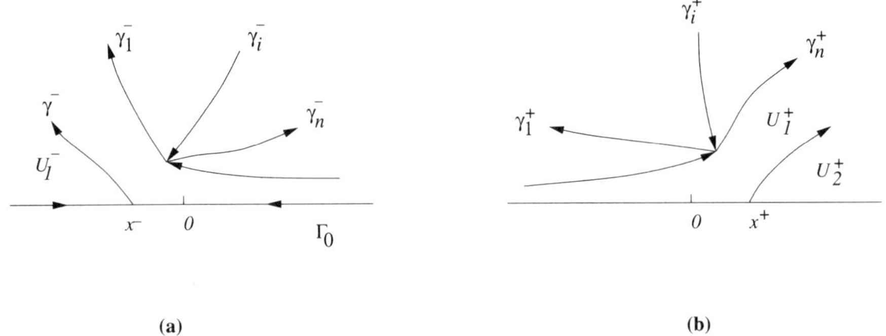

FIGURE 5.3.2. Sketch illustrating the proof of Assertion (1) in Theorem 5.3.7: (a) orbits at $t^- < t_0$, and (b) orbits at $t^+ > t_0$.

Since $v_1^1(0) > 0$, the flow of $v^0 + \varepsilon v^1$ crosses γ_n transversally and enters into V_2 (resp. the flow of $u^0 - \varepsilon u^1$ crosses γ_1 transversally and enters into V_1). Therefore, we have

$$\gamma^+(\varepsilon) \subset V_2, \quad \gamma^-(\varepsilon) \subset V_1; \tag{5.3.17}$$

see Figure 5.3.2. Obviously,

$$\lim_{\varepsilon \to 0} \gamma^-(\varepsilon) = \gamma_1, \quad \lim_{\varepsilon \to 0} \gamma^+(\varepsilon) = \gamma_n.$$

Moreover, there are n orbits $\gamma_i^+ = \gamma_i^+(\varepsilon)$ of $v^0 + \varepsilon v^1$ in $V - V_1$ (resp. n orbits $\gamma_i^- = \gamma_i^-(\varepsilon)$ of $v^0 - \varepsilon v^1$ in $V - V_2$), which are connected to singular points of $v^0 + \varepsilon v^1$ (resp. $v^0 - \varepsilon v^1$) such that

$$\lim_{\varepsilon \to 0} \gamma_i^+(\varepsilon) = \gamma_i, \qquad \lim_{\varepsilon \to 0} \gamma_i^-(\varepsilon) = \gamma_i,$$

where $V \subset \overset{\circ}{M}$ is a neighborhood of $\bar{x} = 0$.

Let U_1^+ (resp. U_1^-) be the domain enclosed by ∂M and γ^+ (resp. by ∂M and γ^-), and $U_2^+ = V - U_1^+$ (resp. $U_2^- = V - U_1^-$). We infer from (5.3.17) that

$$\gamma_i^+ \subset U_1^+, \quad \gamma_i^- \subset U_2^- \quad (1 \leq i \leq n). \tag{5.3.18}$$

We know that under topological equivalence, the orbits of $u^0 + \varepsilon u^1$ connected to a singular boundary point are mapped to the orbits of $u^0 - \varepsilon v^1$ connected to a boundary singular point preserving orientation. Hence, if $u^0 + \varepsilon u^1$ and $v^0 - \varepsilon v^1$

were topologically equivalent locally at $\bar{x} \in \Gamma$, then the restriction of $u^0+\varepsilon v^1$ to U_1^+ would have to be topologically equivalent to $v^0-\varepsilon v^1$ in U_1^-. This is in contradiction to (5.3.18). Thus, Assertion 1 of Theorem 5.3.7 is proven.

Assertion 2 of Theorem 5.3.7 is a corollary of Assertion 1. Indeed, if $\bar{x} \in \partial M$ is a unique singular point of u^0, which has the same index as $\operatorname{ind}(v^0, 0)$ on ∂M, then the structural bifurcation of u locally at $(\bar{x}, t_0)$ implies the structural bifurcation in its global structure.

The proof of Theorem 5.3.7 is complete. □

5.3.4. Structural bifurcations near curved boundaries. We now generalize the main bifurcation theorems on the flat boundary case in the previous subsection to the curved boundary case in this subsection.

Consider structural bifurcation near a ∂-singular point $\bar{x} \in \partial M$ of $u(x,t)$ on the general C^{r+1} $(r \geq 2)$ boundary ∂M. Let $\bar{x} \in \partial M$ and (x_1, x_2) be an orthogonal coordinate system with origin $\bar{x}$, the x_1-axis tangent to ∂M at $\bar{x}$, and the x_2-axis oriented to the inward normal direction.

Let $u \in C^1([0,T], B_0^r(TM))$ $(r \geq 2)$ have the Taylor expansion at t_0 $(0 < t_0 < T)$ as in (5.3.2). In particular, let

$$u^0 = (u_1^0, u_2^0) = u(x, t_0),$$
$$u^1 = (u_1^1, u_1^2) = \frac{\partial u(x,t_0)}{\partial t}.$$

In addition, let n be the number of interior orbits of $u^0(x)$ connected to $\bar{x}$. Then we can restate Assumption 5.3.1 as follows with Condition (5.3.5) replaced by (5.3.20). Condition (5.3.20) is a geometrical condition.

ASSUMPTION 5.3.11. *Let $\bar{x} \in \partial M$ be an isolated degenerate ∂-singular point of $u^0(x) = u(x,t_0)$, $u^0 \in C^{k+1}$ near $\bar{x} \in \Gamma$ for some $k \geq 2$. Assume that*

$$\frac{\partial u^0(\bar{x})}{\partial n} = 0, \tag{5.3.19}$$
$$n \neq 1, \tag{5.3.20}$$
$$\frac{\partial^{k+1} u_1^0(\bar{x})}{\partial^k \tau \partial n} \neq 0, \tag{5.3.21}$$
$$\frac{\partial u_1^1(\bar{x})}{\partial n} \neq 0. \tag{5.3.22}$$

We then have the following structural bifurcation theorems, similar to the previous subsection.

THEOREM 5.3.12. *Let $u \in C^1([0,T], B_0^r(TM))$ satisfy Assumption 5.3.11. Then in a neighborhood $\Gamma \subset \partial M$ of $\bar{x}$, the ∂-singular points of $u(x, t_0 \pm \varepsilon)$ are nondegenerate for any $\varepsilon > 0$ sufficiently small. Moreover,*

(1) *if $n =$ even $(n \geq 0)$, then one of $u(x, t_0 \pm \varepsilon)$ has two ∂-singular points on Γ, and the other one has no ∂-singular point on Γ;*
(2) *if $n =$ odd $(n \geq 3)$, then each of $u(x, t_0 \pm \varepsilon)$ has only one ∂-singular point on Γ.*

THEOREM 5.3.13. (Structural Bifurcation Theorem). *Let $u \in C^1([0,T], B_0^r(TM))$ satisfy Assumption 5.3.11. Then the following assertions hold true:*

(1) *u has a bifurcation in its local structure at $(\bar{x}, t_0)$;*

(2) *if $\bar{x} \in \partial M$ is a unique degenerate ∂-singular point of $u^0(x) = u(x, t_0)$ on ∂M, then $u(x,t)$ has a bifurcation in its global structure at t_0.*

5.3.5. Coordinate transformation. The main ideas and method to prove Theorems 5.3.12–5.3.13 are as follows. First, we introduce a local coordinate transformation, which preserves the divergence-free condition, and maps a neighborhood $\Gamma \subset \partial M$ of $\bar{x}$ to a flat boundary, and show that Assumption 5.3.11 is equivalent to Assumption 5.3.1 for the newly transformed vector field. Then Theorems 5.3.12–5.3.13 follow immediately from Theorems 5.3.6–5.3.7.

In the coordinate system (x_1, x_2) introduced in the beginning of this section, the boundary ∂M can be expressed locally near $\bar{x} = 0 \in \partial M$ by

$$x_1 = f(x_1), \quad f(0) = 0, \quad f'(0) = 0. \tag{5.3.23}$$

We take the local coordinate transformation

$$(\widetilde{x}_1, \widetilde{x}_2) = (x_1, x_2 - f(x_1)). \tag{5.3.24}$$

Obviously, the transformation (5.3.24) takes a neighborhood $U \subset M$ of $\bar{x}$ to a domain $\widetilde{U} \subset \mathbb{R}^2_+ = \{(\widetilde{x}_1, \widetilde{x}_2) \in \mathbb{R}^2 \mid \widetilde{x}_2 \geq 0\}$, and maps the boundary part $U \cap \partial M$ to a neighborhood of $x = 0$ on the $\widetilde{x}_1$-axis.

Let $\varphi : U \to \widetilde{U}$ be the transformation (5.3.24), and let $\varphi^* : C^r(TU) \to C^r(T\widetilde{U})$ be the isomorphism induced by φ. It is easy to see that

$$\varphi^* = D\varphi = \begin{pmatrix} 1 & 0 \\ -f'(\widetilde{x}_1) & 1 \end{pmatrix},$$

and for any $u \in C^r(TU)$,

$$\widetilde{u} = \varphi^* \circ u = \begin{pmatrix} u_1 \\ u_2 - f'(\widetilde{x}_2) u_1 \end{pmatrix}. \tag{5.3.25}$$

LEMMA 5.3.14. *If $u \in C^r(TU)$ is divergence-free, then $\widetilde{u} = \varphi^* \circ u$ is also divergence-free. Moreover, as $u|_{\partial M \cap U} = 0$, then $\widetilde{u}(\widetilde{x}_1, 0) = 0$, and*

$$\begin{cases} \widetilde{u}(\widetilde{x}) = \widetilde{x}_2 \widetilde{v}(\widetilde{x}), \\ \widetilde{v}_2(\widetilde{x}_1, 0) = 0, \quad (\widetilde{v} = (\widetilde{v}_1, \widetilde{v}_2)). \end{cases} \tag{5.3.26}$$

PROOF. We infer from (5.3.26) that

$$\begin{aligned} \frac{\partial \widetilde{u}_1}{\partial \widetilde{x}_1} &= \frac{\partial \widetilde{u}_1}{\partial x_1} \cdot \frac{\partial x_1}{\partial \widetilde{x}_1} + \frac{\partial \widetilde{u}_1}{\partial x_2} \frac{\partial x_2}{\partial \widetilde{x}_1} \\ &= \frac{\partial u_1}{\partial x_1} + \frac{\partial u_1}{\partial x_2} \cdot f'(x_1), \\ \frac{\partial \widetilde{u}_2}{\partial \widetilde{x}_2} &= \frac{\partial \widetilde{u}_2}{\partial x_1} \frac{\partial x_1}{\partial \widetilde{x}_2} + \frac{\partial \widetilde{u}_2}{\partial x_2} \frac{\partial x_2}{\partial \widetilde{x}_2} \\ &= \frac{\partial}{\partial x_2} [u_2 - f'(x_1) u_1] \\ &= \frac{\partial u_2}{\partial x_2} - \frac{\partial u_1}{\partial x_2} f'(x_1). \end{aligned}$$

Hence we have

$$\operatorname{div} \widetilde{u} = \frac{\partial \widetilde{u}_1}{\partial \widetilde{x}_1} + \frac{\partial \widetilde{u}_2}{\partial \widetilde{x}_2} = \frac{\partial u_1}{\partial x_1} + \frac{\partial u_2}{\partial x_2} = 0.$$

Since φ maps the part on ∂M to the $\widetilde{x}_1$-axis, and $u|_{\partial M} = 0$, we have $\widetilde{u}(\widetilde{x}_1, 0) = 0$. Thus $\widetilde{u} = \widetilde{x}_2 \widetilde{v}(\widetilde{x})$. Thanks to the divergence-free condition, it is easy to see that $\widetilde{v}_2(\widetilde{x}_1, 0) = 0$, and the proof is complete. □

5.3.6. Proof of Theorems 5.3.12 and 5.3.13. According to Theorems 5.3.6 and 5.3.7, the proof of these two theorems will be achieved in a few lemmas as follows.

LEMMA 5.3.15. *Let $u \in C^1([0,T], B_0^r(TM))$ satisfy Assumption 5.3.11. Then the vector field $\widetilde{u} = \varphi^* \circ u = \widetilde{x}_2 \widetilde{v}(\widetilde{x}, t)$ satisfies Assumption 5.3.1.*

PROOF. Notice that φ maps $\bar{x} = 0$ to $\widetilde{x} = 0$. Since u and $\widetilde{u}$ are locally topologically equivalent near $x = 0$ and $\widetilde{x} = 0$, (5.3.20) implies that the number n of interior orbits connected to $\widetilde{x} = 0$ of $\widetilde{v}(x, t_0)$ is different from 1. Hence

$$\operatorname{ind}(\widetilde{v}(x, t_0), 0) = -\frac{n}{2} \neq -\frac{1}{2}.$$

By (5.3.26), $\widetilde{u}_1 = u_1$, and φ^* takes the inward normal vector n at $x = 0$ to the normal vector $\widetilde{n} = (0, 1)$ at $\widetilde{x} = 0$. Therefore, $\partial u_1^1(0)/\partial n \neq 0$, which implies that $\partial \widetilde{u}_1^1(0)/\partial x_2 \neq 0$.

By tensor analysis, we know that for a function $f(x)$, under a coordinate transformation $\varphi : U \to \widetilde{U}$, the direction derivative $\partial f/\partial r = \partial f/\partial \widetilde{r}$, where r is a vector and $\widetilde{r} = D\varphi \cdot r$. Actually, we have

$$\begin{aligned}
\frac{\partial f}{\partial \widetilde{r}} &= (\widetilde{r} \cdot \widetilde{\nabla}) f \\
&= (\widetilde{r}_1, \widetilde{r}_2) \begin{pmatrix} \frac{\partial}{\partial \widetilde{x}_1} \\ \frac{\partial}{\partial \widetilde{x}_2} \end{pmatrix} f \\
&= (r_1, r_2) D\varphi^* \cdot (D\varphi^*)^{-1} \begin{pmatrix} \frac{\partial}{\partial x_1} \\ \frac{\partial}{\partial x_2} \end{pmatrix} f \\
&= (r \cdot \nabla) f.
\end{aligned}$$

By (5.3.26) we see that $u_1^0 = \widetilde{u}_1^0$, and $\widetilde{n} = (0, 1)$ at $\widetilde{x} = 0$. Hence we have

$$\frac{\partial^k}{\partial \tau^k} \frac{\partial u_1^0(0)}{\partial n} = \frac{\partial^k}{\partial \widetilde{\tau}^k} \cdot \frac{\partial \widetilde{u}_1^0(0)}{\partial \widetilde{n}} = \frac{\partial^k}{\partial x_1^k} \frac{\partial \widetilde{u}_1^0(0)}{\partial x_2}.$$

The proof of the lemma is complete. □

LEMMA 5.3.16. *A point $x \in \partial M \cap U$ is a ∂-regular point of $u \in B_0^r(TM)$ if and only if $\widetilde{x} = \varphi(x) = (\widetilde{x}_1, 0)$ is a ∂-regular point of $\widetilde{u} = D\varphi \cdot u$.*

PROOF. By Definition 2.2.1, it suffices to show that two vector fields $\partial u/\partial n$ and $\partial \widetilde{u}/\partial \widetilde{x}_2$ have the same singular points on the boundary in the sense of homeomophism, where

$$\frac{\partial u}{\partial n} = (N \cdot \nabla) u = n_1 \frac{\partial u}{\partial x_1} + n_2 \frac{\partial n}{\partial x_2}$$

and the vector field N defined in U with the unit modulus $|N| = 1$ such that the orbits of N are the normal lines λn in U (when U is properly taken, for any $x, y \in U \cap \partial M$, $x \neq y$, the normal lines λn_x and λn_y do not intersect in U).

Equivalently, we only proceed for the two vector fields $\partial\widetilde{u}/\partial\widetilde{n}$ and $\partial\widetilde{u}/\partial\widetilde{x}_2$, where $\partial\widetilde{u}/\partial\widetilde{n}$ is the corresponding transformation of $\partial u/\partial n$, which is expressed by

$$\begin{cases} \dfrac{\partial\widetilde{u}}{\partial\widetilde{n}} = \left(\widetilde{N}\cdot\widetilde{\nabla}\right)\widetilde{u} = \widetilde{n}_1\dfrac{\partial\widetilde{u}}{\partial\widetilde{x}_1} + \widetilde{n}_2\dfrac{\partial\widetilde{u}}{\partial\widetilde{x}_2}, \\ \widetilde{N} = \begin{pmatrix}\widetilde{n}_1\\ \widetilde{n}_2\end{pmatrix} = D\varphi\circ N = \begin{pmatrix}1 & 0\\ -f' & 1\end{pmatrix}\begin{pmatrix}n_1\\ n_2\end{pmatrix} = \begin{pmatrix}n_1\\ n_2 - f'n_1\end{pmatrix}. \end{cases} \tag{5.3.27}$$

From (5.3.26), we see that

$$\widetilde{u}(\widetilde{x}_1, 0) = 0, \quad \forall\, \widetilde{x}_1 \in \partial\mathbb{R}^2_+ \cap \widetilde{U}. \tag{5.3.28}$$

By (5.3.27) and (5.3.28), we deduce that

$$\frac{\partial\widetilde{u}}{\partial\widetilde{n}} = (n_2 - f'(x_1)n_1)\frac{\partial\widetilde{u}}{\partial\widetilde{x}_2}, \qquad \text{on } \partial\mathbb{R}^2_+ \cap \widetilde{U}. \tag{5.3.29}$$

By (5.3.23) and $n_2 \neq 0$ for $x \in \partial M$ near $\bar{x}$, we derive this lemma. The proof is complete. □

LEMMA 5.3.17. *A point $x \in \partial M \cap U$ is a ∂-saddle point of $u \in B^r_0(TM)$ if and only if $\widetilde{x} = \varphi(x) \in \partial\mathbb{R}^2_+ \cap \widetilde{U}$ is a ∂-saddle point of $\widetilde{u} = D\varphi\cdot u$.*

PROOF. It suffices to prove that $\partial\widetilde{u}/\partial\widetilde{n}$ and $\partial\widetilde{u}/\partial\widetilde{x}_2$ have the same nondegenerate singular points on the boundary

$$\Gamma = \partial\mathbb{R}^2_+ \cap \widetilde{U} = \{(\widetilde{x}_1, 0) \mid |\widetilde{x}_1| < \delta\}.$$

Namely, both Jacobian determinants

$$\det\begin{pmatrix} \dfrac{\partial^2\widetilde{u}_1}{\partial\widetilde{x}_1\partial\widetilde{n}} & \dfrac{\partial^2\widetilde{u}_1}{\partial\widetilde{x}_2\partial\widetilde{n}} \\ \dfrac{\partial^2\widetilde{u}_2}{\partial\widetilde{x}_1\partial\widetilde{n}} & \dfrac{\partial^2\widetilde{u}_2}{\partial\widetilde{x}_2\partial\widetilde{n}} \end{pmatrix} \quad \text{and} \quad \det\begin{pmatrix} \dfrac{\partial^2\widetilde{u}_1}{\partial\widetilde{x}_1\partial\widetilde{x}_2} & \dfrac{\partial^2\widetilde{u}_1}{\partial\widetilde{x}_2^2} \\ \dfrac{\partial^2\widetilde{u}_2}{\partial\widetilde{x}_1\partial\widetilde{x}_2} & \dfrac{\partial^2\widetilde{u}_2}{\partial\widetilde{x}_2^2} \end{pmatrix}$$

have the same nonzero points on Γ.

From (5.3.26), we see that

$$\frac{\partial\widetilde{u}_2(\widetilde{x}_1, 0)}{\partial x_2} = 0, \quad \forall\, |x_1| < \delta, \tag{5.3.30}$$

for some $\delta > 0$. By (5.3.29) and (5.3.30) one deduces that

$$\frac{\partial^2\widetilde{u}_2(\widetilde{x}_1, 0)}{\partial\widetilde{x}_1\partial\widetilde{n}} = 0; \quad \frac{\partial^2\widetilde{u}_2(\widetilde{x}_1, 0)}{\partial\widetilde{x}_1\partial\widetilde{x}_2} = 0.$$

Hence, we only need to prove that

$$\frac{\partial^2\widetilde{u}_1(\widetilde{x}_1, 0)}{\partial\widetilde{x}_1\partial\widetilde{n}} \neq 0 \text{ if and only if } \frac{\partial^2\widetilde{u}_1(\widetilde{x}_1, 0)}{\partial\widetilde{x}_1\partial\widetilde{x}_2} \neq 0, \tag{5.3.31}$$

$$\frac{\partial^2\widetilde{u}_2(\widetilde{x}_1, 0)}{\partial\widetilde{x}_2\partial\widetilde{n}} \neq 0 \text{ if and only if } \frac{\partial^2\widetilde{u}_2(\widetilde{x}_1, 0)}{\partial\widetilde{x}_2^2} \neq 0. \tag{5.3.32}$$

From (5.3.29) and (5.3.30), we immediately derive (5.3.32).

Thanks to (5.3.26), for any integer $k \geq 0$,

$$\frac{\partial^k\widetilde{u}_1(\widetilde{x}_1, 0)}{\partial\widetilde{x}_1 k} = 0. \tag{5.3.33}$$

By assumption, $(\widetilde{x}_1, 0)$ is a singular point of $\partial\widetilde{u}/\partial x_2$ (a ∂-saddle point of $\widetilde{u}$), i.e.,

$$\frac{\partial \widetilde{u}_1(\widetilde{x}_1, 0)}{\partial x_2} = 0. \tag{5.3.34}$$

From (5.3.33) and (5.3.34), it follows that

$$\begin{aligned} \frac{\partial^2 \widetilde{u}_1(\widetilde{x}_1, 0)}{\partial \widetilde{x}_1 \partial \widetilde{n}} &= \frac{\partial}{\partial \widetilde{x}_1}\left[\widetilde{n}_1 \frac{\partial \widetilde{u}_1}{\partial \widetilde{x}_1} + \widetilde{n}_2 \frac{\partial \widetilde{u}_1}{\partial \widetilde{x}_2}\right]\Bigg|_{\widetilde{x}=(\widetilde{x}_1,0)} \\ &= \widetilde{n}_2 \frac{\partial \widetilde{u}_1(\widetilde{x}_1, 0)}{\partial \widetilde{x}_1 \partial \widetilde{x}_2}. \end{aligned}$$

By (5.3.27), $\widetilde{n}_2 = n_2 - f'(x_1)n_1 \neq 0$ for $x = (x_1, x_2) \in \partial M$ near $\bar{x}$ $(x = 0)$. Thus we derive (5.3.32), and the proof is complete. □

5.4. Boundary Layer Separations of Incompressible Flows I

5.4.1. Bifurcation time and location.

In this section, we study the relationship between the bifurcation points of solutions of the following Navier-Stokes equations and the given data, e.g., the initial values, the Reynolds number, etc.

Consider the Navier-Stokes equations on M given by

$$\begin{cases} \dfrac{\partial u}{\partial t} + (u \cdot \nabla)u - R^{-1}\Delta u + \nabla p = \lambda f, \\ \operatorname{div} u = 0, \\ u|_{\partial M} = 0, \\ u(x, 0) = \varphi(x), \end{cases} \tag{5.4.1}$$

with $\varphi|_{\partial M} = 0$.

The main theorem in this section is as follows.

THEOREM 5.4.1 (Separation Equation). *If $(\bar{x}, t_0)$ $(\bar{x} \in \partial M,\ t_0 \geq 0)$ is a bifurcation point of the solution u of (5.4.1), then*

$$\frac{\partial \varphi_\tau(\bar{x})}{\partial n} = \int_0^{t_0} \left\{R^{-1}[\nabla \times \Delta u + k\Delta u \cdot \tau] + \lambda(\nabla \times f + kf_\tau)\right\}(\bar{x}, t)dt, \tag{5.4.2}$$

where $\nabla \times \Delta u = -\frac{\partial}{\partial n}(\Delta u \cdot \tau) + \frac{\partial}{\partial \tau}(\Delta u \cdot n)$ is the vorticity of Δu, and $k(\bar{x})$ is the curvature of ∂M at $\bar{x}$.

We remark here that (5.4.2) links the separation location and time with the Reynolds number, the external forcing, boundary curvature, and the initial velocity field. It will useful in future theoretical and engineering studies of the boundary layer separation. For simplicity, we call (5.4.2) the *separation equation* of the Navier-Stokes equation.

PROOF. For simplicity, we prove only the case where $f = 0$. From the Navier–Stokes equations (5.4.1), we have

$$\begin{aligned} &\int_0^{t_0} \frac{d}{dt}\left[\frac{\partial u_\tau(\bar{x}, t)}{\partial n}\right] dt \\ &= \int_0^{t_0} \left[R^{-1}\frac{\partial \Delta u \cdot \tau}{\partial n} - \frac{\partial}{\partial n}\frac{\partial p}{\partial \tau} - \frac{\partial (u \cdot \nabla)u \cdot \tau}{\partial n}\right](\bar{x}, t)dt. \end{aligned} \tag{5.4.3}$$

By assumption, if $(\bar{x}, t_0)$ is a bifurcation point of u, then $\bar{x}$ is a ∂-singular point of u. Hence, it follows from (5.4.3) that

$$\frac{\partial \varphi_\tau(\bar{x})}{\partial n} = -\int_0^{t_0} \left[R^{-1} \frac{\partial \Delta u \cdot \tau}{\partial n} - \frac{\partial}{\partial n} \frac{\partial p}{\partial \tau} - \frac{\partial (u \cdot \nabla) u \cdot \tau}{\partial n} \right] dt. \tag{5.4.4}$$

Observe that

$$\begin{aligned} (u \cdot \nabla) u \cdot \tau &= \left(u_\tau \frac{\partial u}{\partial \tau} + u_n \frac{\partial u}{\partial n} \right) \cdot \tau \\ &= u_\tau \frac{\partial u_\tau}{\partial \tau} - k u_\tau^2 + u_n \frac{\partial u_\tau}{\partial n} - u_n u \cdot \frac{\partial \tau}{\partial n}, \end{aligned}$$

which implies, using the Dirichlet boundary condition and the incompressibility $\partial u_n / \partial n = -\partial u_\tau / \partial \tau = 0$ on Γ, that

$$\frac{\partial (u \cdot \nabla) u \cdot \tau}{\partial n} = 0 \qquad \text{on } \partial M. \tag{5.4.5}$$

For simplicity, let (x_1, x_2) be the coordinating system with $\bar{x} \in \partial M$ at the origin, the x_1-axis pointing to the tangential direction, and the x_2-axis pointing to the normal direction. Then at the origin, $\tau = (\tau_1, \tau_2) = (1, 0)$ and $n = (n_1, n_2) = (0, 1)$. Hence

$$\begin{aligned} \frac{\partial}{\partial n} \frac{\partial p}{\partial \tau} &= n_1 \frac{\partial}{\partial x_1} \left(\tau_1 \frac{\partial p}{\partial x_1} + \tau_2 \frac{\partial p}{\partial x_2} \right) + n_2 \frac{\partial}{\partial x_2} \left(\tau_1 \frac{\partial p}{\partial x_1} + \tau_2 \frac{\partial p}{\partial x_2} \right) \\ &= \frac{\partial^2 p}{\partial x_1 \partial x_2} + \frac{\partial \tau_1}{\partial x_2} \cdot \frac{\partial p}{\partial x_1} + \frac{\partial \tau_2}{\partial x_2} \cdot \frac{\partial p}{\partial x_2} \\ &= \frac{\partial^2 p}{\partial x_1 \partial x_2}. \end{aligned} \tag{5.4.6}$$

Notice that

$$\frac{\partial n}{\partial \tau}(\bar{x}) = (k(\bar{x}), 0),$$

where $k(\bar{x})$ is the curvature of ∂M at $\bar{x}$. Hence, we have

$$\begin{aligned} \frac{\partial}{\partial \tau} \frac{\partial p}{\partial n} &= \frac{\partial^2 p}{\partial x_1 \partial x_2} + \frac{\partial n_1}{\partial x_1} \cdot \frac{\partial p}{\partial x_1} + \frac{\partial n_2}{\partial x_1} \cdot \frac{\partial p}{\partial x_2} \\ &= \frac{\partial^2 p}{\partial x_1 \partial x_2} + \frac{\partial n_1}{\partial \tau} \cdot \frac{\partial p}{\partial \tau} + \frac{\partial n_2}{\partial \tau} \cdot \frac{\partial p}{\partial n} \\ &= \frac{\partial}{\partial n} \frac{\partial p}{\partial \tau} - k(\bar{x}) \frac{\partial p}{\partial \tau}. \end{aligned} \tag{5.4.7}$$

Therefore, (5.4.4) implies that

$$\frac{\partial \varphi_\tau(\bar{x})}{\partial n} = -\int_0^{t_0} \left[R^{-1} \frac{\partial}{\partial n} (\Delta u \cdot \tau) - \frac{\partial}{\partial \tau} \frac{\partial p}{\partial n} - k(\bar{x}) \frac{\partial p}{\partial \tau} \right] dt. \tag{5.4.8}$$

Finally, the theorem follows from the following observation:

$$\frac{\partial p}{\partial \tau} = R^{-1} \Delta u \cdot \tau, \quad \frac{\partial p}{\partial n} = R^{-1} \Delta u \cdot n, \qquad \text{on } \partial M.$$

The proof is complete. □

REMARK 5.4.2. In equation (5.4.2), n is the outward normal vector, and (τ, n) forms a right-hand frame. Thus, the curvature $k(\bar{x}) > 0$ (resp. $k(\bar{x}) < 0$) implies that the boundary ∂M is convex (resp. concave) at $\bar{x}$.

REMARK 5.4.3. Equation (5.4.2) provides a relationship between the bifurcation point $(\bar{x}, t_0)$ $(\bar{x} \in \partial M,\ t_0 \geq 0)$ and the given data, i.e., the initial value φ, the external forcing f and the Reynolds number R.

It is easy to see that the formula (5.4.2) is also a sufficient condition for $\bar{x}$ being a ∂-singular point of the solution u of (5.4.1). As seen in the proof of the theorem, a singular point with integer index must be a bifurcation point. Thus, we immediately derive a sufficient condition of structural bifurcation of solutions for the Navier-Stokes equations with the Dirichlet boundary conditions (5.4.1) as follows.

THEOREM 5.4.4. *Assume that* $\frac{\partial \varphi_\tau}{\partial n} \neq 0$ *on* Γ, *where* $\Gamma \subset \partial M$ *is a connected component of* ∂M. *If* $(\bar{x}, t_0)$ *with* $\bar{x} \in \Gamma$, $t_0 > 0$ *satisfies formula (5.4.2) and*

$$
\begin{cases} \dfrac{\partial u_\tau(x,t)}{\partial n} \neq 0 \quad \forall\, x \in \Gamma \text{ and } 0 \leq t < t_0, \\ \left[R^{-1}(\nabla \times \Delta u + k\Delta u \cdot \tau) + \lambda(\nabla \times f + k f_\tau)\right](\bar{x}, t_0) \neq 0, \end{cases} \tag{5.4.9}
$$

then $(\bar{x}, t_0)$ *must be a bifurcation point of the solution* u *of (5.4.1) in its global structure.*

REMARK 5.4.5. If we assume $\Gamma \subset \partial M$ is an open (connected) portion of ∂M, then $(\bar{x}, t_0)$ must be a bifurcation point of the solution u of (5.4.1) in its *local* structure.

REMARK 5.4.6. An important question is to establish an asymptotic relationship between the bifurcation time t_0 and the Reynolds number R for the Navier-Stokes equations (5.4.1). In the case where $f = 0$, based on (5.4.2), we conjecture that

$$
t_0 \sim cR, \tag{5.4.10}
$$

where c is a constant independent of the Reynolds number. To prove this conjecture, we need to derive some *a priori* estimates for the solutions of the Navier-Stokes equations in terms of the Reynolds number R.

5.4.2. Structural bifurcation driven by forcing. In this section, we study structural bifurcation of the solutions of the Navier-Stokes equations (5.4.1) with the (homogeneous) Dirichlet boundary conditions generated by the initial data and forcing.

For this purpose, let $\varphi \in B_0^{2+\alpha}(TM)$, let $f \in D^\alpha(TM)$ $(0 < \alpha < 1)$, and let $u \in C^1([0,\infty), B_0^{2+\alpha}(TM))$ be a solution of (5.4.1).

THEOREM 5.4.7. *Let* $\Gamma \subset \partial M$ *be a connected component of* ∂M, *and*

$$
\frac{\partial f_\tau}{\partial n} \cdot \frac{\partial \varphi_\tau}{\partial n} < 0, \qquad \text{on } \Gamma.
$$

Then, there is a constant $\lambda_0 > 0$, *depending on* $\|f\|_{C^1}, \|\varphi\|_{C^3}$ *and the Reynolds number* R, *such that for any* $\lambda > \lambda_0$, u *has a bifurcation at* $(\bar{x}, t_0)$ *in its global structure with* $t_0 > 0$ *and* $\bar{x} \in \Gamma$. *Consequently,* $\bar{x}$ *is a* ∂*-singular point of* u^0.

Furthermore, if $\bar{x}$ *is an isolated* ∂*-singular point of* $u(\cdot, t_0)$, *then the number of orbits connected to* $\bar{x}$ *in* $\overset{\circ}{M}$ *is even.*

PROOF. For simplicity, let the orientation of φ be the same as the tangent vector τ on Γ. The solution u of (5.4.1) can be Taylor expanded at $t = 0$ as follows:

$$u(x,t) = \varphi + t\lambda f - t(\varphi \cdot \nabla)\varphi + tR^{-1}\Delta\varphi - t\nabla p_0 + g(x,t,\lambda). \tag{5.4.11}$$

Here g is given by [1]

$$g(x,t;\lambda) = \sum_{k=2}^{\infty} \frac{t^k}{k!} \frac{\partial^k u(x,0,\lambda)}{\partial t^k}.$$

By definition, we have

$$\frac{\partial u(x,0,\lambda)}{\partial t} = \lambda f - (\varphi \cdot \nabla)\varphi + R^{-1}\Delta\varphi - \nabla p_0,$$

which implies that

$$\frac{\partial u(\bar{x},0,\lambda)}{\partial t} = c_1\lambda + c_0,$$

for some constants c_0 and c_1. By induction, it is easy to see that $\frac{\partial^k u(\bar{x},0,\lambda)}{\partial t^k}$ is a polynomial of λ of degree $k-1$ for $k \geq 2$. Hence, we obtain that

$$|g(\bar{x},t,\lambda)| \leq t \sum_{k=2}^{\infty} \frac{(t\lambda)^{k-1} C_k}{k!},$$

which implies that for any constant K, we have

$$g(\bar{x},t,\lambda) \longrightarrow 0 \quad \text{as } t\lambda = K, t \to 0, \lambda \to \infty. \tag{5.4.12}$$

By assumption, from (5.4.11), we derive that on Γ,

$$\begin{aligned} \frac{\partial u}{\partial n} \cdot \tau = &\left|\frac{\partial \varphi_\tau}{\partial n}\right| - t\lambda \left|\frac{\partial f_\tau}{\partial n}\right| \\ &+ t\frac{\partial}{\partial n}\left[(\varphi \cdot \nabla)\varphi - R^{-1}\Delta\varphi + \nabla p_0\right] \cdot \tau + \frac{\partial g}{\partial n} \cdot \tau. \end{aligned} \tag{5.4.13}$$

Since $u|_{\partial M} = 0$ and div $u = 0$, we have

$$\frac{\partial u}{\partial n} \cdot n = \frac{\partial u_n}{\partial n} = -\frac{\partial u_\tau}{\partial \tau} = 0.$$

Hence, a zero point of $\partial u_\tau / \partial n$ on Γ is a ∂-singular point of u. We infer from (5.4.12) and (5.4.13) that there exists a $\lambda_0 > 0$ depending on the modules $\|f\|_{C^1}$, $\|\varphi\|_{C^3}$ and the Reynolds number R, such that, for any $\lambda > \lambda_0$, there is $t_0 > 0$ satisfying

$$\begin{cases} \dfrac{\partial u_\tau(x,t)}{\partial n} > 0, & \forall\, 0 \leq t < t_0 \text{ and } x \in \Gamma, \\ \dfrac{\partial u_\tau(\bar{x},t_0)}{\partial n} = 0, & \text{for some } \bar{x} \in \Gamma, \end{cases} \tag{5.4.14}$$

and

$$\begin{aligned} &\frac{\partial}{\partial t} \frac{\partial u_\tau(\bar{x},t_0)}{\partial n} \\ &= -\lambda \left|\frac{\partial f_\tau(\bar{x})}{\partial n}\right| + \frac{\partial}{\partial n}\left[(\varphi \cdot \nabla)\varphi - R^{-1}\Delta\varphi + \nabla p_0 - g'(\bar{x},t_0,\lambda)\right] \cdot \tau \\ &< 0. \end{aligned} \tag{5.4.15}$$

---

[1] This Taylor expansion is ensured by the time-analyticity of u for positive time. Hence, without loss of generality, we can assume u is also time-analytic at $t = 0$; otherwise, we need only to consider the expansion at $t = \delta$ for an arbitrary small $\delta > 0$.

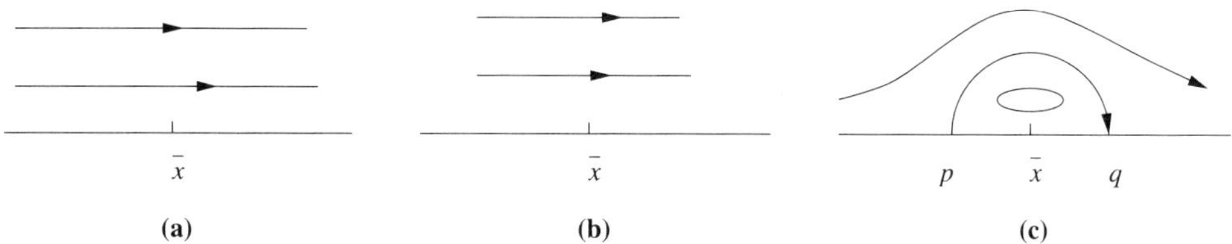

FIGURE 5.5.1. Boundary layer separation of shear flow.

Then by Theorem 5.3.13 and (5.4.14) and (5.4.15), it is easy to show that t_0 is a structural bifurcation point in its global structure.

By the homotopy invariance of index sum of vector fields in an open set, as $\bar{x} \in \Gamma$ is an isolated singular point of $u(\cdot, t_0)$, we have

$$\text{ind}(u(\cdot, t_0), \bar{x}) = \sum_i \text{ind}(u(\cdot, t_0 - \varepsilon), z_i), \quad z_i \in U$$

for any $\varepsilon > 0$ sufficiently small, where $U \subset M$ is a neighborhood of $\bar{x} \in \Gamma$ and $u(\cdot, t_0)$ has no singular point in $U \cap \overset{\circ}{M}$. Since $u(\cdot, t_0 - \varepsilon)$ has no boundary singular points in Γ, the index sum of $\sum_i \text{ind}(u(\cdot, t_0 - \varepsilon), z_i)$ must be an integer.

The proof is complete. □

5.5. Boundary Layer Separations of Incompressible Flows II

5.5.1. Structural bifurcation with integer indices. By Theorem 5.1.5, the index $\text{ind}(u, \bar{x}) = -n/2$ for $u \in D^r(TM)$ and $\bar{x} \in \partial M$ implies that there are n orbits of u in $\overset{\circ}{M}$ connected to $\bar{x}$. For a vector field $u \in B_0^r(TM)$ and a ∂-singular point $\bar{x} \in \partial M$, we define the index

$$\text{ind}(u, \bar{x}) = -\frac{n}{2}, \quad n \geq 0 \text{ an integer}, \, \bar{x} \in \partial M, \, u \in B_0^r(TM),$$

which amounts to saying that there are n orbits of u in $\overset{\circ}{M}$ connected to $\bar{x} \in \partial M$.

Let $u \in C^1([0, T], B_0^r(TM))$ be a solution of (5.4.1). From the structural stability theorems and the structural bifurcation theorems, we know that $(\bar{x}, t_0)$ $(\bar{x} \in \partial M)$ being a bifurcation point of u implies that $\text{ind}(u(\cdot, t_0), \bar{x}) \neq -\frac{1}{2}$. In fluid dynamics, the structural bifurcation often occurs in the case with integer index as demonstrated in the theorems in the previous sections. Actually, in the real world, we can only observe the structural bifurcation of fluid flows in which the index is either 0 or -1. The structural bifurcation from index zero is well known as the boundary layer separation. We proceed now with some schematic pictures characterizing the structural bifurcation when the index is either 0 or -1.

First, we consider the case where $\text{ind}(u(\cdot, t_0), \bar{x}) = 0$. In this case, the bifurcation occurs as shown in Figure 5.5.1. Figure 5.5.1(a) shows the typical shear flow, and u has no singular point near $\bar{x}$ for $t < t_0$. The flow pattern for $u^0 = u(\cdot, t_0)$ given in Figure 5.5.1(b) has an isolated degenerate ∂-singular point $\bar{x} \in \partial M$. When $t > t_0$, the flow pattern given by Figure 5.5.1(c) illustrates that $u(x, t)$ bifurcates from $\bar{x} \in \partial M$ one vortex or multi-vortices. Although solutions with either a single vortex or multiple vortices may occur, the case given by Figure 5.5.1(c) with one vortex is generic.

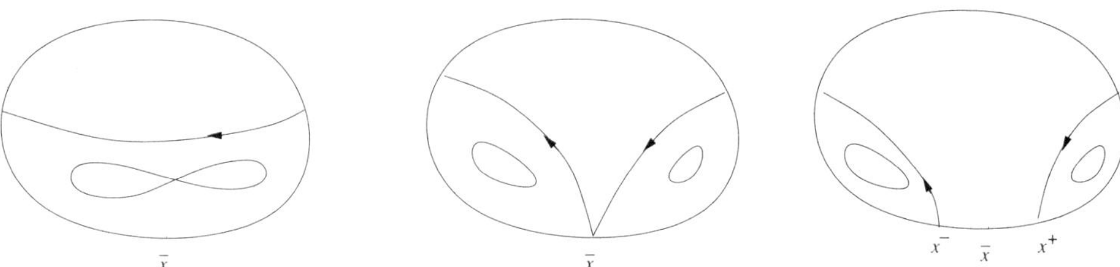

FIGURE 5.5.2. A type of structural bifurcation for the case of $\text{ind}(u(\cdot, t_0), \bar{x}) = -1$.

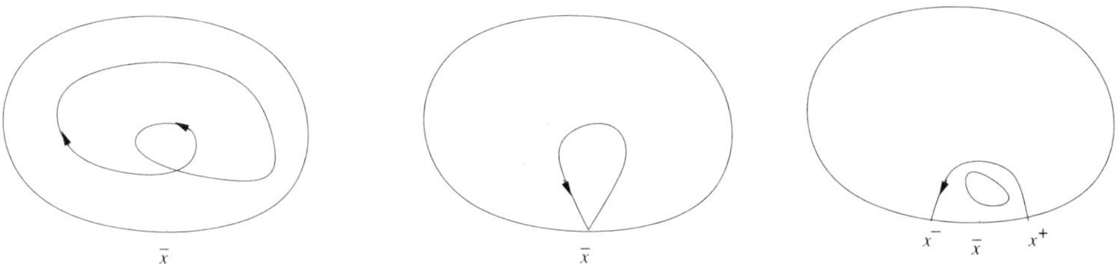

FIGURE 5.5.3. Another type of structural bifurcation for the case of $\text{ind}(u(\cdot, t_0), \bar{x}) = -1$.

For the case where $\text{ind}(u(\cdot, t_0), \bar{x}) = -1$, there are two types of flow transitions, which are shown respectively in Figure 5.5.2 and Figure 5.5.3.

The above two cases can be classified in the following theorem.

THEOREM 5.5.1. *Let $u \in C^1([0, T), X)$, $\bar{x} \in \partial M$ be a singular (or ∂-singular) point of u, and $0 < t_0 < T$. We have the following assertions:*

(1) *If $\text{ind}(u(\cdot, t_0), \bar{x}) = 0$, the type of structural bifurcation of u at $(\bar{x}, t_0)$ is unique and is called the boundary layer separation, which is shown as in Figure 5.5.1. In other words, there are some vortices bifurcated from $\bar{x} \in \partial M$, which are enclosed by orbit lines connecting to both bifurcated boundary saddle points x^- and x^+.*
(2) *If $\text{ind}(u(\cdot, t_0), \bar{x}) = -1$, there are only two types of structural bifurcation of u at $(\bar{x}, t_0)$, which are respectively shown as in Figure 5.5.2 and Figure 5.5.3.*

REMARK 5.5.2. The types of structural bifurcation mentioned in the above theorem are defined in terms of connecting the two bifurcated boundary saddle points, not in the sense of topological equivalence.

The proof of the above theorem is a direct application of the stability theorem of orbit lines, Theorem 2.1.15 in Section 2.1.

5.5.2. Determination of boundary layer separation. In the bifurcation theorems in Section 2, structural bifurcation of solutions of the Navier-Stokes equations occurs at a degenerate singular point with integer index for the velocity field at the critical time instant t_0. In this subsection, we prove necessary and sufficient kinematic conditions for the boundary layer separation case with zero index.

Let $X = D^r(TM)$ or $B^r(TM)$ or $B^r_0(TM)$, let $u \in C^1([0, T], X)$, and let $\bar{x} \in \partial M$ be a degenerate singular (or ∂-singular) point of u at $t = t_0$ $(0 < t_0 <$

T). Let $u^0(x) = u(x, t_0)$ and $u^1(x) = \partial u(x, t_0)/\partial t$. We start with the following assumptions:

$$\frac{\partial^2 u_\tau^0(\bar{x})}{\partial \tau^2} \neq 0, \quad \frac{\partial u_\tau^0(\bar{x})}{\partial n} \neq 0, \quad u_\tau^1(\bar{x}) \neq 0, \text{ if } X = D^r(TM), \tag{5.5.1}$$

$$\frac{\partial^2 u_\tau^0(\bar{x})}{\partial \tau^2} \neq 0, \quad \frac{\partial^2 u_\tau^0(\bar{x})}{\partial n^2} \neq 0, \quad u_\tau^1(\bar{x}) \neq 0, \text{ if } X = B^r(TM), \tag{5.5.2}$$

$$\frac{\partial^3 u_\tau^0(\bar{x})}{\partial \tau^2 \partial n} \neq 0, \quad \frac{\partial^2 u_\tau^0(\bar{x})}{\partial n^2} \neq 0, \quad \frac{\partial u_\tau^1(\bar{x})}{\partial n} \neq 0, \text{ if } X = B_0^r(TM). \tag{5.5.3}$$

REMARK 5.5.3. By assumption, $\bar{x} \in \partial M$ is a degenerate singular (or ∂-singular) point of $u \in C^1([0,T], X)$ at $t = 0$. Thus, we have

$$\frac{\partial u_\tau^0(\bar{x})}{\partial \tau} = 0 \qquad \left(\text{or } \frac{\partial^2 u_\tau^0(\bar{x})}{\partial \tau \partial n} = 0\right).$$

Hence, the above conditions are generic for a bifurcation point $(\bar{x}, t_0)$ of $u \in C^1([0,T], X)$ with $(\bar{x}) \in \partial M$ and $0 < t_0 < T$.

The following theorem determines the type of structural bifurcations.

THEOREM 5.5.4. *Let $u \in C^1([0,T], X)$, let n be an inward normal vector and let (τ, n) form a right-hand frame. Let the corresponding condition of (5.5.1), or (5.5.2), or (5.5.3) hold true. Then we have the assertions:*

(1) *When the following corresponding condition is satisfied*

$$\begin{cases} \mathit{Sign}\, \frac{\partial^2}{\partial \tau^2} u_\tau^0(\bar{x}) = \mathit{Sign}\, \frac{\partial}{\partial n} u_\tau^0(\bar{x}), & \text{if } X = D^r(TM), \\ \mathit{Sign}\, \frac{\partial^2}{\partial \tau^2} u_\tau^0(\bar{x}) = \mathit{Sign}\, \frac{\partial^2}{\partial n^2} u_\tau^0(\bar{x}), & \text{if } X = B^r(TM), \\ \mathit{Sign}\, \frac{\partial^2}{\partial \tau^2} \frac{\partial u_\tau^0(\bar{x})}{\partial n} = \mathit{Sign}\, \frac{\partial^2 u_\tau(\bar{x})}{\partial n^2}, & \text{if } X = B_0^r(TM), \end{cases} \tag{5.5.4}$$

then the structural bifurcation of u at $(\bar{x}, t_0)$ is of the type $ind(u^0, \bar{x}) = 0$, i.e., the boundary layer separation, and the vortex separated from $\bar{x} \in \partial M$ is unique as shown in Figure 5.5.1(c).

(2) *When the corresponding condition in (5.5.4) is not valid, then the structural bifurcation of u is of the type $ind(u^0, \bar{x}) = -1$; the portrait of flows is as shown in Figure 5.5.2 or Figure 5.5.3.*

PROOF. We prove only the cases where the vector fields $u \in C^1([0,T], D^r(TM))$ or $u \in C^1([0,T], B^r(TM))$; the third case can be proved in the same fashion.

Without loss of generality, we assume that $\bar{x} \in \partial M$ has a flat neighborhood $\Gamma \subset \partial M$. We take a coordinate system (x_1, x_2) with $\bar{x}$ at the origin, $\Gamma = \{(x_1, 0) \mid |x_1| < \delta\}$, the x_2-axis oriented toward the inward normal direction. By assumption,

$$\frac{\partial u_\tau}{\partial \tau} = \frac{\partial u_1}{\partial x_1}, \qquad \frac{\partial u_\tau}{\partial n} = \frac{\partial u_1}{\partial x_2},$$

and so on.

1. The case of $u \in C^1([0,T], D^r(TM))$. From (5.5.1), the vector fields u^0 and u^1 have the Taylor expansion as follows:

$$\begin{cases} u_1^0(x) = C_1 x_2 + C_2 x_1^2 + C_3 x_1 x_2 + o(|x|^2), \\ u_2^0(x) = -2C_2 x_1 x_2 + x_2 \cdot o(|x|), \end{cases} \tag{5.5.5}$$

with $C_1, C_2 \neq 0$, and

$$\begin{cases} u_1^1(x) = \beta + O(|x|), \\ u_2^1(x) = x_2 \cdot O(|x|), \end{cases} \tag{5.5.6}$$

with $\beta \neq 0$. Without loss of generality, we assume that $C_2 > 0$ and $\beta < 0$. Obviously, $u(x, t_0 - \varepsilon) = u^0(x) - \varepsilon u^1(x) + o(|\varepsilon|)$ has no singular point on $\Gamma = \{(x_1, 0) \mid |x_1| < \delta\}$ for $\delta > 0$ and $\varepsilon > 0$ sufficiently small, and $u(x, t_0 + \varepsilon) = u^0(x) + \varepsilon u^1(x) + o(|\varepsilon|)$ has exactly two singular points on Γ near $\bar{x}$ $(x = 0)$. Hence, we have that $\text{ind}(u^0, \bar{x}) = \text{integer}$.

When $\text{ind}(u^0, \bar{x}) = -n$, there are exactly $2n$ orbits of u^0 in $\overset{\circ}{M}$ connecting to $\bar{x}$. Hence, $u_1^0(x)$ has at least $2n$ zero points for each $x_2 > 0$ sufficiently small because the sign of $u_1^0(x)$ changes near $\bar{x}$ $(x = 0)$ at least n times. Hence, as $\text{Sign}\, C_1 = \text{Sign}\, C_2$, i.e., $C_1 > 0$, from (5.5.5) we infer that $u_1^0(x)$ has no zero point near $x_1 = 0$ for any $x_2 > 0$ sufficiently small. Thus, we verify that $\text{ind}(u^0, \bar{x}) = 0$.

From (5.5.5) and (5.5.6), we see that the interior singular points $(\widetilde{x}_1, \widetilde{x}_2)$ of $u(x, t_0 + \varepsilon)$ with $\widetilde{x}_2 > 0$ satisfies the equation

$$\begin{cases} C_1 x_2 + C_2 x_1^2 + C_3 x_1 x_2 + \varepsilon\beta + o(|\varepsilon|, |x_2|, |x_1|^2) = 0, \\ -2C_2 x_1 + \varepsilon \cdot O(|x|) + o(|\varepsilon|, |x|) = 0. \end{cases} \tag{5.5.7}$$

From the implicit function theorem, it follows that the solution $(\widetilde{x}_1, \widetilde{x}_2)$ of (5.5.7) with $\widetilde{x}_2 > 0$ sufficiently small, if it exists, is unique for any $\varepsilon > 0$ sufficiently small. And the existence of (5.5.7) can be derived by the invariance of the index sum in a neighborhood of $\bar{x}$ and Theorem 5.2.6. Thus, the first conclusion is proven.

When $C_1 < 0$, it is easy to see that $u_1^0(x) = 0$, i.e., the equation below has exactly two solutions near $x_1 = 0$ for any $x_2 > 0$ sufficiently small:

$$C_2 x_1^2 + C_1 x_2 + C_3 x_1 x_2 + o(|x|^2) = 0, \quad C_2 > 0, C_1 < 0.$$

Moreover, the domain

$$D_\varepsilon = \{x \mid u_1^0(x) < 0, |x| < \varepsilon\} \neq \phi, \quad \forall\, \varepsilon > 0 \text{ sufficiently small.}$$

It implies that there are exactly two orbits of u^0 in $\overset{\circ}{M}$ connecting to $\bar{x}$ $(x = 0)$. Hence, $\text{ind}(u^0, \bar{x}) = -1$. The conclusion for $u \in C^1([0, T], D^r(TM))$ is proven.

2. The case of $u \in C^1([0, T], B^r(TM))$. By (5.5.2), we obtain that

$$\begin{cases} u_1^0(x) = \alpha_1 x_1^2 + \alpha_2 x_2^2 + o(|x|^2), \\ u_2^0(x) = -2\alpha_1 x_1 x_2 + x_2 \cdot o(|x|), \end{cases} \tag{5.5.8}$$

and

$$\begin{cases} u_1^1(x) = \beta_0 + \beta_1 x_1 + O(|x|), \\ u_2^1(x) = -\beta_1 x_2 + x_2 \cdot O(|x|). \end{cases} \tag{5.5.9}$$

Assume that $\alpha_1 < 0$, $\beta_0 < 0$. In the same fashion as above, we can verify that

$$\text{ind}(u^0, \bar{x}) = \begin{cases} 0 & \text{if } \alpha_2 > 0, \\ -1 & \text{if } \alpha_2 < 0. \end{cases}$$

We only have to prove that the equations below have only one solution near $x = 0$ for any $\varepsilon > 0$ sufficiently small:

$$
\begin{cases}
\alpha_1 x_1^2 + \alpha_2 x_2^2 + \varepsilon\beta_0 + o(|\varepsilon|, |x|^3) = 0, \\
-2\alpha_1 x_1 - \varepsilon\beta_1 + \varepsilon \cdot O(|x|) + O(|x|^2) = 0, \\
x_2 > 0,\ \alpha_1 > 0,\ \alpha_2 > 0,\ \beta < 0.
\end{cases} \tag{5.5.10}
$$

Equation (5.5.10) is equivalent to the equation

$$
\begin{cases}
\alpha_2 x_2^2 + \varepsilon\beta_0 + \alpha_1 g^2(x_2) + o(|\varepsilon|, |x_2|^3) = 0, \\
g(x_2) = -\dfrac{\beta_1}{2\alpha_1}\varepsilon + \varepsilon \cdot O(|x_2|) + O(|x_2|^2), \\
x_2 > 0,\ \alpha_2 > 0,\ \beta < 0.
\end{cases} \tag{5.5.11}
$$

The existence and uniqueness of solutions of (5.5.11) is obvious. Thus, this theorem is proven. □

5.5.3. Adverse pressure gradient. In this subsection, we show the existence of the adverse pressure gradient at the neighborhood of the bifurcation point $(\bar{x}, t_0)$ by utilizing the relation between the vorticity and pressure in the solutions of the Navier Stokes equation and applying the Hopf Lemma for the subharmonic function.

Consider the 2D Navier-Stokes equations (5.4.1). For simplicity, we take the external forcing $f = 0$. Namely, we consider

$$
\begin{cases}
\dfrac{\partial u}{\partial t} + (u \cdot \nabla)u - R^{-1}\Delta u + \nabla p = 0, \\
\operatorname{div} u = 0, \\
u|_{\partial M} = 0, \\
u(x, 0) = \varphi(x),
\end{cases} \tag{5.5.12}
$$

with $\varphi|_{\partial M} = 0$.

In the vorticity formulation, the first equation in (5.5.12) can be rewritten as

$$
\frac{\partial \omega}{\partial t} + (u \cdot \nabla)\omega - R^{-1}\Delta\omega = 0, \tag{5.5.13}
$$

where the vorticity function ω is defined by

$$
\omega = -\frac{\partial u_1}{\partial x_2} + \frac{\partial u_2}{\partial x_1}.
$$

For simplicity of presentation and for comparison with numerical results in Section 6.1, we assume that a section of the vertical line $x = 1$, denoted by Γ_0, is included in ∂M with its normal direction going rightward and the possible separation point $\bar{x}$ lies on the boundary section Γ_0. The unit normal and tangential vectors on Γ_0 are given as $n = (1, 0)$, $\tau = (0, 1)$, respectively, which determines the normal and tangential derivatives to be $\frac{\partial}{\partial n} = \partial_{x_1}$, $\frac{\partial}{\partial \tau} = \partial_{x_2}$.

Regarding the boundary layer, the pure shear flow was presented around the boundary section Γ_0 before the critical time t_0. Without loss of generality, and for the sake of the consistency with the discussion of the numerical example in Section 6.1, it is assumed that such shear flow is downward. Mathematically speaking, this amounts to assuming that there exists a neighborhood δ_1 of $\bar{x}$ such that

$$
\omega \geq 0, \quad \forall x \in \delta_1, \quad t \leq t_0. \tag{5.5.14}
$$

The following lemma makes connections between the main assumption, Assumption 5.3.1, for structure bifurcation and the so-called vorticity "crisis" conditions. The proof of the lemma is trivial, and we omit the details.

LEMMA 5.5.5. (1) *Assumption 5.3.1 with $k = 2$ for u is equivalent to the following conditions on the vorticity ω:*

$$\omega = 0\,, \quad \frac{\partial \omega}{\partial \tau} = 0, \quad \frac{\partial^2 \omega}{\partial \tau^2} \neq 0, \quad \frac{\partial \omega}{\partial t} \neq 0 \qquad at\ (\bar{x}, t_0). \tag{5.5.15}$$

(2) *If (5.5.13) holds true, i.e., only downward shear flow is present around Γ^0 for $t \leq t_0$, then Assumption 5.3.1 with $k = 2$ is equivalent to the following:*

$$\omega = 0\,, \quad \frac{\partial \omega}{\partial \tau} = 0, \quad \frac{\partial^2 \omega}{\partial \tau^2} > 0, \quad \frac{\partial \omega}{\partial t} < 0 \qquad at\ (\bar{x}, t_0). \tag{5.5.16}$$

The main result of this section is the following theorem, showing the adverse pressure gradient at the separation point.

THEOREM 5.5.6 (Vorticity Criss and Adverse Pressure Gradient). *Let (u, p) be the solution of the 2D Navier-Stokes equations (5.5.12). Assume that there exist a boundary point $\bar{x} \in \partial M$ and a critical time t_0 such that (5.5.14) and (5.5.16) hold true. Then we have:*

(1) *u has structural bifurcation in its local structure and boundary layer separation at the boundary point $\bar{x}$ as t crosses t_0. Meanwhile, the separation process is shown exactly in Figure 5.5.1.*
(2) *An adverse pressure gradient in the tangent direction is present at $\bar{x}$, i.e.,*

$$\frac{\partial p}{\partial \tau} = \frac{\partial p}{\partial y} < 0 \qquad at \quad (\bar{x}, t_0). \tag{5.5.17}$$

PROOF. It suffices to prove Assertion (2). By (5.5.13), we have

$$\triangle \omega = R\Big(\partial_t \omega + (u \cdot \nabla)\omega\Big) = R\partial_t \omega < 0 \qquad \text{at } (\bar{x}, t_0), \tag{5.5.18}$$

where the first step is based on the no-penetration, no-slip boundary condition for the velocity field. Thus, we arrive at the conclusion that the vorticity field keeps rigorously subharmonic in a neighborhood δ_2 of $\bar{x}$:

$$\triangle \omega < 0\,, \qquad \text{in} \qquad \delta_2. \tag{5.5.19}$$

Let $\partial^* = \delta_1 \cap \delta_2$; then the strong maximum principle for elliptic equations implies that

$$\frac{\partial \omega}{\partial n}(\bar{x}, t_0) < 0. \tag{5.5.20}$$

On the other hand, it is easy to see that

$$\triangle u \cdot \tau = \frac{\partial \omega}{\partial n}, \quad \text{on } \Gamma^0,$$

which implies that

$$\frac{\partial p}{\partial \tau} = \frac{\partial p}{\partial y} = R^{-1}\frac{\partial \omega}{\partial n}, \qquad \text{on } \Gamma^0.$$

The proof is complete. □

5.6. Structural Bifurcation near Interior Singular Points

5.6.1. Characterization of degenerate singularities with nonzero Jacobian. The structural stability theorem, Theorem 2.1.2, suggests studying the structure of divergence-free vector fields near degenerate singular points. To this end, we now introduce some lemmas characterizing degenerate interior singularities with nonzero Jacobians that are useful for studying interior structural bifurcation.

LEMMA 5.6.1. *Let $u \in D^r(TM)$ $(r \geq 1)$, and let $x_0 \in \overset{\circ}{M}$ be an isolated singular point of u. If the index $ind(u, x_0) \neq 1, 0, -1$, then the Jacobian matrix*

$$Du(x_0) = 0. \tag{5.6.1}$$

PROOF. By Theorem 5.1.5, the index of an interior singular point of a divergence-free vector field is determined by the $2n$ $(n \geq 0)$ orbits connected to x_0, i.e., $\text{ind}(u, x_0) = 1 - n$. Hence, by assumption, $n \geq 3$.

Let γ be an orbit of u connected to x_0. Let (x_1, x_2) be the orthogonal coordinate system with x_0 as its origin, and with its x_1-axis tangent to γ at x_0. Then u can be expressed locally by

$$\begin{cases} u(x) = \begin{pmatrix} a & b \\ c & -a \end{pmatrix} \begin{pmatrix} x_1 \\ x_2 \end{pmatrix} + o(|x|), \\ Du(x_0) = Du(0) = \begin{pmatrix} a & b \\ c & -a \end{pmatrix}. \end{cases} \tag{5.6.2}$$

We shall prove that $a = b = c = 0$ in several steps as follows.

Step 1. We show that $a = c = 0$. By definition, the x_1-axis is tangent to γ at $x_0 = 0$, which yields

$$\lim_{\substack{x \in \gamma \\ x \to 0}} \frac{u_2(x)}{u_1(x)} = 0. \tag{5.6.3}$$

In addition, for $(x_1, x_2) \in \gamma$, $x_2 = o(|x_1|)$. Hence, we infer from (5.6.2) and (5.6.3) that

$$\begin{aligned} \lim_{\substack{x \in \gamma \\ x \to 0}} \frac{u_2(x)}{u_1(x)} &= \lim_{\substack{x \in \gamma \\ x \to 0}} \frac{cx_1 - ax_2 + o(|x|)}{ax_1 + bx_2 + o(|x|)} \\ &= \lim_{\substack{x \in \gamma \\ x_1 \to 0}} \frac{cx_1 + o(|x_1|)}{ax_1 + o(|x_1|)} \\ &= \frac{c}{a} \\ &= 0. \end{aligned}$$

Hence, $c = 0$. By $\text{ind}(u, x_0) \neq 1, -1$, the singular point x_0 of u is degenerate. Therefore $c = 0$ implies that $a = 0$.

Hence, when $\text{ind}(u, x_0) \neq 1, -1$, u can be expressed near $x_0 = 0$ as

$$\begin{cases} u_1(x) = bx_2 + o(|x|), \\ u_2(x) = o(|x|). \end{cases} \tag{5.6.4}$$

Step 2. Consider the case when there is another orbit γ_1 of u connected to x_0, and the angle between γ_1 and γ is θ different from 0 and π. Then, by (5.6.4), we

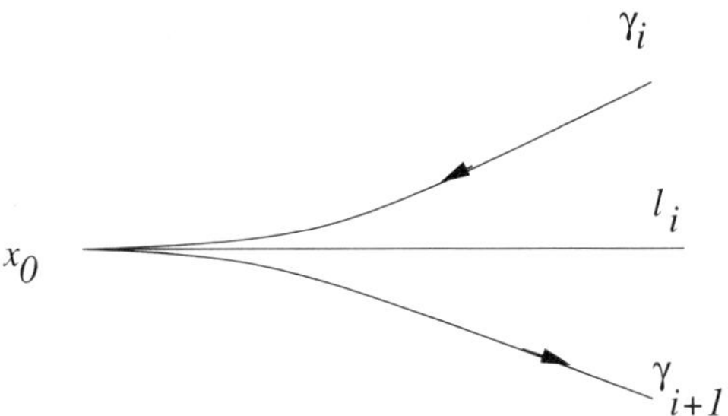

FIGURE 5.6.1. Schematic illustrating the proof of Lemma 5.6.1.

deduce that

$$\lim_{\substack{x\in\gamma_1 \\ x\to 0}} \frac{u_2(x)}{u_1(x)} = \lim_{\substack{x\in\gamma_1 \\ x\to 0}} \frac{o(|x|)}{bx_2 + o(|x|)} = \tan\theta \neq 0,$$

which yields that $b = 0$. Hence, (5.6.1) holds true in this case.

Step 3. Consider the case when all orbits connected to x_0 are tangent to γ at x_0. Let $O \in M$ be a sufficiently small neighborhood of x_0, let F_i $(1 \leq i \leq 2n)$ be the domains in O enclosed by the orbits connected to x_0, and let θ_i be the angle of the boundary of F_i at x_0. It is easy to see that in each F_i with $\theta_i = 0$, there exists at least a curvilinear segment ℓ_i, with x_0 being its end point, such that $u_1(x) = 0$, $x \in \ell_i$; see Figure 5.6.1. Hence, there are at least $2(n-1)$ curvilinear segments in O with x_0 as their common end point where $u_1 = 0$. On the other hand, by the implicit function theorem, if $b \neq 0$ in (5.6.4), then there is a unique curve $L \subset O$ with $x_0 \in L$ such that $u_1(x) = 0$, $x \in L$, i.e., there are only two line segments $L = \ell_1 \cup \ell_2$ in O along which $u_1 = 0$; hence, if $n \geq 3$, it follows that $b = 0$ and (5.6.1) holds true.

This completes the proof of this lemma. □

LEMMA 5.6.2. *Let $u \in D^r(TM)$ $(r \geq 1)$, and let $x_0 \in \overset{\circ}{M}$ be an isolated singular point of u. If the index $ind(u, x_0) = 0$, and the angle θ between the two orbits connected to x_0 is different from 0, then (5.6.1) holds true.*

PROOF. By Step 2 in the proof of Lemma 5.6.1, it suffices to prove (5.6.1) when $\theta = \pi$. In this case, the two orbits γ_1 and γ_2 connected to x_0 form a curve Γ with the x_1-axis tangent to Γ at x_0. By Theorem 5.1.5, it is obvious that for any $x_2 > 0$ sufficiently small, we have

$$\operatorname{Sign} u_1(0, x_2) = \operatorname{Sign} u_1(0, -x_2),$$

which, together with (5.6.4), yields that $b = 0$. This proof is complete. □

From Lemma 5.6.1, we see that a degenerate singular point $x_0 \in \overset{\circ}{M}$ of $u \in D^r(TM)$ $(r \geq 1)$ with non-zero Jacobian $Du(x_0) \neq 0$ can only be one of the following three cases:

(1) a degenerate center,
(2) a degenerate saddle such that the 4 orbits connected to x_0 are tangent to each other at x_0, and
(3) a point with $\operatorname{ind}(u, x_0) = 0$ such that the angle between the two orbits connected to x_0 is zero.

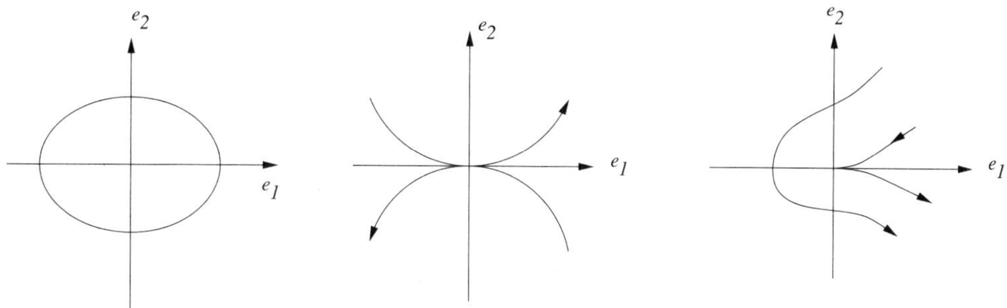

FIGURE 5.6.2. (a) The case with index 1, (b) the case with index −1, and (c) the case with index 0.

We now take a further examination of these cases. Let $x_0 \in \overset{\circ}{M}$ be an isolated degenerate singular point of $u \in D^r(TM)$ $(r \geq 1)$ with non-zero Jacobian: $Du(x_0) \neq 0$. Since $Du(x_0)$ is a degenerate matrix, $Du(x)$ has an eigenvector e_1 satisfying

$$Du(x_0)e_1 = 0, \qquad |e_1| = 1. \tag{5.6.5}$$

Let e_2 be a unit vector, orthogonal to e_1, and satisfies that

$$Du(x_0)e_2 = \alpha e_1, \tag{5.6.6}$$

for some constant $\alpha \neq 0$.

For simplicity, we always take the orthogonal coordinate system (x_1, x_2) with the origin at x_0, the x_1-axis and the x_2-axis pointing respectively in the e_1 and e_2 directions. In this case, the matrix $Du(x_0)$ and the vectors e_1, e_2 can be written as follows:

$$\begin{cases} Du(x_0) = Du(0) = \begin{pmatrix} 0 & \alpha \\ 0 & 0 \end{pmatrix}, \\ e_1 = (1,0), \ \ e_2 = (0,1), \end{cases} \tag{5.6.7}$$

where $\alpha \neq 0$. Geometrically, e_1 and e_2 can be illustrated as in Figure 5.6.2.

5.6.2. Index and kinematic conditions. We now make connections between the index of u at x_0 and different orders of u in its Taylor expansion near x_0. Let $x_0 \in \overset{\circ}{M}$ be an isolated degenerate singular point of u, and

$$Du(x_0) \neq 0, \tag{5.6.8}$$

$$\frac{\partial^m (u(x_0) \cdot e_2)}{\partial e_1^m} = \begin{cases} 0, & 1 \leq m < n, \\ \neq 0, & m = n. \end{cases} \tag{5.6.9}$$

Under conditions (5.6.8) and (5.6.9), the vector field $u(x)$ has the Taylor expansion, by (5.6.7), as follows:

$$u(x) = \begin{cases} \alpha x_2 + f(x_1) + x_2 g_1(x), \\ \beta x_1^n - x_2 f'(x_1) + x_2^2 g_2(x) + o(|x_1|^n), \end{cases} \tag{5.6.10}$$

where $\alpha, \beta \neq 0$, $f(x_1) = o(|x_1|)$ and $g_i(0) = 0$ $(i = 1, 2)$. Let $k = \deg f$ be defined by

$$\lim_{z \to 0} \frac{f(z)}{z^k} = \lambda \neq 0, \qquad k \leq \infty.$$

LEMMA 5.6.3. *Let $x_0 \in \overset{\circ}{M}$ be an isolated degenerate singular point of u satisfying (5.6.8) and (5.6.9).*

(1) *If $2k > n+1$, then*

$$ind(u, x_0) = \begin{cases} 0 & \text{if } n = \text{even}, \\ -1 & \text{if } n = \text{odd and } \alpha \cdot \beta > 0 \text{ in (5.6.10)}, \\ 1 & \text{if } n = \text{odd and } \alpha \cdot \beta < 0; \end{cases}$$

(2) *If either $2k < n+1$, or $2k = n+1$ and $\alpha\beta \neq -k\lambda^2$, then $ind(u, x_0) = -1$.*

PROOF. PROOF OF ASSERTION (1). Let

$$u_t(x) = \begin{cases} \alpha x_2 + t[f(x_1) + x_2 g_1(x)], \\ \beta x_1^n + t[-x_2 f'(x_1) + x_2^2 g_2(x) + o(|x_1|^n)], \end{cases}$$

where $0 \leq t \leq 1$. Since $2k - 1 > n$, it is easy to see that there exists a neighborhood $U \subset M$ of x_0 $(= 0)$, such that $u_t(x)$ has only a singular point $x = 0$ in U for all $t \in [0, 1]$. By the homotopy invariance of the index, we derive that

$$\text{ind}(u_0, x_0) = \text{ind}(u_1, x_0) = \text{ind}(u, x_0). \tag{5.6.11}$$

In a neighborhood of $x = 0$, orbits of $u_0 = (\alpha x_2, \beta x_1^n)$ are given by the following equations:

$$\frac{\alpha}{2} x_2^2 - \frac{\beta}{n+1} x_1^{n+1} = C, \qquad 0 \leq |C| < \delta. \tag{5.6.12}$$

Obviously, we can see from (5.6.12) that if $n =$ even, the flow of u_0 in a neighborhood of $x = 0$ is as shown in Figure 5.6.2(c). If $n =$ odd, when $\alpha \cdot \beta > 0$ (resp. $\alpha \cdot \beta < 0$), the flows of u_0 look as shown in Figure 5.6.2(b) (resp. as shown in Figure 5.6.2(a)). Thus, we derive from (5.6.11) this claim.

PROOF OF ASSERTION (2). We take $\varepsilon > 0$ sufficiently small, and consider singular points of the following vector field near $x = 0$:

$$u_\varepsilon = \begin{cases} \alpha x_2 + f(x_1) + x_2 g_1(x), \\ \beta x_1^n - x_2 f'(x_1) + x_2^2 g_2(x) + o(|x_1|^n) - \varepsilon. \end{cases}$$

By assumption, f can be expressed near $x = 0$ by

$$f(x_1) = \lambda x_1^k + o(|x_1|^k), \qquad \lambda \neq 0,\ 1 < k \leq \frac{n+1}{2}.$$

Thus, singular points of u_ε in a small neighborhood of $x = 0$ satisfy the equation below:

$$\begin{cases} x_2 = -\frac{\lambda}{\alpha} x_1^k + o\left(|x_1|^k\right), \\ \beta x_1^n + \frac{1}{\alpha} k\lambda^2 x_1^{2k-1} = \varepsilon + o\left(|x_1|^{2k-1}\right). \end{cases} \tag{5.6.13}$$

Obviously, when $2k - 1 < n$, or $2k - 1 = n$ and $\alpha\beta \neq -k\lambda^2$, (5.6.13) has a unique solution

$$x_\varepsilon \sim \left(C\varepsilon^{\frac{1}{2k-1}}, -\frac{\lambda}{\alpha} C^k \varepsilon^{\frac{k}{2k-1}} \right),$$

where

$$C = \begin{cases} \alpha k^{-1}\lambda^{-2} & \text{if } 2k-1<n, \\ \alpha(\alpha\beta+\lambda^2)^{-1} & \text{if } 2k-1=n \text{ and } \alpha\beta\neq k\lambda^2. \end{cases}$$

It is easy to check that

$$\text{Sign}\det Du_\varepsilon(x_\varepsilon) = -1, \tag{5.6.14}$$

for any $\varepsilon>0$ sufficiently small. By the invariance of index sums in a small domain with a perturbation, we infer from (5.6.14) that $\text{ind}(u,x_0)=-1$. The proof of this lemma is complete. □

5.6.3. Structural bifurcation near interior singular points with index zero. Let $u\in C^1([0,T],D^r(TM))$ $(r\geq 1)$ be a one-parameter family of divergence-free vector fields. As in the case for structural bifurcation near boundary singular points, consider the Taylor expansion of $u(x,t)$ at t_0 $(0<t_0<T)$:

$$\begin{cases} u(x,t)=u^0(x)+(t-t_0)u^1(x)+o(|t-t_0|), \\ u^0(x)=u(x,t_0), \\ u^1(x)=\frac{\partial}{\partial t}u(x,t_0). \end{cases} \tag{5.6.15}$$

Assumption 5.6.4. *Let $x_0\in \overset{\circ}{M}$ be an isolated degenerate singular point of $u^0(x)$. Suppose that*

$$\text{ind}(u^0,x_0)=0, \tag{5.6.16}$$

$$Du^0(x_0)\neq 0, \tag{5.6.17}$$

$$u^1(x_0)\cdot e_2\neq 0, \tag{5.6.18}$$

where e_2 is the unit vector defined as in (5.6.6).

Assumption 5.6.5. *Under the conditions of Assumption 5.6.4, we also assume that $u^0\in C^n$ near $x_0\in\overset{\circ}{M}$ for some $n\geq 2$, and*

$$\frac{\partial^k(u^0(x_0)\cdot e_2)}{\partial e_1^k} = \begin{cases} 0 & \text{for } 1\leq k<n=\text{even}, \\ \neq 0 & \text{for } k=n=\text{even}. \end{cases} \tag{5.6.19}$$

Remark 5.6.6. Conditions (5.6.16) and (5.6.17) imply, by Lemma 5.6.2, that the flows of u^0 near x_0 are as shown in Figure 5.6.2(c); i.e., both orbits of u^0 connected to x_0 are tangent to each other at x_0, and the eigenvector e_1 of $Du(x_0)$ is their common tangent vector. We shall see later that conditions (5.6.16) and (5.6.17) are generic for the interior structural bifurcation.

Remark 5.6.7. In view of fluid mechanics applications, condition (5.6.18) is equivalent to nonzero acceleration of the flow in the orthogonal direction to the eigenvector e_1 of $Du(x_0)$. This is a natural condition for the structural bifurcation.

Remark 5.6.8. Condition (5.6.19) is a technical condition which, by Lemma 5.6.3, ensures the regularity of the bifurcated singular points of $u(x,t)$ from (x_0,t_0). In addition, from (5.6.7), we can see that the integer number n satisfying (5.6.19) must be $n\geq 2$.

The interior structural bifurcation of $u(x,t)$ near a singular point with index zero is described by the following theorems.

THEOREM 5.6.9 (Interior Structural Bifurcation Theorem). *Let a one-parameter family of divergence-free vector fields* $u \in C^1([0,T], D^r(TM))$ $(r \geq 1)$ *satisfy Assumption 5.6.4. Then*

(1) *the vector field* u *has a bifurcation in its local structure at* (x_0, t_0). *More precisely,* $u(x,t)$ *has no singular point in a small neighborhood of* x_0 *for any* $t < t_0$ *(or* $t > t_0$*) sufficiently close to* t_0, *and* $u(x,t)$ *bifurcates at least two singular points from* x_0 *as* $t > t_0$ *(or* $t < t_0$*), and*

(2) *if* $x_0 \in \overset{\circ}{M}$ *is a unique singular point with index zero of* u^0, *then* $u(x,t)$ *has a bifurcation in its global structure at* $t = t_0$.

THEOREM 5.6.10. *Let* $u \in C^1([0,T], D^r(TM))$ $(r \geq 1)$ *satisfy Assumption 5.6.5. Then,* $u(x,t)$ *bifurcates from* (x_0, t_0) *exactly two nondegenerate singular points, one being a center and the other being a saddle.*

REMARK 5.6.11. Both Theorems 5.6.9 and 5.6.10 study structural bifurcation of u near an interior point x_0 with $\text{ind}(u^0, x_0) = 0$. When $\text{ind}(u^0, x_0)$ is different from zero, interior structural bifurcation may not occur as we shall see in Examples 5.6.15 and 5.6.16. In addition, Theorem 5.6.14 shows that the bifurcation given in Theorems 5.6.9 and 5.6.10 is generic.

This is quite different from the structural bifurcation near a boundary singular point. We have shown in previous sections that under suitable necessary conditions, u will always have a structural bifurcation near a boundary singular point with index different from $-1/2$.

PROOF OF THEOREM 5.6.9. To investigate the structural bifurcation of $u(x,t)$ at t_0, by the Taylor expansion (5.6.15) and condition (5.6.18), it suffices to consider only the topological structure of the first-order approximation $u^0 \pm \varepsilon u^1$ of (5.6.15) for $\varepsilon > 0$ sufficiently small.

By Lemma 5.6.2, let γ_1 and γ_2 be the two orbits of $u^0(x)$ connected to $x_0 \in \overset{\circ}{M}$. The eigenvector e_1 of $Du^0(x_0)$ is a common tangent vector of γ_1 and γ_2 at x_0; see Figure 5.6.3(a).

Both orbits γ_1 and γ_2 divide a neighborhood of x_0 into open domains I and II as shown in Figure 5.6.3(a). Since the angles between e_1 and the vectors of u^0 on γ_1 and γ_2 vary from 0 to π, and by assumption the angle θ between e_1 with $u^1(x_0)$ satisfies $0 < \theta < \pi$, there exist curves ℓ_1 in domain I and curve ℓ_2 in domain II connected to x_0 such that u^1 are parallel to u^0 on ℓ_1 and ℓ_2 (see Figure 5.6.3(b)). Obviously, the singular points of $u^0 \pm \varepsilon u^1$ are only on the curves as ℓ_1 and ℓ_2. Without loss of generality, we assume that u^0 and u^1 have a reverse orientation on ℓ_1 and ℓ_2, i.e., u^0 and $-\varepsilon u^1$ have the same orientation on ℓ_1 and ℓ_2. By condition (5.6.18), it follows that $u^0 - \varepsilon u^1$ has no singular points in ℓ_1 and ℓ_2, and therefore has no singular points in a neighborhood of x_0. Because $x_0 \in \overset{\circ}{M}$ is an isolated singular point of u^0, the values $|u^0(x)|$ are variant from 0 to a $\delta > 0$ sufficiently small on ℓ_1 and ℓ_2, i.e.,

$$
\begin{cases}
0 < |u^0(x)| < \delta \quad \forall\, x \in \ell_1 \cup \ell_2, \\
\sup\limits_{\ell_1 \cup \ell_2} |u^0| = \delta, \\
\inf\limits_{\ell_1 \cup \ell_2} |u^0| = 0.
\end{cases}
\tag{5.6.20}
$$

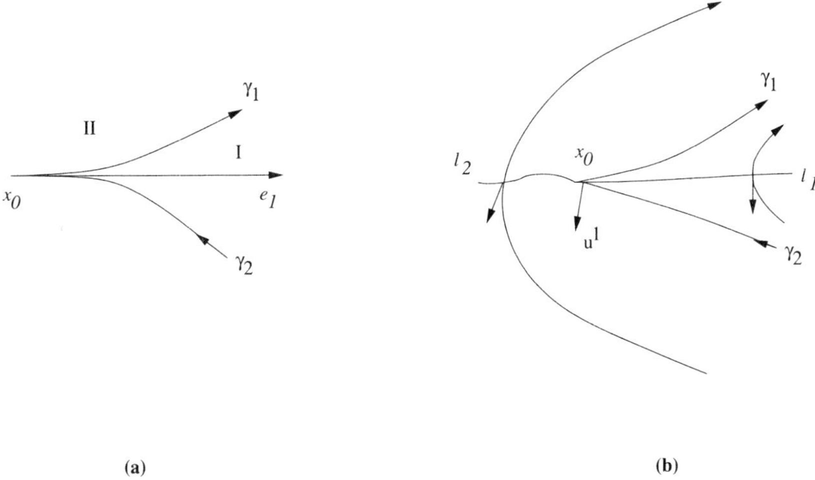

FIGURE 5.6.3. Schematic illustrating the proof of Theorem 5.6.9.

It follows from (5.6.20) that there is an $\varepsilon_0 > 0$ sufficiently small such that, for any $0 < \varepsilon < \varepsilon_0$, the vector field $u^0 + \varepsilon u^1$ has at least a singular point on each of ℓ_1 and ℓ_2; i.e., $u^0 + \varepsilon u^1$ has at least two singular points near x_0. Thus, $u^0 + \varepsilon u^1$ and $u^0 - \varepsilon u^1$ are not topologically equivalent in a neighborhood of x_0. The first assertion is proved.

Obviously, if x_0 is a unique singular point of $u^0(x)$ with index zero, then $u^0+\varepsilon u^1$ and $u^0 - \varepsilon u^1$ are not topologically equivalent on M. Hence, $u(x,t)$ has a globally structural bifurcation at $t = t_0$. This theorem is proved. □

PROOF OF THEOREM 5.6.10. By Assumption 5.6.5 and Lemma 5.6.3, the vector field $u^0(x)$ has the Taylor expansion at x_0 $(x = 0)$ as follows:

$$u^0(x) = \begin{cases} \alpha x_2 + f(x_1) + x_2 g_1(x), \\ \beta x_1^{2m} - x_2 f'(x_1) + x_2 g_2(x) + o(|x_1|^{2m}), \end{cases} \tag{5.6.21}$$

where $\alpha \neq 0$, $\beta \neq 0$, and

$$\begin{cases} f(x_1) = o\left(|x_1|^{m+\frac{1}{2}}\right), \\ f'(x_1) = o\left(|x_1|^{m-\frac{1}{2}}\right), \\ g_i(0) = 0 \qquad\qquad (i = 1, 2). \end{cases} \tag{5.6.22}$$

By (5.6.18), we have

$$u^1(x) = \begin{cases} \lambda_1 + h_1(x), \\ \lambda_2 + h_2(x), \end{cases} \tag{5.6.23}$$

where $\lambda_2 \neq 0$ and $h_i(x) = O(|x|)$, $i = 1, 2$.

By Theorem 5.6.9, one of $u^0 \pm \varepsilon u^1$ has no singular points, and another has at least two singular points near x_0 for all $\varepsilon > 0$ sufficiently small. We assume that $u^0 - \varepsilon u^1$ has singular points, i.e., $\beta > 0$ in (5.6.21) and $\lambda_2 > 0$ in (5.6.23). We need to prove that the equations below have exactly two solutions

$$\alpha x_2 + x_2 g_1(x) = \lambda_1 \varepsilon + \varepsilon h_1(x) - f(x_1), \tag{5.6.24}$$

$$\beta x_1^{2m} - x_2 f'(x_1) + x_2^2 g_2(x) = o\left(|x_1|^{2m}\right) = \lambda_2 \varepsilon + \varepsilon h_2(x). \tag{5.6.25}$$

By the Implicit Function Theorem, we derive from (5.6.22) and (5.6.24) that

$$\begin{cases} x_2 = \alpha^{-1}\lambda_1\varepsilon - f(x_1) + G(\varepsilon, x_1), \\ G(\varepsilon, x_1) = o\left(|\varepsilon|, |x_1|^{m+\frac{1}{2}}\right). \end{cases} \tag{5.6.26}$$

Putting (5.6.26) in (5.6.25), we get the algebraic equation

$$\beta x_1^{2m} = \lambda_2 \varepsilon + \alpha^{-1}\lambda_1 \varepsilon f'(x_1) + \varepsilon \cdot O(|x|) + o(|\varepsilon|, |x_1|^{2m}), \tag{5.6.27}$$

where $\beta, \lambda_2 > 0$. It is clear that for any $\varepsilon > 0$ sufficiently small, the equation (5.6.27) has exactly two solutions:

$$x_1 = \pm(\beta^{-1}\lambda_2)^{\frac{1}{2m}} \varepsilon^{\frac{1}{2m}} + o\left(\varepsilon^{\frac{1}{2m}}\right).$$

Thus, we deduce that the vector field $u^0 - \varepsilon u^1$ has exactly two singular points $x(\varepsilon) = (x_1(\varepsilon), x_2(\varepsilon))$ as follows:

$$\begin{cases} x_1^{\pm}(\varepsilon) = \pm(\beta^{-1}\lambda_2)^{\frac{1}{2m}} \varepsilon^{\frac{1}{2m}} + o\left(\varepsilon^{\frac{1}{2m}}\right), \\ x_2^{\pm}(\varepsilon) = \alpha^{-1}\lambda_1\varepsilon + o\left(\varepsilon, |x_1^{\pm}|^{m+\frac{1}{2}}\right). \end{cases} \tag{5.6.28}$$

Finally, we shall show that $x^{\pm}(\varepsilon)$ are nondegenerate for all $\varepsilon > 0$ sufficiently small. By $\operatorname{div} u = 0$, we have

$$\begin{aligned} \det D(u^0 - \varepsilon u^1)_{x = x(\varepsilon)} &= -\left(\frac{\partial(u_1^0 - \varepsilon u_1^1)}{\partial x_1}\right)^2 - \frac{\partial(u_1^0 - \varepsilon u_1^1)}{\partial x_2} \cdot \frac{\partial(u_2^0 - \varepsilon u_2^1)}{\partial x_1} \\ &= \text{(by (5.6.22) and (5.6.28))} \\ &= \pm 2m\alpha\beta(\beta^{-1}\lambda_2)^{\frac{2m-1}{2m}} \varepsilon^{\frac{2m-1}{2m}} + o\left(\varepsilon^{\frac{2m-1}{2m}}\right), \end{aligned}$$

which yields that

$$\det D(u^0 - \varepsilon u^1) = \begin{cases} > 0 & \text{if } x = x^-, \\ < 0 & \text{if } x = x^+. \end{cases}$$

Thus, we prove that $u^0 - \varepsilon u^1$ has exactly two singular points x^+ and x^- for any $\varepsilon > 0$ sufficiently small; x^- is a center, x^+ is a saddle, and they are nondegenerate. This proof is complete. □

5.6.4. Kinematic theory for interior separation of fluid flows. We start with a typical example, illustrating how structural bifurcation occurs in the interior of fluid flows.

Let $u \in C^1([0,T], D^r(TM))$, and let $x_0 \in \overset{\circ}{M}$ be an isolated singular point of $u^0(x) = u(x, t_0)$, $0 < t_0 < T$.

EXAMPLE 5.6.12. Consider the case when the index of $u^0(x)$ at the singular point $x_0 \in \overset{\circ}{M}$ is zero, and the Jacobian matrix at x_0 is nonzero, i.e.,

$$\operatorname{ind}(u^0, x^0) = 0, \qquad Du^0(x_0) \neq 0.$$

The structural bifurcation occurs as shown in Figure 5.6.4; it corresponds to interior separation phenomena in fluid mechanics. When $t = t_0 - \varepsilon$ with $\varepsilon > 0$ small, the flow

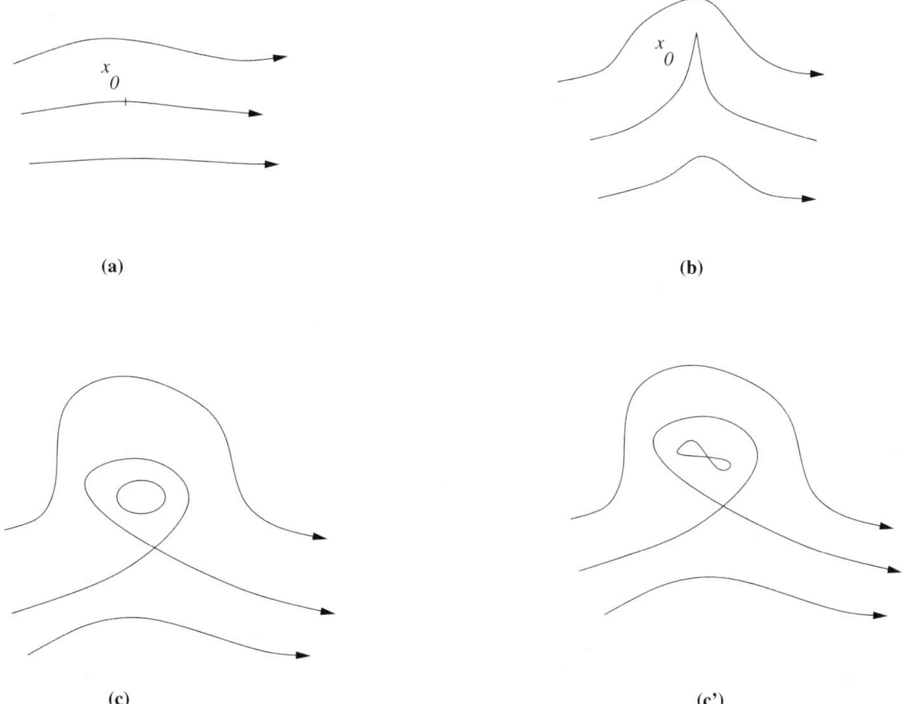

FIGURE 5.6.4. Schematic illustrating interior structural bifurcation.

of $u(x, t_0 - \varepsilon)$ given by Figure 5.6.4(a) exhibits no singular points in a neighborhood of x_0. At $t = t_0$, $u^0 = u(x, t_0)$ is given by Figure 5.6.4(b), which has an isolated singular point $x_0 \in \overset{\circ}{M}$ with index zero. When $t = t_0 + \varepsilon$ for $\varepsilon > 0$ small, $u(x, t_0 + \varepsilon)$ is given by either Figure 5.6.4(c) or Figure 5.6.4(c$'$) or even more complicated circulation patterns in the back flow region. As we shall see in Theorem 5.6.14, the flow pattern given by Figure 5.6.4(c) is generic. In other words, the flow transition from Figure 5.6.4(a) to Figure 5.6.4(b), and then to Figure 5.6.4(c), or vice versa, is, in general, the pattern transition obtained both experimentally and numerically.

We now address interior flow separation from a rigorously analytic point of view.

THEOREM 5.6.13. *Let $u \in C^1([0, T], D^1(TM))$ satisfy Assumption 5.6.4. Then*

(1) *there must be some centers of u separated from $x_0 \in \overset{\circ}{M}$ as shown schematically in either Figure 5.6.4(c) or Figure 5.6.4(c$'$);*
(2) *the centers (back flows) are enclosed by a closed orbit line $\gamma(t)$ consisting of orbits of $u(\cdot, t)$, and $\gamma(t)$ converges/shrinks to x_0 as $t \to t_0$; and*
(3) *if Assumption 5.6.5 is satisfied, then the center separated from $x_0 \in \overset{\circ}{M}$ is unique, as shown in Figure 5.6.4(c).*

The centers in Figures 5.6.4(c) and (c$'$) correspond, in a real fluid, to isolated vortices or, in the case of the figure-eight shaped centers, to pairs of co-rotating vortices. This theorem is a direct corollary of Theorems 5.6.9 and 5.6.10 and Theorem 2.1.15.

5.6.5. Genericity of structural bifurcation with index zero. In the following, we shall show that the type of structural bifurcation as shown in Figure 5.6.4(c), i.e., one center interior separation, is generic in the interior structural bifurcation. This is remarkably different from the structural bifurcation near the boundary addressed in the previous sections. It also explains why interior separation to multiple centers and the interior flow separation from the singularities with nonzero index are seldom observed in fluid motions.

Let $x_0 \in \overset{\circ}{M}$ and $0 < t_0 < T$ be given. We define a topological space $B \subset C'([0,T], D^2(TM))$ as follows:

$$B = \left\{u \in C^1([0,T], D^2(TM)) \mid u^0(x_0) = 0, \det Du^0(x_0) = 0, u^0 = u(\cdot, t_0)\right\},$$

with the topology of $C^1([0,T], D^2(TM))$. Obviously, the space B contains all vector fields in $C^1([0,T], D^2(TM))$ that have a bifurcation in their local structure at (x_0, t_0). It is easy to see that the set

$$B_0 = \left\{u \in B \,\middle|\, Du^0(x_0) \neq 0, \ \frac{\partial^2 (u^0(x_0) \cdot e_2)}{\partial e_1^2} \neq 0, \ u^1(x_0) \cdot e_2 \neq 0\right\}$$

is open and dense in B, where e_1 and e_2 are as in (5.6.5) and (5.6.6), and $u^1(x) = \frac{\partial}{\partial t} u(x, t_0)$. The following genericity theorem of structural bifurcation immediately follows from Lemma 5.6.3.

THEOREM 5.6.14 (Genericity of Structural Bifurcation). *For any $u \in B_0$, u has a bifurcation in its local structure at (x_0, t_0). More precisely, u bifurcates from (x_0, t_0) exactly two nondegenerate singular points, one of which is a center and the other of which is a saddle, as shown in Figure 5.6.4(a–c). Moreover, the set B_0 is open and dense in the topological space B, which contains all vector fields in $C^1([0,T], D^2(TM))$ having a locally structural bifurcation at (x_0, t_0).*

PROOF. If $u \in B_0$, then $u^0 \in D^2(TM)$, and u^0 has the Taylor expansion (5.6.10) with $n = 2$, and

$$\deg f = k \geq 2 \qquad (\text{by } f \in C^2 \text{ and } f(z) = o(|z|).$$

Hence, by Lemma 5.6.3, we have

$$\operatorname{ind}(u^0, x_0) = 0.$$

Theorem 5.6.14 follows from Theorems 5.6.9–5.6.10. The proof is complete. □

5.6.6. Examples of no structural bifurcation. We know that for a locally structurally stable singular point of a vector field $v \in D^r(TM)$, its index obeys

$$\operatorname{ind}(v, x_0) = \begin{cases} -\frac{1}{2}, & x_0 \in \partial M, \\ -1 \text{ or } +1, & x \in \overset{\circ}{M}. \end{cases}$$

For a vector field $u(x,t) = u^0(x) + (t - t_0)u^1(x) + o(|t - t_0|)$, the structural bifurcation theorem near a boundary singular point amounts to saying that if x_0 is a boundary singular point with $\operatorname{ind}(u^0, x_0) \neq -\frac{1}{2}$ and $u^1(x_0) \neq 0$ for $x_0 \in \partial M$, then $u(x,t)$ has a bifurcation in its local structure at (x_0, t_0). However, for an interior singular point $x_0 \in \overset{\circ}{M}$ of u^0 with $\operatorname{ind}(u^0, x_0) \neq 1$ or -1, the vector field $u(x,t)$ may have no

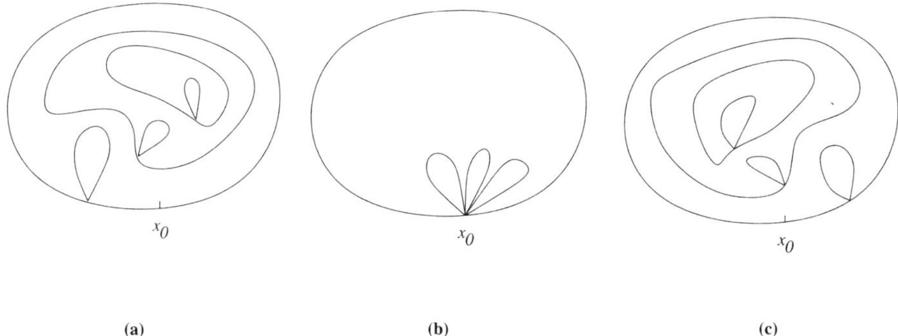

FIGURE 5.6.5. Schematic illustrating lack of structural bifurcation as given in Example 5.6.15.

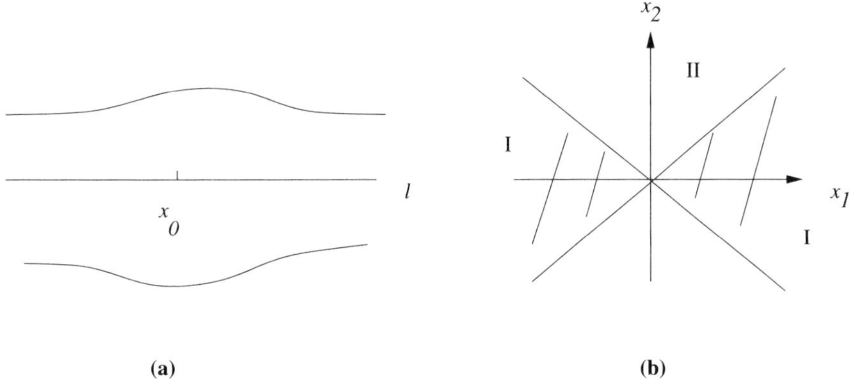

FIGURE 5.6.6. An example showing lack of structural bifurcation for the case with zero Jacobian.

structural bifurcation near (x_0, t_0). In the following, we give two examples to show this.

EXAMPLE 5.6.15. Figure 5.6.5(a–c) illustrates a structural evolution of a vector field $u(x,t)$ near $x_0 \in \overset{\circ}{M}$ as time t crosses t_0, where $\text{ind}(u^0, x_0) = -n$ $(n > 1)$, $u^0 = u(x, t_0)$, and $u^1(x_0) = \partial u(x_0, t)/\partial t \neq 0$.

In Figure 5.6.5, we see that the vector fields $u(x, t_0 - \varepsilon)$ given by (a) and $u(x, t_0 - \varepsilon)$ given by (c) are topologically equivalent for all $\varepsilon > 0$ small. Hence, $u(x,t)$ has no structural bifurcation at (x_0, t_0).

EXAMPLE 5.6.16. Let $\text{ind}(u^0, x_0) = 0$, let $Du^0(x_0) = 0$, and let the structure of u^0 near x_0 is illustrated by Figure 5.6.6(a).

Let the x_1-axis of the coordinate system in Figure 5.6.6(b) be tangent to the orbit line ℓ in (a) at x_0. The angles between $u^0(x)$ and the x_1-axis near x_0 vary in the shadow domain I in (b). Hence, if the angle between $u^1(x_0)$ and the x_1-axis is in the domain II in (b), then $u^1(x)$ is transversal to $u^0(x)$ near x_0, which implies that the vector field $u(x,t) = u^0(x) + (t - t_0)u^1 + o(|t - t_0|^2)$ has no structural bifurcation near (x_0, t_0).

5.7. Genericity of Structural Bifurcations

5.7.1. Abstract theorems on Banach spaces. Let X be a Banach space, and let $X_1 \subset X$ be an open and dense set. We denote by $R \subset C^r([0,T],X)(r \geq 0)$ the set that for any $x \in R$, there is an open and dense set $I \subset [0,T]$ such that $x(t) \in X_1$ for $t \in I$.

A subset of a topological space is residual if it contains a countable intersection of open dense sets. A topological space is a Baire space if every residual subset is dense. It is well known that a complete metric space is a Baire space; hence, $C^r([0,T],X)$ is a Baire space.

THEOREM 5.7.1. *The subset $R \subset C^r([0,T],X)$ is residual, and therefore dense.*

PROOF. We first prove the theorem for the case of $T < \infty$. Without loss of generality, we assume $T = 1$.

Note that X_1 is open and dense in X. If $x \notin R$, then the set

$$J_x = \{t \in [0,1] \quad | \quad x(t) \notin X_1\}$$

is closed and has non-zero Lebesque measure.

Let

$$R_n = \{x \in C^r([0,1],X) \quad | \quad |J_x| < \frac{1}{n}\},$$

where $|J_x|$ stands for the Lebesque measure of J_x. It is clear that

$$R = \bigcap_{n=1}^{\infty} R_n.$$

Hence, we only need to prove that $R_n (n \geq 1)$ are open and dense in $C^r([0,1],X)$. Let $A_n = C^r([0,1],X) \setminus R_n$. Then

$$A_n = \{x \in C^r([0,1],X) \quad | \quad |J_x| \geq \frac{1}{n}\}.$$

It suffices then to prove that A_n is closed and nowhere dense.

Let $x_k \in A_n$ and $x_k(t) \to \bar{x}(t)(k \to \infty)$ in $C^r([0,1],X)$. Then $x_k(t_0) \to \bar{x}(t_0)$ in $X, \forall t_0 \in [0,1]$. Since $X \setminus X_1$ is closed, if $x_k(t_0) \in X/X_1$, then $\bar{x}(t_0) \in X/X_1$, which implies

$$|J_{x_0}| \geq \frac{1}{n},$$

and therefore $\bar{x} \in A_n$. Hence, A_n is closed.

Let $O \subset C^r([0,1],X)$ be an open set such that $A_n \cap O$ is dense in O. Then for any $t_0 \in [0,1]$, the section

$$O_n(t_0) = \{x(t_0) \in X \quad | \quad x \in A_n \cap O\}$$

is dense in the section

$$O(t_0) = \{x(t_0) \in X \quad | \quad x \in O\}.$$

Since $|J_x| \geq \frac{1}{n}$ for any $x \in A_n$, for any $n+1$ elements $x_1, \cdots, x_{n+1} \in O \cap A_n$, there exists a subset $J_0 \in [0,1]$ with mes $J_0 > 0$ and at least there are two elements x_i, x_j $(1 \leq i,j \leq n+1)$ such that, for and $t \in J_0$, $x_i(t), x_j(t) \in X/X_1$. It implies that there exist an open subset $O_1 \subset O$ and a $t_1 \in [0,1]$ such that, for any $x \in O_1 \cap A_n$, $x(t_1) \in X/X_1$. Hence, the section $O_1(t_1) \cap A_n$ is dense in $O_1(t_1)$, a contradiction to the statement that X/X_1 is nowhere dense in X (due to $O_1(t_1) \cap A_n \subset X/X_1$).

When $T = \infty$, we take

$$R_n = \left\{ x \in C^r([0,\infty], X) \quad | \quad |(J_x \cap [k, k+1))| < \frac{1}{n}, \quad \forall k = 1, 2, \cdots \right\}.$$

We can prove, in the same fashion as before, that R_n are open and dense in $C^r([0,\infty], X)$, and $R = \cap_{n=1}^{\infty} R_n$.

The proof is complete. □

For hydrodynamic equations, we are interested in the following type of spaces. Let Z and X be two Banach spaces, and X can be embedded in Z. Let

$$\begin{aligned}
L^2([0,T), X) &= \left\{ x(t) \in X \quad | \quad \int_0^T \|x(t)\|_X^2 dt < \infty \right\}, \\
L^2_{x_0}([0,T), X) &= \left\{ x \in L^2([0,T], X) \quad | \quad x(0) = \bar{x} \right\}, \\
H^1([0,T), Z) &= \left\{ \frac{dx}{dt} \in z \quad | \quad \int_0^T \|\frac{dx}{dt}\|_Z^2 dt < \infty \right\}.
\end{aligned}$$

For $\bar{x} \in X$, let

$$\begin{aligned}
R(\bar{x}) = \{ x \in L^2_{x_0}([0,T), X) \cap H^1([0,T), Z) \quad | \quad & \text{there is an open} \\
& \text{and dense set } I \subset [0,T) \text{ s.t. } x(t) \in X_1 \quad \forall t \in I \},
\end{aligned}$$

where $X_1 \subset X$ is open and dense.

Then the following theorem can be proved in the same fashion as that of Theorem 5.7.1; we omit the details:

THEOREM 5.7.2. *The subset $R(\bar{x}) \subset L^2_{x_0}([0,T), X) \cap H^1([0,T), Z)$ is residual, and therefore dense.*

REMARK 5.7.3. The conclusion that R is residual in $C^r([0,T), X)$ (or $R(\bar{x})$ is residual in $L^2_{x_0}([0,T), X) \cap H^1([0,T), Z)$) cannot be strengthened; i.e., one can not conclude that R is open and dense in $C^r([0,T], X)$ (or $R(\bar{x})$ is open and dense in $L^2_{x_0}([0,T), X) \cap H^1([0,T), Z)$) only from the condition that X_1 is open and dense in X. In fact, we have the following counterexample.

EXAMPLE 5.7.4. Let $X = \mathbb{R}^2$, and let $S \subset X$ be a set defined by

$$S = \bigcup_{n=0}^{\infty} \widetilde{S}_n + S_2 + \bigcup_{n=0}^{\infty} B_n,$$

where

$$\begin{aligned}
\widetilde{S}_n &= \left\{ x \in \mathbb{R}^2 \quad \middle| \quad \|x\| = 2 - \frac{1}{2^n} \right\}, \\
S_2 &= \left\{ x \in \mathbb{R}^2 \quad | \quad \|x\| = 2 \right\}, \\
B_n &= \left\{ (\alpha_k, r) \quad \middle| \quad a_k = \frac{2k\pi}{n} \quad \forall 1 \le k \le n; \quad 2 - \frac{1}{2^n} \le r \le 2 - \frac{1}{2^{n+1}} \right\},
\end{aligned}$$

where (α, r) is the polar coordinate in $\mathbb{R}^2$. It is easy to see that for $X_1 = X \setminus S$, the set $R \subset C^r([0,T), X)$ is not open.

5.7.2. Genericity of structural bifurcation of the solutions of the Navier-Stokes equations. We consider the following problem:

$$\frac{\partial u}{\partial t} + (u \cdot \nabla)u = \mu\Delta u - \nabla p + f \qquad \forall x \in M,\ 0 < t < T, \tag{5.7.1}$$

$$\text{div } u = 0, \tag{5.7.2}$$

$$u(x,0) = \varphi(x), \tag{5.7.3}$$

with the corresponding boundary conditions given in (5.3.4–5.3.7).

Let X be either $B^r(TM)$, or $B_0^r(TM)$, or $B_\varphi^r(TM)$, or $H^r(TM)(r \geq 2)$, and let $X_1 \subset X$ be the set of all structurally stable vector fields in X. By the structural stability theorems in Chapter 2, X_1 is open and dense in X.

We introduce the usual Sobolev spaces of the Navier-Stokes equations

$$V_X^k = \text{ the closure of } X \text{ in } W^{k,2}(TM).$$

By Sobolev embedding theorems, the embedding

$$\begin{aligned} &L^2((0,T), V_X^k) \cap H^1((0,T), V_X^{k-2}) \\ &\qquad \to L^2((0,T), C^{k-2+\alpha}(TM)) \cap C((0,T), C^{k-3+\alpha}(TM)) \end{aligned}$$

is continuous, where $0 < \alpha < 1$ and $k \geq 3$ ($k \geq 4$ if $X = B_0^r(TM)$, or $X = B_\phi^r(TM)$).

We say that $u \in L^2((0,T), V_X^k) \cap H^1((0,T), V_X^{k-2})$ has bifurcation property if there is an open and dense set $I \subset (0,T)$ such that $u(\cdot,t) \in X_1$ for any $t \in I$, i.e., $u(\cdot,t)$ is structurally stable for any $t \in I$.

THEOREM 5.7.5. *There is a residual set $F \subset L^2((0,T), W^{k-2,2}(TM))$ such that, for any $f \in F$, the solution $u \in L^2((0,T), V_X^k) \cap H^1((0,T), V_X^{k-2})$ of (5.7.1–5.7.3) with the corresponding boundary conditions has the bifurcation property.*

PROOF. Let

$$\begin{aligned} Z &= L_\varphi^2([0,T), V_X^k) \cap H^1([0,T), V_X^{k-2}), \\ G^k &= \{\nabla p | p \in W^{k+1,2}(M)\}. \end{aligned}$$

We define a mapping A by

$$\begin{aligned} &A : Z \times L^2((0,T), G^{k-2}) \to L^2([0,T), W^{k-2,2}(TM)), \\ &A(u,p) = \frac{\partial u}{\partial t} + (u \cdot \nabla)u - \mu\Delta u + \nabla p. \end{aligned}$$

By the theory of the Navier-Stokes equations, it is well known that the mapping A is one to one and onto. For any $(u_0, p_0) \in z \times L^2([0,T), G^{k-2})$, the derivative operator $A'(u_0, p_0)$ corresponds to the following linear equations:

$$\begin{cases} \dfrac{\partial u}{\partial t} + (u_0 \cdot \nabla)u + (u \cdot \nabla)u_0 = \mu\Delta u - \nabla p + f, \\ \text{div } u = 0, \\ u(x,0) = \varphi(x), \end{cases} \tag{5.7.4}$$

with corresponding boundary conditions given in (5.3.4–5.3.7).

It is known that for any $f \in L^2([0,T), W^{k-2,2}(TM))$, there exists a unique solution in $Z \times L^2([0,T), G^{k-2})$, for (5.7.4) with the corresponding boundary conditions. Therefore, the derivative operator

$$A'(u_0, p_0) : Z \times L^2([0,T), G^{k-2}) \to L^2([0,T), W^{k-2,2}(TM))$$

is one to one and onto for any $(u_0, p_0) \in Z \times L^2([0,T), G^{k-2})$. By the linear operator theory in Banach spaces, we know that $A'(u_0, p_0)$ are isomorphic. By the inverse function theorem, the mapping A is a local homeomorphism.

By Theorem 5.7.2, the set

$$R_\varphi = \{u \in Z \quad | \quad u(\cdot, t) \in X_1 \text{ for } t \in \text{ an open and dense set in } (0,T)\}$$

is residual in Z. Hence

$$F = A(R_\varphi \times L^2([0,T), G^{k-2}))$$

is a residual set in $L^2([0,T), W^{k-2,2}(TM))$.

The proof is complete. □

REMARK 5.7.6. The genericity of structural bifurcation of one-parameter families of general vector fields was considered by Sotomayor [**77, 99, 98**]. A basic result obtained by Sotomayor is that there is a residual subset of one-parameter families of vector fields $R \subset C^\infty([0,T], C^r(TM))$ such that any $v \in R$ is Kupka-Smale for any $t \in [0,T]$ except a countable set $\{t_i\} \subset [0,T]$. □

Notes for Chapter 5

The material in this chapter is based on [**29, 30, 28, 57, 59, 63, 61**]. The classical boundary layer theory can be found in Goldstein [**31**], H. Schlichting [**92**], and A. Chorin and J. Marsden [**11**]. Also, we refer readers to a recent textbook by O. Oleinik and V. N. Samokhin [**76**] and articles by J. Bona [**8**], Weinan E [**17**], W. E and B. Engquist [**18**], and R. Temam and X. Wang [**104**] for the mathematical analysis of the Prandtl equation, an approximation of the Navier-Stokes equations for the boundary layer analysis.

CHAPTER 6

Two Examples

To guide the theoretical studies as well as to identify some key issues, it is important to supplement rigorous analysis with numerical simulations. We present in this chapter numerical simulations for 1) a wind-driven, double-gyre ocean circulation model, and 2) driven cavity flow and boundary-layer separation, focusing on transitions of different flow structures as predicted by the theory developed in early chapters.

6.1. Fluid Flow Maps and Double-Gyre Ocean Circulation

6.1.1. An index formula and flow maps. In this section, we apply the aforementioned topological results to the dynamics of two-dimensional flows. Large-scale planetary flows, atmospheric and oceanic, are characterized by fast rotation (small Rossby number) and strong stratification (small vertical-to-horizontal aspect ratio), at least in the midlatitudes; see Ghil and Childress [**27**] and Pedlosky [**80**]. Therefore, many features of large-scale atmospheric and oceanic flows are two-dimensional.

In this section, we present an application of the general theory developed in previous chapters to study flow maps of both general 2-D fluid flows and geophysical fluid flows. In the atmospheric flow context, flow maps are related to weather maps.

Let $v \in D^r(TM)$. Since v is divergence-free, it is proved that an interior nondegenerate singular point must be either a center or a saddle point, and a nondegenerate singular point on the boundary must be a saddle point. The indices and physical meanings of these singular points are as follows.

First, when the eigenvalues of the matrix A are purely imaginary numbers, the interior singularity $x = 0$ is a center of the vector field $v = Ax + o(|x|)$ with index 1:

$$\text{ind } (v, 0) = 1.$$

In fluid mechanics, $x = 0$ corresponds to the center of a local circulation.

Second, when the eigenvalues λ_1 and λ_2 of A are real numbers, we have $\lambda_1 = -\lambda_2 > 0$. In this case, $x = 0$ is a saddle point with index -1:

$$\text{ind } (v, 0) = -1.$$

Third, when $x = 0 \in \partial M$, it has to be a saddle point of the vector field v. In this case, according to the definition given in Section 2, the index is $-1/2$:

$$\text{ind } (v, 0) = -\frac{1}{2}.$$

In large scale oceanic flows (resp. a general fluid flows), saddle points on the boundary are normally caused by a jet flowing toward or away from the coastal line or the boundary.

We now use the index theorem to establish a global formula on the numbers of islands, mountains, saddle points, and centers of divergence-free vector fields. To this end, assume that $M \subset \mathbb{R}^2$ (or $M \subset S^2$) is a compacted manifold with boundary with k holes. For the oceanic motion, the k holes represent k islands, and for the atmospheric motion, they represent the horizontal sections of k mountains. It is easy to see that the Euler characteristic of M is

$$\chi(M) = 1 - k.$$

Since degenerate singular points are often unstable, *we assume that* $v \in D^r(TM)$ $(r \geq 1)$ *is a regular vector field on* M.

We let

$$\begin{aligned} C = &\ \text{the number of centers of } v, \\ S = &\ \text{the number of interior saddle points of } v, \\ B = &\ \text{the number of boundary saddle points of } v. \end{aligned}$$

Then by the index theorem, Theorem 1.2.8, we have

$$C - S - \frac{1}{2}B = 1 - k, \tag{6.1.1}$$

which yields the following method for flow maps.

METHOD 6.1.1. *Let* $v \in D^r(TM)$.

(1) Interior centers correspond to circulations, which are observable; boundary saddles correspond to jets moving toward or away from the boundary, which are also observable. By the index formula, the number of interior saddles can be determined.
(2) The structural classification theorem and the global structural theorem ensure that the global structure of the flow pattern is then uniquely determined by saddle connections. Flow maps can be obtained.
(3) On the other hand, given a divergence-free vector field, it is extremely easy to verify the structural stability conditions that can be directly used in verifying the structural stability of the flow pattern. Moreover, stable divergence-free vector fields are dense and open in the space of all regular divergence-free vector fields.

6.1.2. Global oceanic flows. For the global ocean, if we neglect small islands and consider only the continents, we may consider the ocean occupies a connected region on the surface of the earth S^2_a, excluding 6 isolated continents: Eurasia, Africa, North America, South America, Antarctica, and Australia. This global oceanic domain is isomorphic to a planar region having 5 holes. In this case, formula (6.1.1) becomes

$$C - S - \frac{1}{2}B = -4. \tag{6.1.2}$$

Since the set of all incompressibly structurally stable vector fields is open and dense in $D^r(TM)$, in most cases, the global ocean flow pattern that one observes is structurally stable. By the structural classfication theorem, the flow pattern is divided into several circulation regions (circle cells and circle bands), which are separated by saddle connections (separatrix).

Again, by the structural stability theorem, we notice that all interior saddles of the structurally stable ocean flow are self-connected, and each boundary saddle is connected to saddles on the same connected component of ∂M. Therefore, it is easy to see that if we know the number, position and orientation of the centers, and the numbers and positions of both interior and boundary saddles, then we can, in principle, obtain the topological structure of saddle connections, and then determine completely the global topological structure of the oceanic flow pattern.

In fact, the interior centers correspond to circulations; their position, number and orientation are observable. Boundary saddles correspond to jets moving toward or away from the boundary, and their position and number are also observable. Then the number of interior saddles is determined by (6.1.2). Under the condition of interior saddles being self-connected, the number, position and orientation of centers and boundary saddles can help us to determine the position of interior saddles. Hence, our theory is useful in drawing the weather map and the map of the large-scale ocean currents.

The method and theory we used to analyze the global oceanic flow pattern can also be applied to determine flow patterns in local regions of the atmosphere and the ocean. In the next section, we shall study the double-gyre ocean circulation.

6.1.3. Wind-Driven, Double-Gyre Ocean Circulations. We now study the double-gyre phenomena of large-scale ocean circulation, a typical phenomena in the northern midlatitude ocean basins. The double-gyre phenomena here refers to the two gyre motions (circulations) observed in the ocean basins: one is the sub-polar gyre, the other is the sub-tropical gyre. These gyres have a typical horizontal scale of about one thousand kilometers. Several main features of the double-gyre ocean circulation have been identified by analyzing the observational data as well as by numerical simulations; see, for instance, Speich and Ghil [**101**], Jiang, Jin and Ghil [**40**], and Speich, Dijkstra and Ghil [**100**]. First, these gyres are dominant and persistent; second, they represent typical seasonal and interannual oscillations of the large-scale ocean; and third, they transfer potential energy. Thus, the study of the double-gyre motion will provide a better understanding of the predictability, and possibly, a better long-term prediction on the dynamics of the ocean.

6.1.3.1. *The model.* The model we adapt in our studies is the following β-plane, wind-driven, double-gyre, large-scale ocean circulation:

$$\begin{cases} \dfrac{\partial v}{\partial t} - \dfrac{1}{\mathrm{Re}}\Delta v + v\cdot\nabla v + \dfrac{1}{\varepsilon}[yk\times v + \nabla p] = \dfrac{\alpha_r}{\varepsilon\pi}\tau, \\ \operatorname{div} v = 0, \end{cases} \tag{6.1.3}$$

where $v = (u, w)$ is the (horizontal) velocity field, p is the surface pressure, and k is the unit vector in the vertical direction so $k \times v = (-w, u)$. The spatial coordinate system is denoted by (x, y) with the x-axis in the east-west direction and y-axis in the north-south direction. The two-dimensional model can be derived from a three-dimensional one by integrating in the vertical direction.

We proceed with notations and comments used in (6.1.3).

(1) The corresponding dimensional form of (6.1.3) reads

$$\begin{cases} \dfrac{\partial v}{\partial t} - A\Delta v + v\cdot\nabla v + \beta yk\times v + \nabla p = \alpha_r\dfrac{\tau_{dim}}{\rho_0 h}, \\ \operatorname{div} v = 0. \end{cases} \tag{6.1.4}$$

Here, βy represents the first order Coriolis parameter, and A represents the effective turbulent viscosity coefficient. The nondimensional parameter α_r

is introduced here to study the effects of the wind stress to the dynamics; its value can be allowed to range between 0 and 1.8 (equivalent to taking the wind stress from abnormally low to abnormally high, as in [**40**]). The term $\tau_{dim}/\rho_0 h$ is dictated by the Ekman pumping on the surface Ekman boundary layer, and is also related to the leading order Coriolis term $f_0 k \times v$ in the quasi-geostrophic asymptotics; see Pedlosky [**80**].

(2) The flow is driven by the wind force, which is given in the dimensional form as a sinusoidal function

$$\tau_{dim} = \tau_0(-\cos(\pi y/L), 0),$$

where τ_0 is the amplitude of the wind force. The nondimensional form of the wind force used in (6.1.3) is $\tau = (-\cos(\pi y), 0)$.

(3) The typical (horizontal) velocity U is calculated via the Sverdrup equation as

$$U = \frac{\tau_0 \pi}{\beta L \rho_0 H},$$

where L is the typical horizontal length, H is the typical depth of the ocean layer, and ρ_0 is the typical surface density.

(4) The parameters ε and Re are the Rossby number and the Reynolds number, given by

$$\varepsilon = \frac{U}{\beta L^2}, \qquad Re = \frac{LU}{A},$$

where A is the effective turbulent viscosity coefficient. The values of the parameters used in this work are:

$$\begin{aligned} &L = 10^6 m, \qquad H = 500m, \\ &f_0 = 5 \times 10^{-5}\ s^{-1}, \qquad \beta = 2 \times 10^{-11}\ m^{-1} \cdot s^{-1}, \\ &\tau_0 = 0.1\ N \cdot m^{-2}, \qquad \rho_0 = 1000\ kg \cdot m^{-3}, \\ &U = 10^{-2}\pi\ m \cdot s^{-1}, \qquad \varepsilon = 5 \times 10^{-4}\pi. \end{aligned}$$

(5) The dimensional time and velocity are calculated by

$$t_{dim} = \frac{1}{\varepsilon \beta L} t = \frac{10^8}{\pi} t\ s, \qquad v_{dim} = Uv.$$

We take the non-dimensional spatial domain M and the non-dimensional depth h of the ocean to be

$$M = (0,1) \times (0,2) \subset \mathbb{R}^2, \qquad h = 1. \tag{6.1.5}$$

The non-dimensional form of the boundary conditions and the initial conditions are

$$\begin{cases} v \cdot n = 0, \quad \dfrac{\partial (v \cdot \tau)}{\partial n} = 0 \text{ on } \partial M, \\ v|_{t=0} = 0 \qquad \text{on } M. \end{cases} \tag{6.1.6}$$

6.1.3.2. *Structural analysis.* The system (6.1.3) is solved numerically using a spectral-projection method; see [**93**] for details. In our numerical experiments, we use $\alpha_r = .95$ with the Reynolds numbers to be , respectively, 5, 10, 20 and 30.

At each time instant, since the number k in (6.1.1) is zero, we have

$$C - S - \frac{1}{2}B = 1. \tag{6.1.7}$$

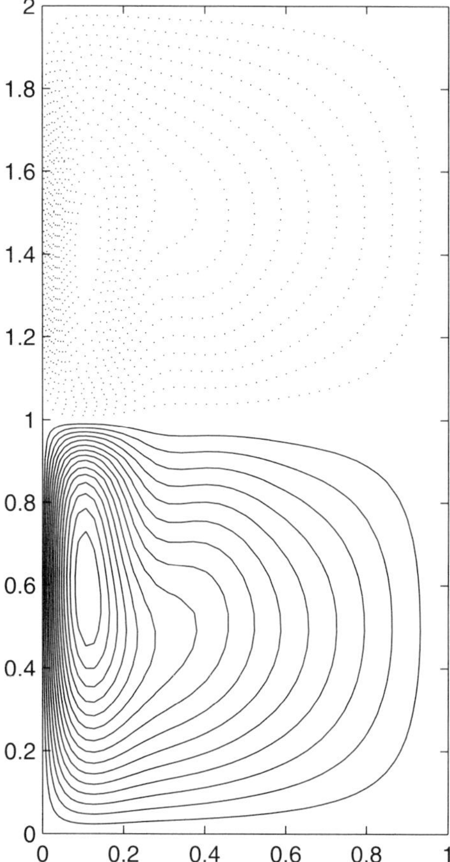

FIGURE 6.1.1. Snapshot of streamlines of the unique steady state for $Re = 5$.

If there is no singular point of v on the boundary, then the number of centers equals the number of interior saddle points plus 1. Hence, the simplest structure of the flow in this case is a circular motion with one center in the interior as the only singular point.

For the large-scale motion of the Atlantic Ocean, it is observed that there is a strong zonal jet, which is partially caused by the rotation of the earth. Because of this jet, if v is the nondegenerate surface velocity field of the ocean flow, then there exist two saddle points on the boundary with one on the east coast of North America and the other on the west coast of Europe. In this case, (6.1.7) shows that the number of centers equals the number of interior saddle points plus 2.

Case Re= 5: The flow, driven by the wind stress, generates two circular gyres, and then approaches a steady state of the QG equations. This steady state has the topological structure as shown in Figure 6.1.1, which is structurally stable. In view of the structure classification, the phase structure of this pattern consists of two circle cells (two gyres) and the saddle connections.

Case Re= 10: The structure of the flow at Re=10 behaves similarly to the flow at Re=5; see Figure 6.1.2.

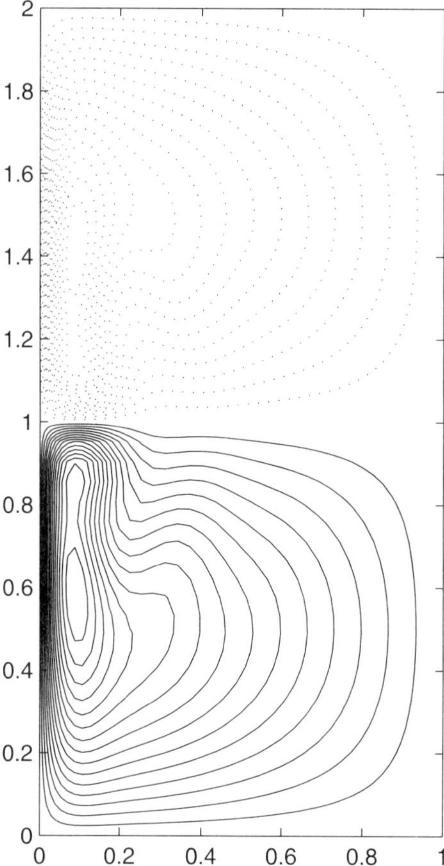

FIGURE 6.1.2. Snapshot of streamlines of the unique steady state for $Re = 10$.

Case Re= 20: The flow also approaches a steady state as indicated by the energy history (figure not shown here). However, the steady state in this case is topologically equivalent to the schematic pattern in Figure 6.1.3. The flow pattern has 6 centers, 4 interior saddles, and 2 boundary saddles, and the index formula holds true in this case. It is easy to see that the flow pattern is structurally stable by the structural stability theorem.

When the Reynolds number is small, the flow converges quickly to a unique steady state, which is structurally stable. Meanwhile, as the Reynolds number changes but remains small, the structural bifurcation would take place. For example, there is one bifurcation value of the Reynolds number between 5 and 20.

Case Re= 30: The energy history (not shown here) shows a quasi-periodic motion; the corresponding dominant frequencies indicate interseasonal and interannual changes of the double-gyre phenomena. The energy spectrum and the phase portrait also suggest that the flow is quasi-periodic, although the dominant frequencies are superimposed with turbulent small-scale motions.

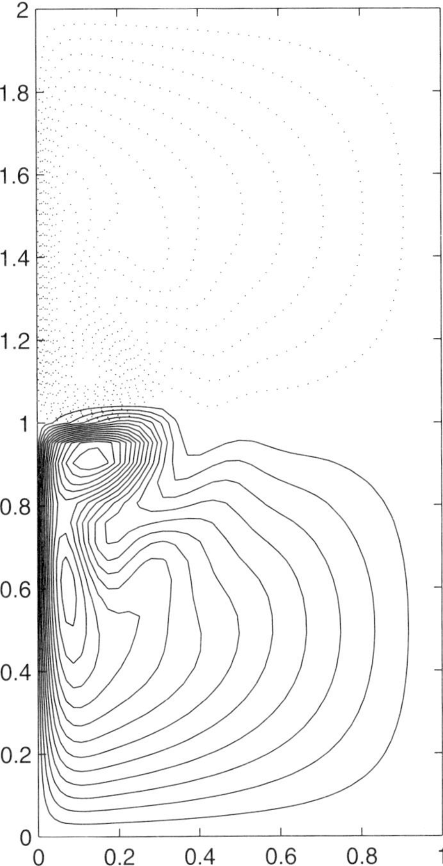

FIGURE 6.1.3. Snapshot of streamlines of the unique steady state for $Re = 20$.

From the viewpoint of structural analysis, the flow patterns in the first 0.87 years, with the pattern at $t = 0.87$ years given in Figure 6.1.4, indicate several structural transitions/bifurcations between structurally stable patterns, and a symmetry breaking of the structure between the sub-polar and sub-tropical cells. In addition, if we generate flow patterns for time long enough as done in see Figures 4.8–4.10 in [**93**] to simulate the long-term dynamics, we can observe from these snapshots the following typical features:

- the typical gyre circulation patterns appear frequently;
- several structural transitions/bifurcations between structurally stable patterns appear during this time interval as well as in the beginning period;
- there is a symmetry breaking of the structure between the sub-polar and sub-tropical cells;
- mixing phenomena is a typical feature;
- as the mixing forces the sub-gyre circulation to detach, the anti-cyclonic anomalies do appear in both the sub-polar and sub-tropical circulations.

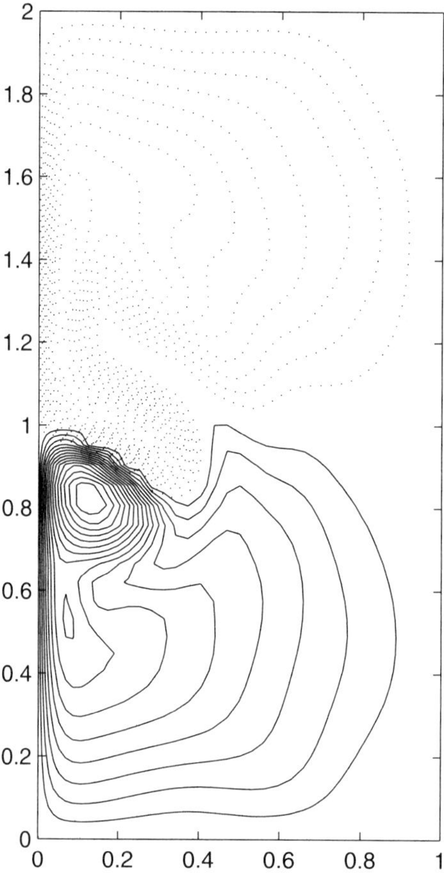

FIGURE 6.1.4. Snapshot of streamlines for $Re = 30$, at time $t = 0.87$ years.

6.2. Boundary Layer Separation on Driven Cavity Flow

In this section, we give a numerical verification of the classification criterion for structural bifurcation in incompressible flow by presenting an example of smoothly-started driven cavity flow at Reynolds number $R = 10^5$.

Let $M = [0,1] \times [0,1]$ with coordinate system (x, y). Consider the cavity flow in a non-dimensional square M that is driven by a slip velocity of parabolic type on the top boundary. The flow is governed by the two-dimensional incompressible Navier-Stokes equations

$$\begin{cases} u_t + (u \cdot \nabla)u + \nabla p = \dfrac{1}{R}\Delta u, \\ \text{div } u = 0, \end{cases} \tag{6.2.1}$$

where $u = (u, v)$ is the velocity, p is the pressure, and R is the Reynolds number. The boundary ∂M is composed of four sections: the left, right and bottom boundary sections are denoted by Γ_r and the top section is denoted by Γ_t. The following

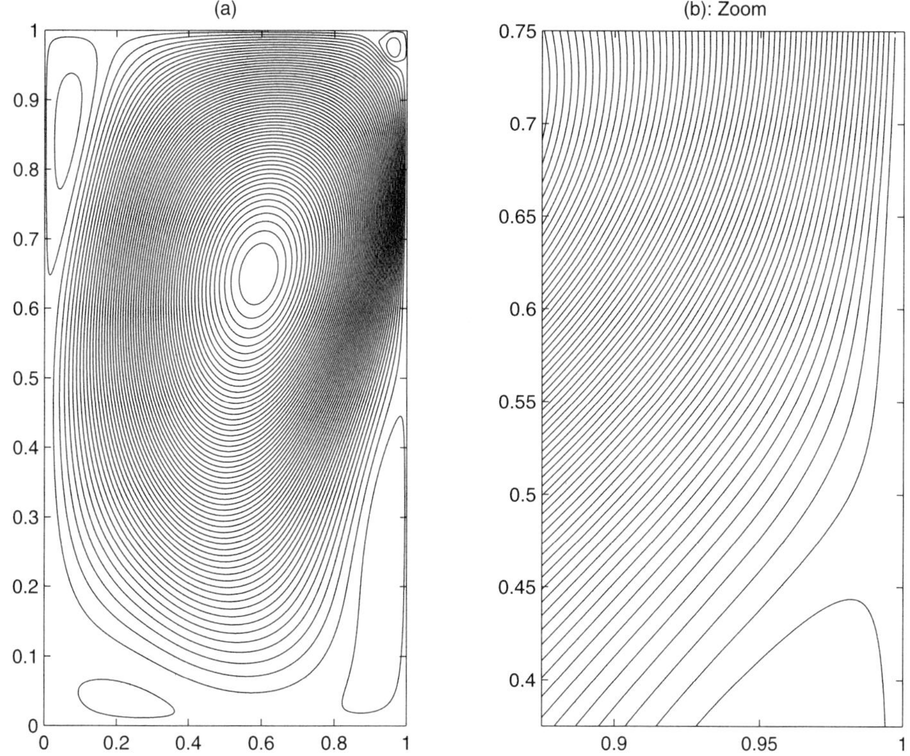

FIGURE 6.2.1. (a): Contour plot of the stream function at the time $t = 1$ over the whole square cavity $[0, 1]^2$. (b): Zoom plot of the stream function in the region $[\frac{7}{8}, 1] \times [\frac{3}{8}, \frac{3}{4}]$ near the mid-portion of the right boundary.

Dirichlet boundary condition for the velocity field is imposed:

$$\begin{cases} u = 0, & \text{on } \Gamma_r, \\ u = u_s = (16x^2(1-x)^2, 0), & \text{on } \Gamma_t, \end{cases} \tag{6.2.2}$$

where u_s denotes the slip velocity on the top boundary that drives the cavity flow. In other words, the no-penetration, no-slip boundary conditions are imposed on the left, right and bottom boundaries and a no-penetration, slip velocity is prescribed at the top boundary.

The initial velocity profile is taken as the following:

$$\begin{cases} u(x, y, 0) = -16x^2(1-x)^2(2y - 3y^2), \\ v(x, y, 0) = 32(x - 3x^2 + 2x^3)(y^2 - y^3), \end{cases} \tag{6.2.3}$$

so that the flow is smoothly-started.

The above system is solved numerically using the Essentially Compact Fourth-Order Scheme (EC4), incorporated with the fourth-order Runge-Kutta time discretization.

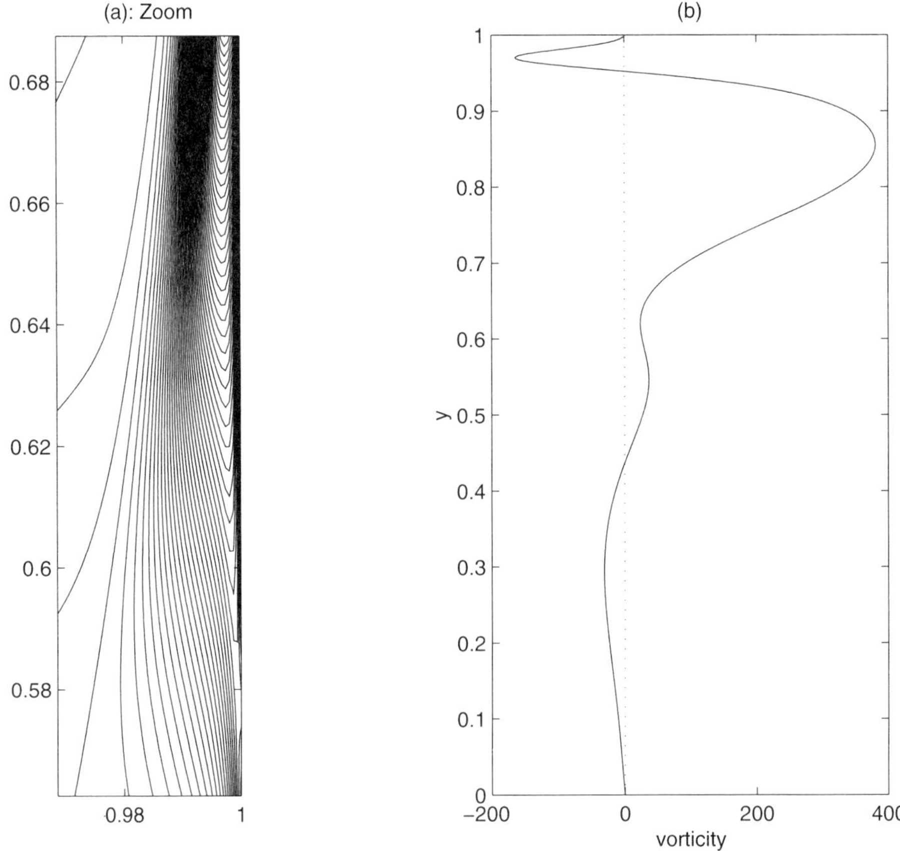

FIGURE 6.2.2. (a): Zoom plot for the vorticity at the time $t = 1$ in the region $[\frac{31}{32}, 1] \times [\frac{9}{16}, \frac{11}{16}]$. (b): The vorticity plot on the right boundary $x = 1$ at $t = 1$.

6.2.1. Structural bifurcation. The driven velocity u_s at the top boundary drives the flow clockwise in the cavity. We call this clockwise rotation the basic circulation. As a result of this circulation, the flow moves downward in the area near the mid-portion of the right boundary. To illustrate more clearly the flow behavior with respect to the issue of structural bifurcation caused by boundary-layer separation, we only concentrate on the mid-portion of the boundary, which is away from the corner area. In other words, the recirculation area around the corner is not taken into consideration. See the contour and zoom plot of the stream function at the time $t = 1$ in Figure 6.2.1, which shows clearly the structure of basic circulation.

As mentioned above, the flow recirculation in the corner areas is neglected since it was caused by corner singularity instead of boundary layer separation. Near the mid-portion of the right boundary, the tangential velocity v is negative, as a result of basic circulation. It is well known from classical fluid mechanics that the combination of the no-slip boundary condition and the basic circulation results in the presence of a thin boundary layer, due to the sharp, initially monotone transition of the velocity from zero tangential velocity on the boundary to a negative value of v in the basic circulation region. The width of this transition layer is proportional to $O(R^{-1/2})$, by formal asymptotic analysis. Accordingly, the vorticity field $\omega = \partial_y u + \partial_x v$ is positive in the thin boundary layer, at least initially. See the vorticity plots near and on the right boundary at $t = 1$ in Figure 6.2.2.

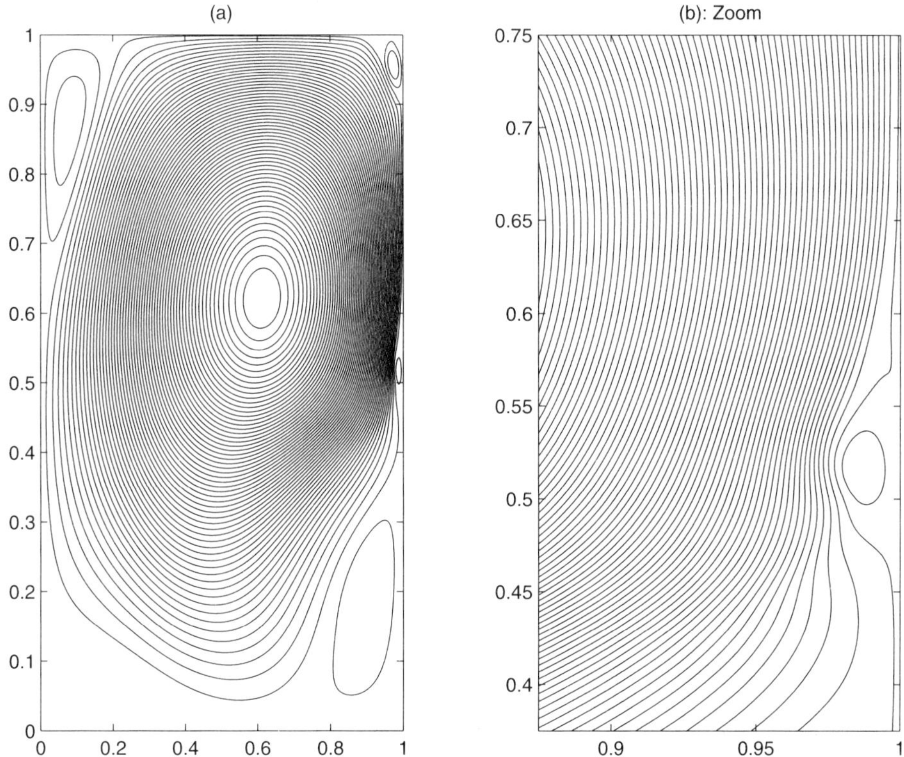

FIGURE 6.2.3. (a): Stream function plot at the time $t = 1.5$ over the whole square cavity $[0, 1]^2$. (b): Zoom plot of the stream function in the region $[\frac{7}{8}, 1] \times [\frac{3}{8}, \frac{3}{4}]$ near the mid-portion of the right boundary.

It is shown in Figure 6.2.2 (b) that the vorticity has positive values in the mid-portion of the right boundary, say, from 0.5 to 0.9 in y. The negative vorticity in the top and bottom portions comes from the recirculation flow at the two corner areas, which is not considered here. The two singular points (zero points) for the vorticity on the right boundary are non-degenerate, i.e, with nonzero tangential derivative, as can be seen in the figure. Thus the structural stability theorems in Chapter 2 indicate that the flow structure keeps stable at the moment. No transition in the flow's structure is going to happen. Moreover, the zoom plot over the region $[\frac{31}{32}, 1] \times [\frac{9}{16}, \frac{11}{16}]$ given in Figure 6.2.2 (a) tells us that the whole vorticity field remains positive in the boundary layer area near the mid-portion of the right boundary at $t = 1$. It in turn verifies that only shear flow is present in the boundary layer before the separation.

THE FIRST BIFURCATION TIME AND BIFURCATION POINT. At the time $t = 1.5$, Figure 6.2.3 presents the stream function plot and Figure 6.2.4 shows the plot of the vorticity profile near and on the right boundary.

The subtle difference of the flow structure between $t = 1$ and $t = 1.5$ can be clearly seen in the above figure. At $t = 1$, Figure 6.2.1 (b) shows that there is no recirculation near the mid-portion of the right boundary. Accordingly, Figure 6.2.2 (b) indicates that the vorticity on the right boundary is positive in the mid-portion. No degenerate singular point for the vorticity forms and the flow structure is stable at that moment. At time $t = 1.5$, recirculation can be obviously observed in the zoom contour plot of Figure 6.2.3 (b). Figure 6.2.4(b) shows that there are two non-degenerate singular points for vorticity ($Y_1^0 = 0.4590$, $Y_2^0 = 0.537$) on the right boundary in the mid-portion. Moreover, a 2-D area with negative vorticity field is illustrated in Figure 6.2.4 (a); the boundary is composed of the boundary section between $(1, Y_1^0)$ and $(1, Y_2^0)$ on Γ_r and a smooth curve (lying inside the interior cavity region) that connects the two singular points. Consequently, it can be concluded that the structural transition occurs between $t = 1$ and $t = 1.5$. Detailed inspection of the numerical results shows that the first singular point for vorticity on the right boundary appears at $T_1^* = 1.0788$, $Y_1^* = 0.6107$. The corresponding plot of boundary vorticity profile at that moment is provided in Figure 6.2.5(a), while the degenerate singular point is located at $y = Y_1^*$; the time history of the vorticity at the boundary point $(1, Y_1^*)$ is given in Figure 6.2.5(b).

By Theorem 5.5.6, the velocity vector field is structurally unstable at that moment. It is indicated by Figure 6.2.5 that the vorticity reaches zero at $(1, Y_1^*, T_1^*)$ as a local minimum in space and decreases in time, i.e, at $(1, Y_1^*, T_1^*)$,

$$\omega = 0, \qquad \frac{\partial \omega}{\partial y} = 0, \qquad \frac{\partial^2 \omega}{\partial^2 y} > 0, \qquad \frac{\partial \omega}{\partial t} < 0,$$

so that condition (5.5.16) in Theorem 5.5.6 is satisfied. That shows that conditions in 5.5.16 hold true. The detailed process of bifurcation, in its local structure in a neighborhood which contains the boundary point $(1,\ Y_1^*)$, is outlined in Theorem 5.5.6 and shown exactly in Figure 5.5.1.

When $t < T_1^*$, there is no singular point for vorticity in the mid-portion of the right boundary, and the local structure of the stream function near that boundary section is equivalent to Figure 5.5.1(a), as can be seen in Figures 6.2.4(b) and 6.2.1(b) at $t = 1$, respectively.

When $t = T_1^*$, there is one degenerate singular point for vorticity, which is a local minimum on Γ_r and decreases in time, as shown in Figure 6.2.5. The local structure of the stream function near the mid-portion of the right boundary is equivalent to Figure 5.5.1(b).

When $t > T_1^*$, there are two isolated singular points for vorticity on Γ_r, and their trajectories are connected. For example, at $t = 1.5$, it appears in Figure 6.2.4(b) that the locations of the two isolated singular points are $Y_1^0 = 0.4590$, $Y_2^0 = 0.537$. One can observe from the zoom plot Figure 6.2.3(b) that the local structure of the stream function near that area is equivalent to Figure 5.5.1(c). The zoom plot of the tangential velocity v, which can be seen in Figure 6.2.6, shows that there is really a recirculation area between Y_1^0 and Y_2^0.

Theorem 5.5.6 assures the occurrence of structural transition of the flow at the time $t = T_1^*$. The phenomenon of boundary-layer separation is a physical explanation of such bifurcation. Furthermore, it is shown in Figures 6.2.7 and 6.2.8

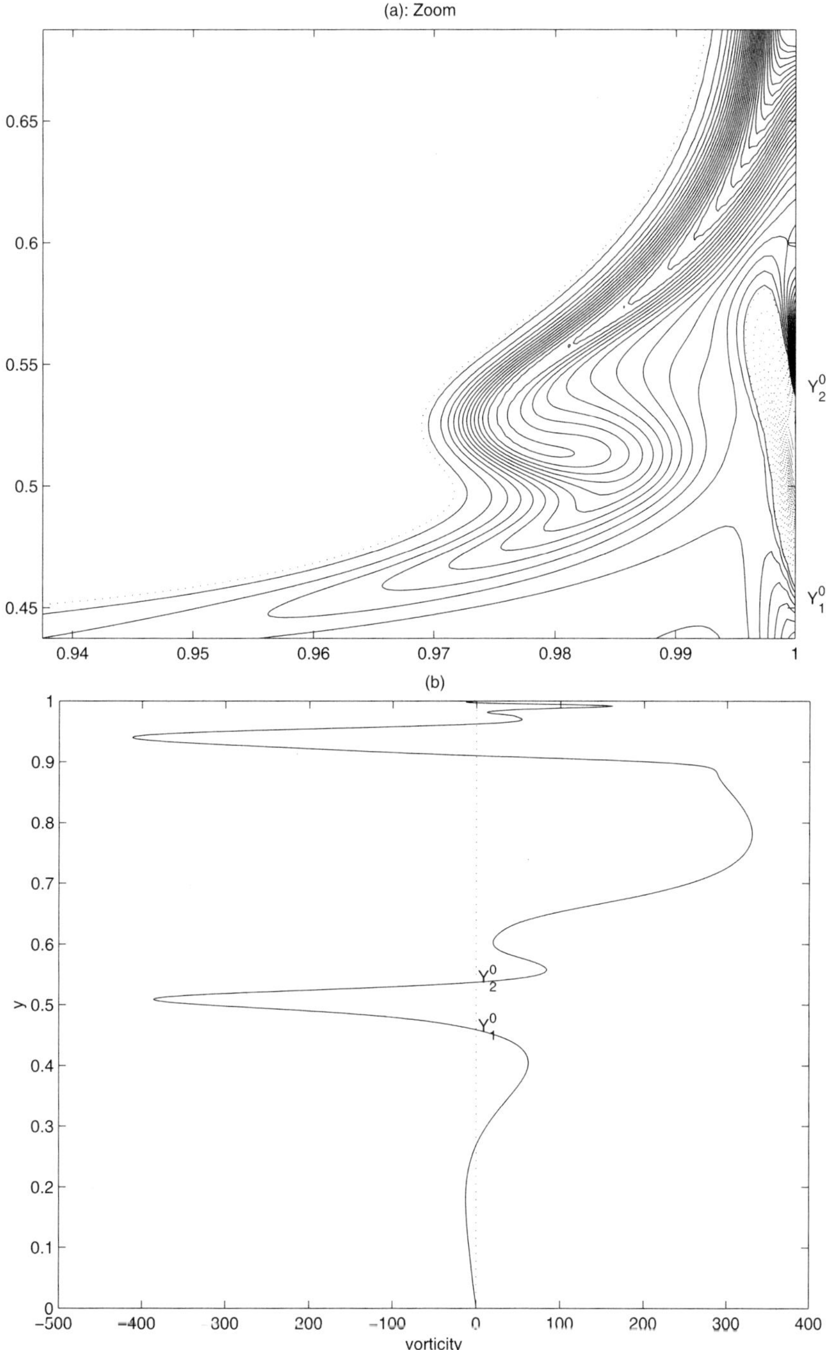

FIGURE 6.2.4. (a): Zoom plot for the vorticity at the time $t = 1.5$ in the region $[\frac{15}{16}, 1] \times [\frac{7}{16}, \frac{11}{16}]$. (b): The vorticity plot on the right boundary $x = 1$ at $t = 1.5$.

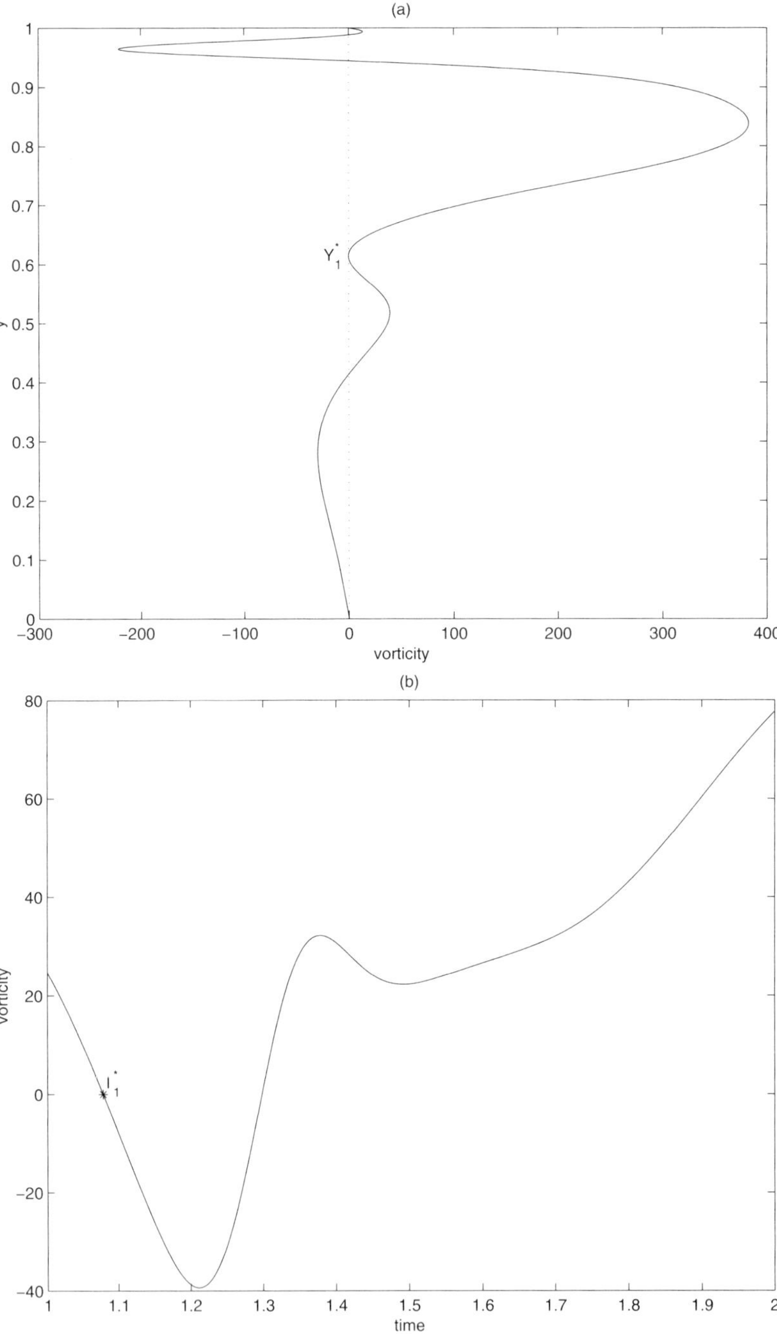

FIGURE 6.2.5. (a): The vorticity on the boundary at $t = T_1^*$. (b): The time history for the vorticity profile on the first bifurcation point $(1, Y_1^*)$ on the right boundary.

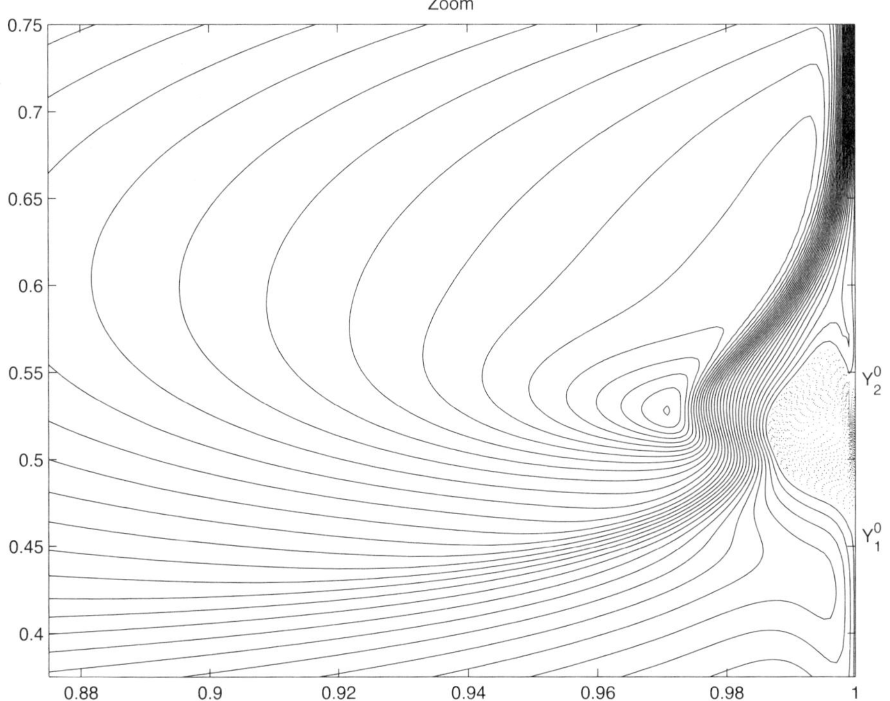

FIGURE 6.2.6. Zoom plot of the vertical velocity near the right boundary at $t = 1.5$. The solid line represents the contour with downward velocity, while the dotted line represents the contour with upward velocity.

that ω keeps positive in a small 2-D neighborhood at the bifurcation point $(1, Y_1^*)$ and it has a negative normal derivative at that point. Consequently, the pressure pushes the flow upward, i.e., opposite to the basic circulation direction, around the bifurcation point. That matches our argument in Chapter 5.

In addition, it is easy to see that

$$\frac{\partial p}{\partial n} = R^{-1}\frac{\partial \omega}{\partial n} = 0, \qquad \text{at } (1, Y_1^*, T_1^*), \tag{6.2.4}$$

which implies that the normal derivative of the pressure vanishes on the first bifurcation point $(1, Y_1^*)$ at the first bifurcation time $t = T_1^*$, since the vorticity field keeps smooth enough around the right boundary, which can be verified by Figure 6.2.5(a) and Figure 6.2.7. As a result, the normal gradient of the pressure stays small around the bifurcation point, and the main contribution of the pressure field in a small neighborhood of $(1, Y_1^*)$ is the backward force for the fluid in the slip direction. Around this time, the boundary layer is still confined in a thin layer around the boundary wall.

SECOND BIFURCATION TIME. As time goes on, the recirculation area expands and moves downward along the right boundary, i.e., in the same direction as the basic circulation. See Figure 6.2.9, which gives the stream function plot at $t = 1.75$.

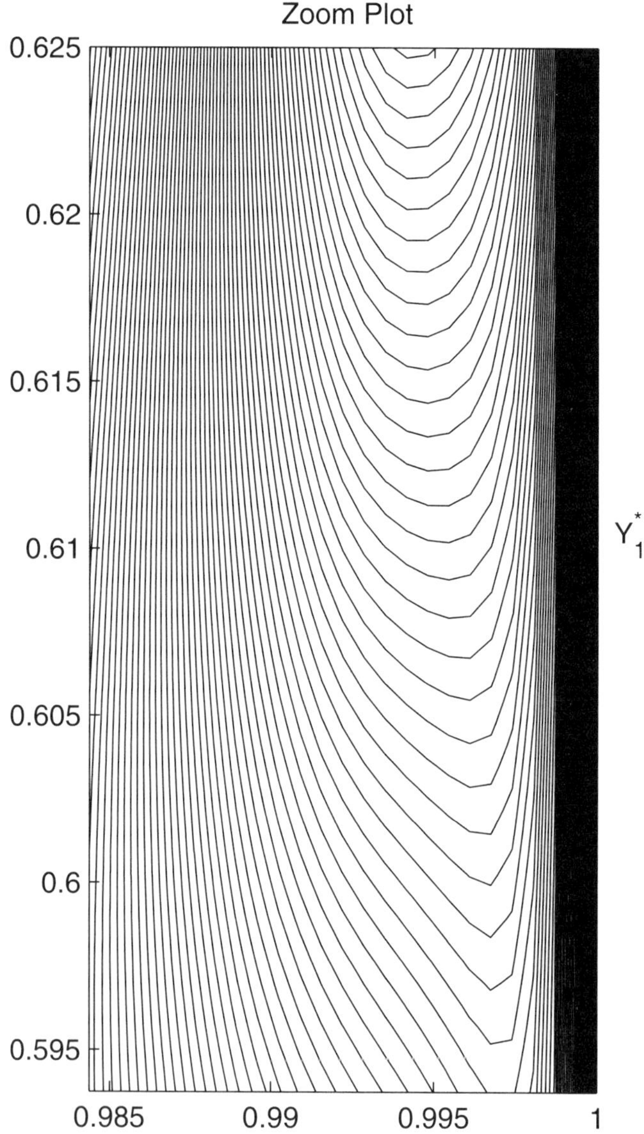

FIGURE 6.2.7. Zoom plot for the vorticity at the first critical time $t = T_1^* = 1.0788$ in the region $[\frac{63}{64}, 1] \times [\frac{19}{32}, \frac{5}{8}]$. The solid line and dotted line represent the contours for the positive and negative levels, respectively.

The process of the recirculation formation described above, including the details of the transition pattern of the flow structure, is repeated after the first one. Figure 6.2.10 presents the stream function plots when $t = 2$ after the second bifurcation time.

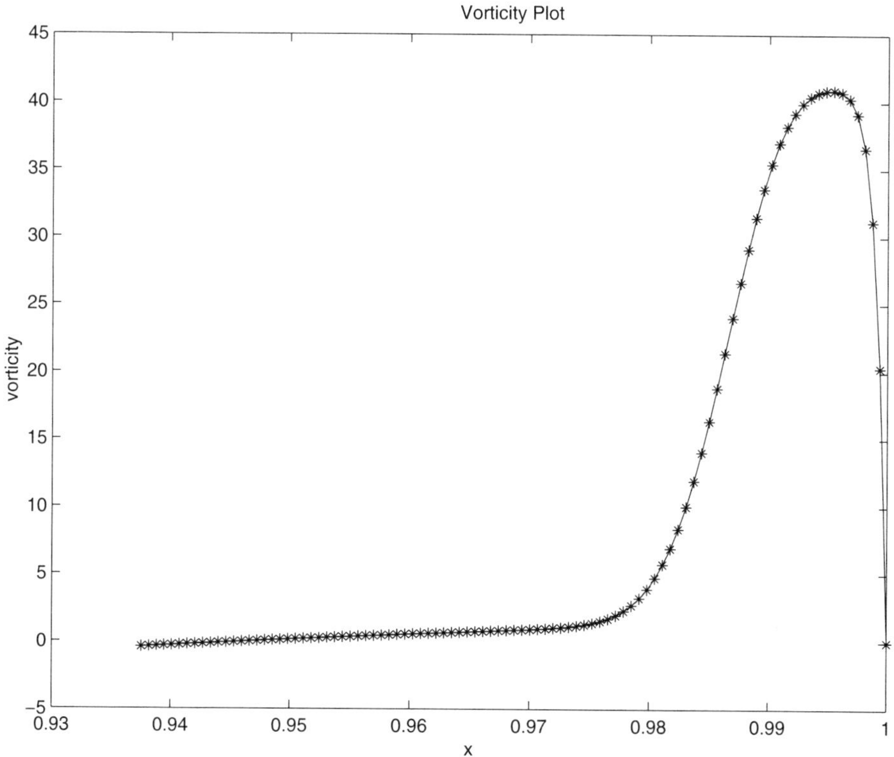

FIGURE 6.2.8. The vorticity plot on the horizontal cut $y = Y_1^* = 0.6107$ near the first bifurcation point, at the first critical time $t = T_1^* = 1.0788$.

The difference in flow structure between $t = 1.75$ and $t = 2$ can be verified by the evolution of the vorticity profile on the right boundary, which can be seen in Figure 6.2.11 at the end of this chapter.

Our numerical results indicate that the second critical time $T_2^* = 1.8086$, and the position of the singular point for the vorticity as $Y_2^* = 0.5182$; see Figure 6.2.12 at the end of this chapter.

Similar to the first bifurcation, the zoom plot for the vorticity near the second bifurcation point at the second critical moment $t = T_2^*$, as presented in Figure 6.2.13 at the end of this chapter, shows that the vorticity field stays positive in a small 2-D neighborhood around the bifurcation point $(1, Y_2^*)$. In fact, the shear flow structure is still preserved in a narrow strip at that moment. There is a thin region filled with backward jet away from the boundary layer. Yet it is strictly kept from the boundary layer at the critical moment. Consequently, a negative normal derivative for the vorticity at the point $(1, Y_2^*)$ can be seen in Figure 6.2.14 at the end of this chapter. That gives a numerical verification of the upward push direction of the pressure around the bifurcation point.

From 6.2.4, we see that the pressure does not exert normal force at the bifurcation point for the first bifurcation time. Yet, such phenomenon is dramatically changed in the second bifurcation time. There is a cusp of the vorticity at the

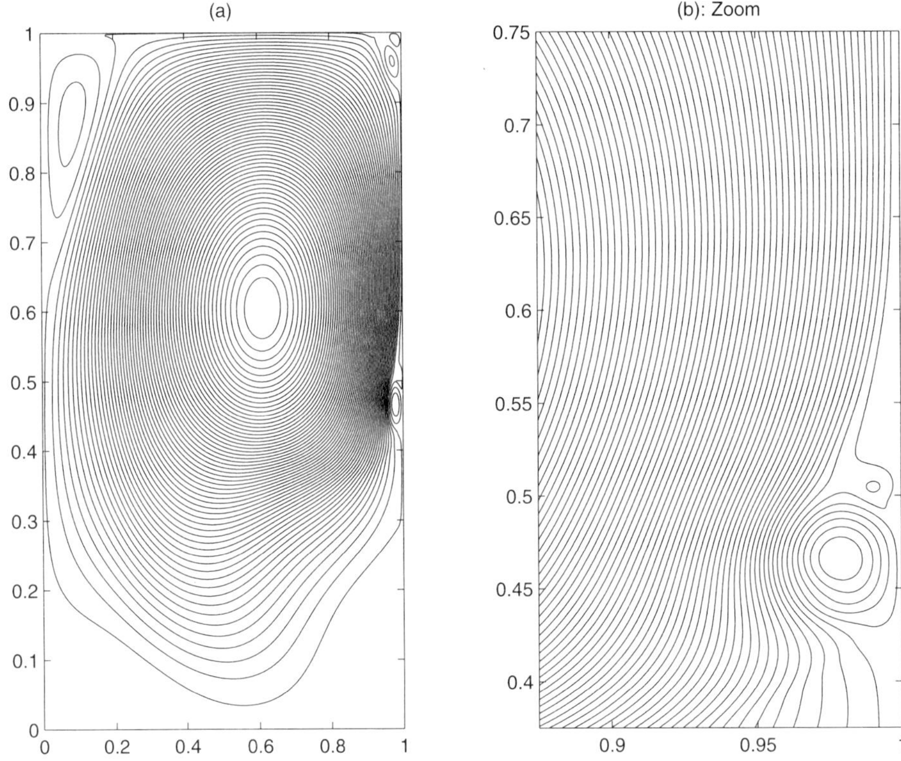

FIGURE 6.2.9. (a): Stream function plot at the time $t = 1.75$ over the whole square cavity $[0,1]^2$. (b): Zoom plot of the stream function in the region $[\frac{7}{8}, 1] \times [\frac{3}{8}, \frac{3}{4}]$.

critical point $(1, Y_2^*)$ in the plot Figure 6.2.12(a) at the end of this chapter. In addition, the tangent gradient of the vorticity changes sign across $(1, Y_2^*)$, which indicates a pair of vortices with opposite signs. That is a common phenomenon at the boundary layer separation. Another peak for the boundary vorticity appears nearby the critical point, as shown in Figure 6.2.12(a) at the end of this chapter. As a result, there is a strong normal pressure gradient which pushes the vortex to shed, due to the large gradient of the vorticity along the boundary.

The second bifurcation time is a crucial moment for boundary-layer separation of incompressible flow. It marks the critical moment for the onset of vortex shedding. After this moment, the vorticity starts to roll up. The difference of roll-up structure can be clearly seen in Figure 6.2.15(a)–(c) at the end of this chapter, which present the zoom plot of vorticity at $t = 1.75$, $t = T_2^* = 1.8086$ and $t = 2$, respectively.

Meanwhile, the "first" bubble moves deeper into the cavity due to the interaction between the boundary layer and the flow in the interior. In other words, the boundary layer that separated from the first bifurcation becomes detached from the boundary. From this point, the evolution of the "first" bubble depends more and more on its interaction with the interior flow.

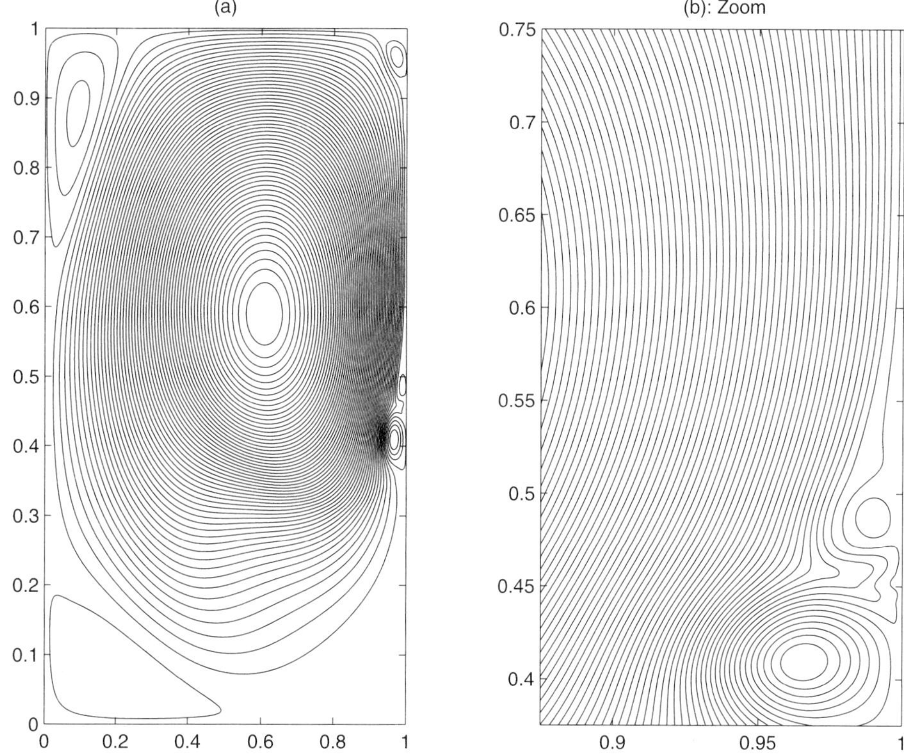

FIGURE 6.2.10. (a): Stream function plot at the time $t = 2$ over the whole square cavity $[0,1]^2$. (b): Zoom plot of the stream function in the region $[\frac{7}{8}, 1] \times [\frac{3}{8}, \frac{3}{4}]$.

MORE BUBBLES AND THE INTERACTION WITH THE INTERIOR. Throughout the time history of the vorticity profile along the right boundary, degenerate singular points appear again and again. Thus, more and more bubbles form along the boundary and move into the interior region afterward. The process of structural transition and recirculation formation and structural transition each time is the same.

Another interesting phenomenon is the structural bifurcation caused by the bubble which moved into the interior region. Theoretical studies in this direction were given in Section 5.6, and numerical studies will be conducted elsewhere. Such an occurrence is illustrated in the sequence of zoom plots of stream function near the right boundary at the later times $t = 3$, 3.5, and 4 as shown in Figure 6.2.16(a)–(c) at the end of this chapter.

Notes for Chapter 6

The material in this chapter is based on [**93, 28**].

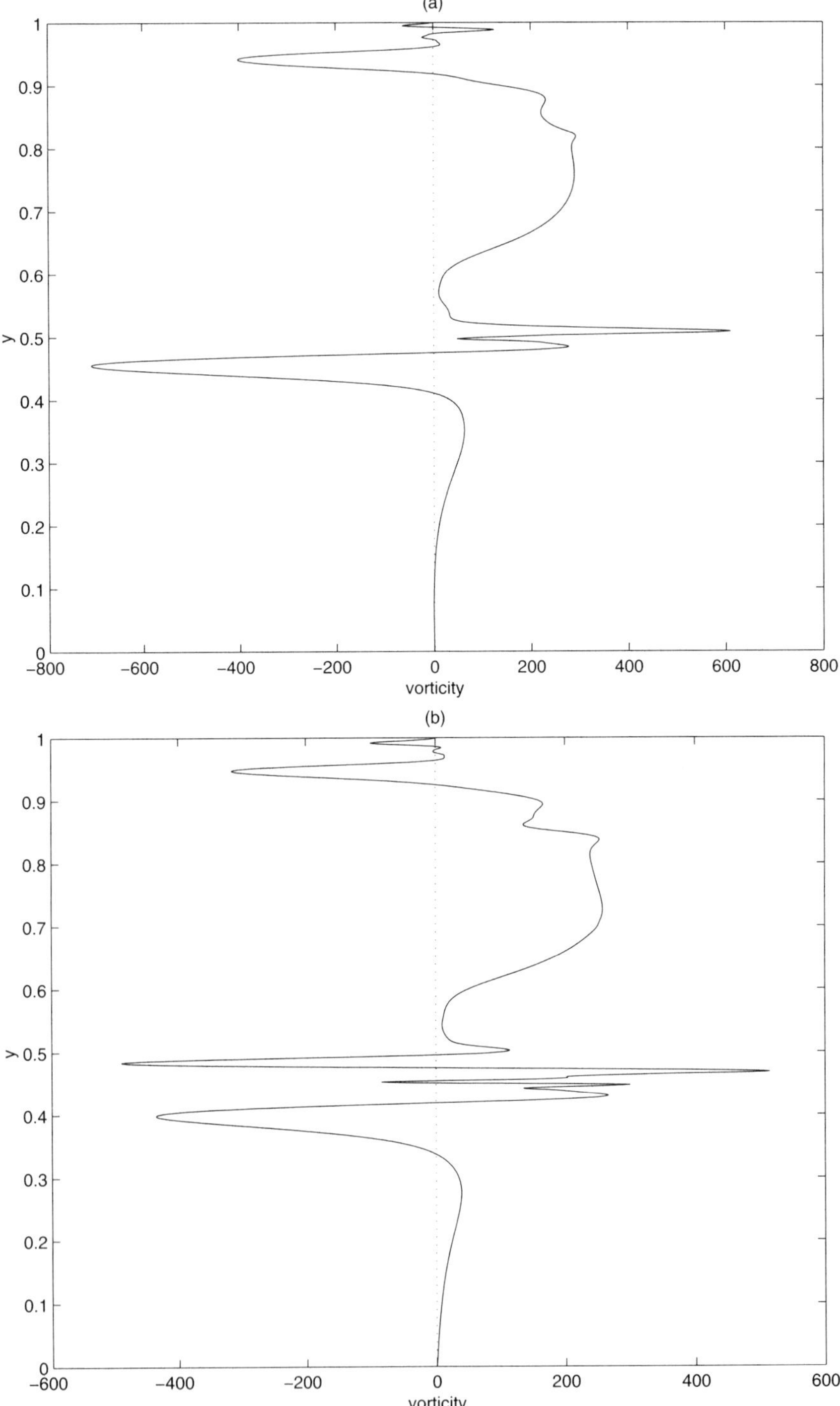

FIGURE 6.2.11. The comparison of the vorticity plots on the right boundary between $t = 1.75$ and $t = 2$.

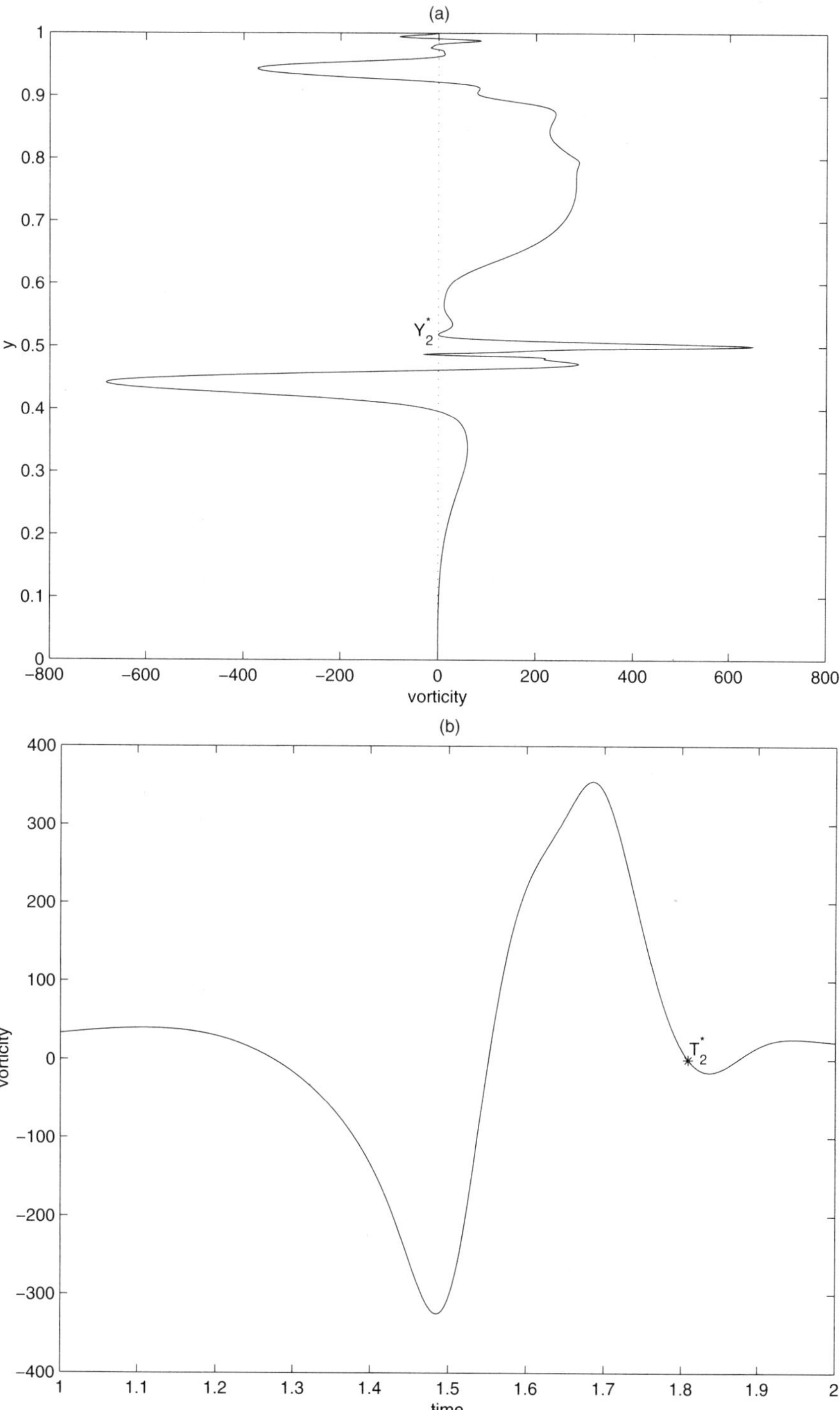

FIGURE 6.2.12. (a): The vorticity on the boundary at the second critical moment $t = T_2^*$. (b): The time history for the vorticity profile on the first bifurcation point $(1, Y_2^*)$ on the right boundary.

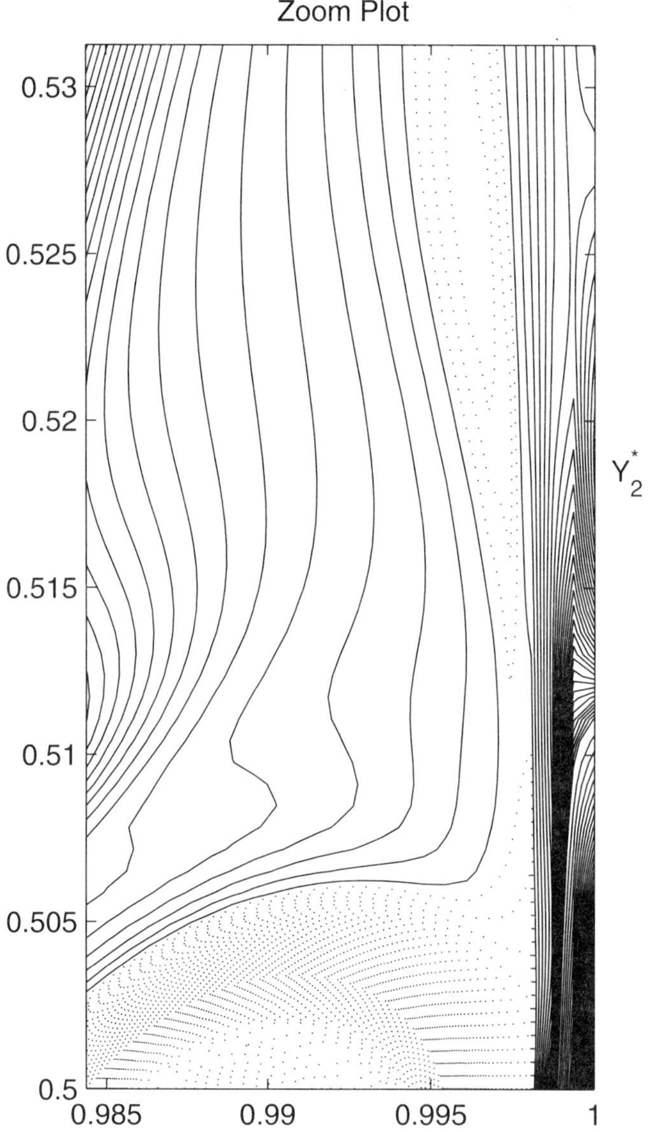

FIGURE 6.2.13. Zoom plot for the vorticity at the second critical time $t = T_2^* = 1.8086$ in the region $[\frac{63}{64}, 1] \times [\frac{1}{2}, \frac{17}{32}]$. The solid line and dotted line represent the contours for the positive and negative levels, respectively.

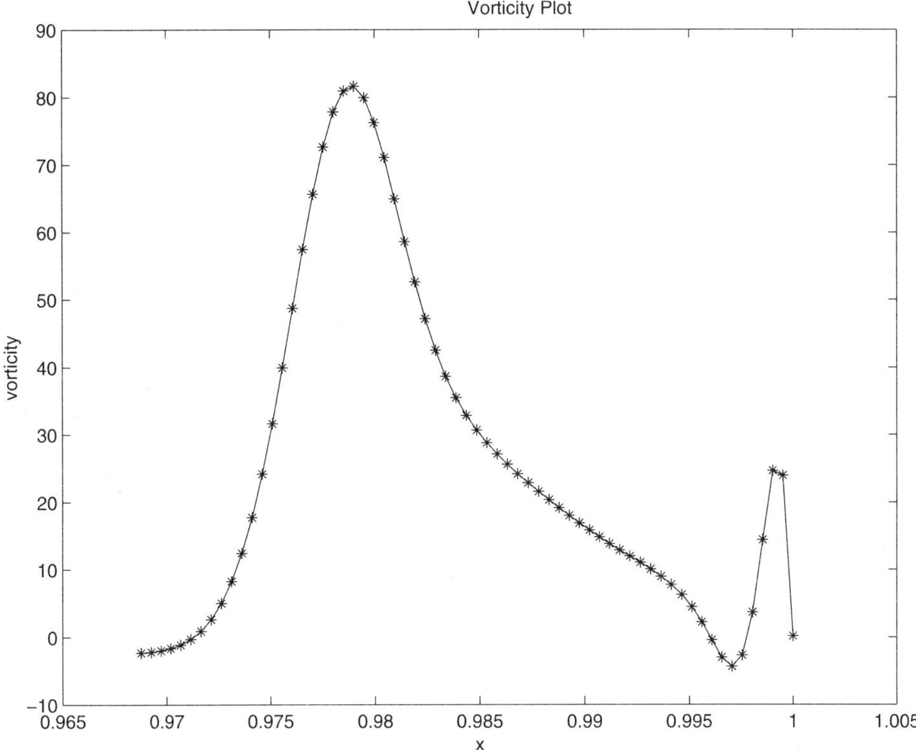

FIGURE 6.2.14. The vorticity plot on the horizontal cut $y = Y_2^* = 0.5182$ near the second bifurcation point, at the second critical time $t = T_2^* = 1.8086$.

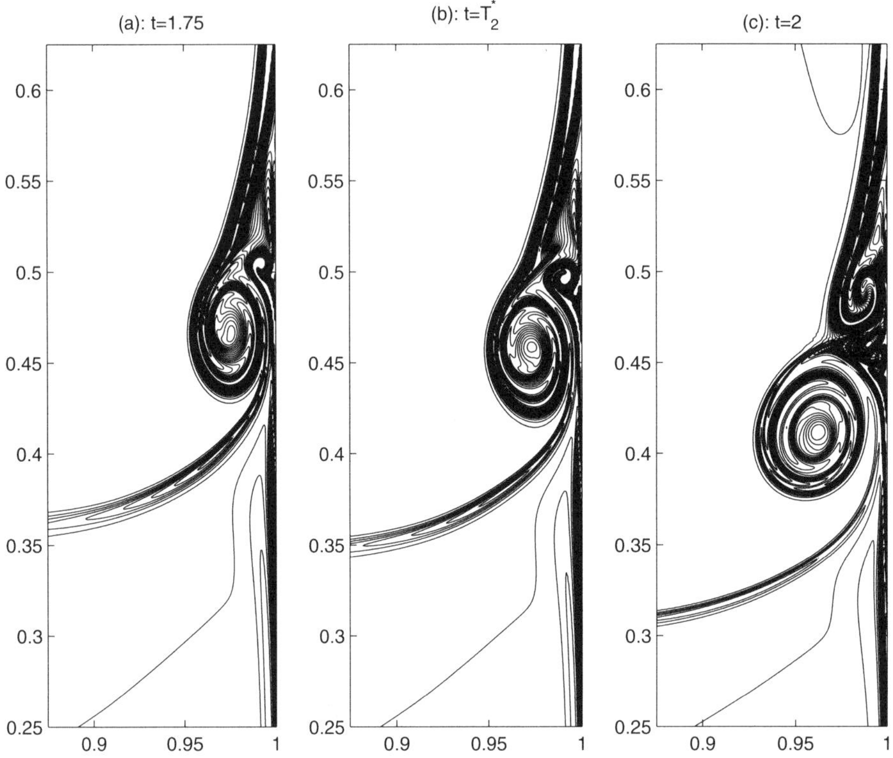

FIGURE 6.2.15. Comparison of zoom plots of vorticity near the right boundary at the time sequence $t = 1.75$, $t = T_2^* = 1.8086$ and $t = 2$.

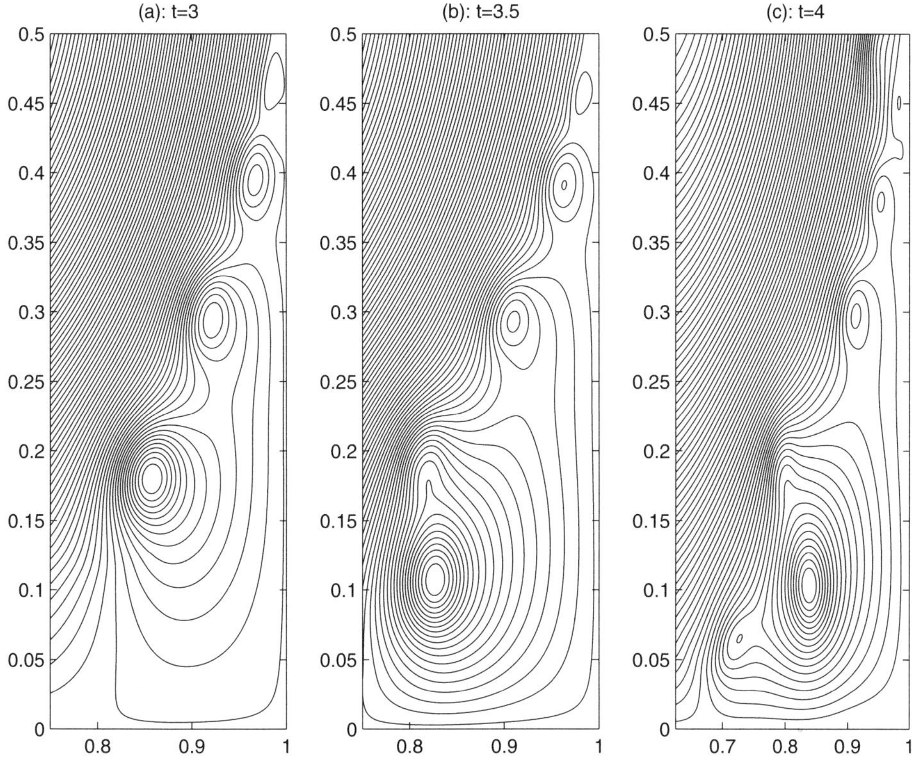

FIGURE 6.2.16. Zoom plot for the stream function near the right boundary at a sequence of time after the second bifurcation: (a) $t = 3.0$,(b) $t = 3.5$, and (c) $t = 4.0$.

Bibliography

[1] R. Abraham and J. Marsden, *Foundations of Mechanics*, Addison–Wesley: Reading, MA, 1978.

[2] S. Agmon, A. Douglis, and L. Nirenberg, *Estimates near the boundary for solutions of elliptic partial differential equations satisfying general boundary conditions. II*, Comm. Pure Appl. Math., 17 (1964), pp. 35–92.

[3] D. V. Anosov and V. Arnold, *Dynamical Systems I*, Springer-Verlag, New York, Heidelberg, Berlin, 1985.

[4] H. Aref, *Chaos in the dynamics of a few vortices—fundamentals and applications*, in Theoretical and applied mechanics (Lyngby, 1984), North-Holland, Amsterdam, 1985, pp. 43–68.

[5] V. Arnold, *Mathematical Methods of Classical Mechanics*, Springer-Verlag, New York, Heidelberg, Berlin, 1978.

[6] G. I. Batchelor, *An Introduction to Fluid Mechanics*, Cambridge University Press, London, 1967.

[7] A. Bensoussan, J.-L. Lions, and G. Papanicolaou, *Asymptotic analysis for periodic structures*, vol. 5 of Studies in Mathematics and its Applications, North-Holland Publishing Co., Amsterdam, 1978.

[8] J. L. Bona and J. Wu, *The zero-viscosity limit of the 2D Navier-Stokes equations*, Stud. Appl. Math., 109 (2002), pp. 265–278.

[9] L. Cattabriga, *Su un problema al contorno relativo al sistema di equazioni di Stokes*, Rend. Sem. Mat. Univ. Padova, 31 (1961), pp. 308–340.

[10] S. Chandrasekhar, *Hydrodynamic and Hydromagnetic Stability*, Dover Publications, Inc., 1981.

[11] A. Chorin and J. Marsden, *A Mathematical Introduction to Fluid Mechanics*, Springer-Verlag, 1997.

[12] P. Constantin, *Three lectures on mathematical fluid mechanics*, in From finite to infinite dimensional dynamical systems (Cambridge, 1995), vol. 19 of NATO Sci. Ser. II Math. Phys. Chem., Kluwer Acad. Publ., Dordrecht, 2001, pp. 145–175.

[13] P. Constantin and C. Foias, *The Navier-Stokes Equations*, Univ. of Chicago Press, Chicago, 1988.

[14] C. R. Doering and J. D. Gibbon, *Applied analysis of the Navier-Stokes equations*, Cambridge Texts in Applied Mathematics, Cambridge University Press, Cambridge, 1995.

[15] P. Drazin and W. Reid, *Hydrodynamic Stability*, Cambridge University Press, 1981.

[16] M. V. Dyke, *Album of Fluid Motion*, Stanford, Calif. : Parabolic Press, 1982.

[17] W. E, *Boundary layer theory and the zero-viscosity limit of the Navier-Stokes equation*, Acta Math. Sin.(Engl. Ser.), 16 (2000), pp. 207–218.

[18] W. E and B. Engquist, *Blow up of solutions of the unsteady Prandtl's equation*, Comm. Pure Appl. Math., (1997), pp. 1287–1293.

[19] C. Foias, M. S. Jolly, I. Kukavica, and E. S. Titi, *The Lorenz equation as a metaphor for the Navier-Stokes equations*, Discrete Contin. Dynam. Systems, 7 (2001), pp. 403–429.

[20] C. Foias, O. Manley, and R. Temam, *Attractors for the Bénard problem: existence and physical bounds on their fractal dimension*, Nonlinear Anal., 11 (1987), pp. 939–967.

[21] C. Foias and J.-C. Saut, *On the smoothness of the nonlinear spectral manifolds associated to the Navier-Stokes equations*, Indiana Univ. Math. J., 33 (1984), pp. 911–926.

[22] ———, *Linearization and normal form of the Navier-Stokes equations with potential forces*, Ann. Inst. H. Poincaré Anal. Non Linéaire, 4 (1987), pp. 1–47.

[23] C. Foias and R. Temam, *Structure of the set of stationary solutions of the Navier-Stokes equations*, Comm. Pure Appl. Math., 30 (1977), pp. 149–164.

[24] S. FRIEDLANDER, *An introduction to the mathematical theory of geophysical fluid dynamics*, vol. 70 of Notas de Matemática [Mathematical Notes], North-Holland Publishing Co., Amsterdam, 1980.

[25] H. FUJITA, *A mathematical analysis of motions of viscous incompressible fluid under leak or slip boundary conditions*, Sūrikaisekikenkyūsho Kōkyūroku, (1994), pp. 199–216. Mathematical fluid mechanics and modeling (Kyoto, 1994).

[26] J.-M. GHIDAGLIA, *Régularité des solutions de certains problèmes aux limites linéaires liés aux équations d'Euler*, Comm. Partial Differential Equations, 9 (1984), pp. 1265–1298.

[27] M. GHIL AND S. CHILDRESS, *Topics in geophysical fluid dynamics: atmospheric dynamics, dynamo theory, and climate dynamics*, vol. 60 of Applied Mathematical Sciences, Springer-Verlag, New York, 1987.

[28] M. GHIL, J.-G. LIU, C. WANG, AND S. WANG, *Boundary-layer separation and adverse pressure gradient for 2-D viscous incompressible flow*, Physica D, 197 (2004), pp. 149–173.

[29] M. GHIL, T. MA, AND S. WANG, *Structural bifurcation of 2-D incompressible flows*, Indiana Univ. Math. J., 50 (2001), pp. 159–180. Dedicated to Professors Ciprian Foias and Roger Temam (Bloomington, IN, 2000).

[30] ———, *Structural bifurcation of 2-D incompressible flows with the Dirichlet boundary conditions and applications to boundary layer separations*, SIAM J. Applied Math., 65:5 (2005), pp. 1576–1596.

[31] S. GOLDSTEIN, *Modern developments in fluid dynamics, Vol I and II*, Dover Publications, New York, 1965.

[32] D. GOTTLIEB, *Vector fields and classical theorems of topology*, Rendiconti del Seminario Matematico e Fisico, Milano, 60 (1990), pp. 193–203.

[33] J. GUCKENHEIMER AND P. J. HOLMES, *Nonlinear oscillations, dynamical systems, and bifurcations of vector fields*, Springer-Verlag, New York, Heidelberg, Berlin, 1983.

[34] J. K. HALE, *Ordinary differential equations*, Robert E. Krieger Publishing Company, Malabar, Florida, 1969.

[35] D. HENRY, *Geometric theory of semilinear parabolic equations*, vol. 840 of Lecture Notes in Mathematics, Springer-Verlag, Berlin, 1981.

[36] M. W. HIRSCH, *Differential topology*, Springer-Verlag, New York, Heidelberg, Berlin, 1976.

[37] D. D. HOLM, J. E. MARSDEN, T. RATIU, AND A. WEINSTEIN, *Nonlinear stability of fluid and plasma equilibria*, Phys. Rep., 123 (1985), p. 116.

[38] H. HOPF, *Abbildungsklassen n-dimensionaler mannigfaltigkeiten*, Math. Annalen, 96 (1926), pp. 225–250.

[39] W. JÄGER, P. LAX, AND C. S. MORAWETZ, *Olga Oleinik (1925–2001)*, Notices Amer. Math. Soc., 50 (2003), pp. 220–223.

[40] S. JIANG, F.-F. JIN, AND M. GHIL, *Multiple equilibria, periodic, and aperiodic solutions in a wind-driven, double-gyre, shallow-water model*, J. Phys. Oceanogr., 25 (1995), pp. 764–786.

[41] C. JONES AND S. WINKLER, *Invariant manifolds and Lagrangian dynamics in the ocean and atmosphere*, in Handbook of dynamical systems, Vol. 2, North-Holland, Amsterdam, 2002, pp. 55–92.

[42] A. KATOK AND B. HASSELBLATT, *Introduction to the Modern Theory of Dynamical Systems*, Cambridge University Press, 1995.

[43] K. KIRCHGÄSSNER, *Bifurcation in nonlinear hydrodynamic stability*, SIAM Rev., 17 (1975), pp. 652–683.

[44] M. KRASNOSELSKII AND P. ZABREIKO, *Geometrical Methods of Nonlinear Analysis*, Springer-Verlag, New York, Heidelberg, Berlin, 1984.

[45] L. D. LANDAU AND E. M. LIFSHITZ, *Fluid Mechanics*, Vol. 6 of Course in Theoretical Physics Series, 2nd ed., Butterworth-Heinemann, Oxford, U.K., 1987.

[46] J. L. LIONS, *Quelques Méthodes de Résolution des Problèmes aux Limites Non Linéaires*, Dunod, Paris, 1969.

[47] J. L. LIONS, R. TEMAM, AND S. WANG, *New formulations of the primitive equations of the atmosphere and applications*, Nonlinearity, 5 (1992), pp. 237–288.

[48] ———, *Models of the coupled atmosphere and ocean (CAO I)*, Computational Mechanics Advance, 1 (1993), pp. 3–54.

[49] P.-L. LIONS, *Mathematical topics in fluid mechanics. Vol. 1*, vol. 3 of Oxford Lecture Series in Mathematics and its Applications, The Clarendon Press, Oxford University Press, New York, 1996. Incompressible models, Oxford Science Publications.

[50] ———, *Mathematical topics in fluid mechanics. Vol. 2*, vol. 10 of Oxford Lecture Series in Mathematics and its Applications, The Clarendon Press, Oxford University Press, New York, 1998. Compressible models, Oxford Science Publications.

[51] T. MA AND S. WANG, *Dynamics of incompressible vector fields*, Appl. Math. Lett., 12 (1999), pp. 39–42.

[52] ———, *The geometry of the stream lines of steady states of the Navier-Stokes equations*, in Nonlinear partial differential equations (Evanston, IL, 1998), vol. 238 of Contemp. Math., Amer. Math. Soc., Providence, RI, 1999, pp. 193–202.

[53] ———, *Dynamics of 2-D incompressible flows*, in Differential equations and computational simulations (Chengdu, 1999), World Sci. Publishing, River Edge, NJ, 2000, pp. 270–276.

[54] ———, *Structural evolution of the Taylor vortices*, M2AN Math. Model. Numer. Anal., 34 (2000), pp. 419–437. Special issue for R. Temam's 60th birthday.

[55] ———, *A generalized Poincaré-Hopf index formula and its applications to 2-D incompressible flows*, Nonlinear Anal. Real World Appl., 2 (2001), pp. 467–482.

[56] ———, *Global structure of 2-D incompressible flows*, Discrete Contin. Dynam. Systems, 7 (2001), pp. 431–445.

[57] ———, *Structure of 2D incompressible flows with the Dirichlet boundary conditions*, Discrete Contin. Dyn. Syst. Ser. B, 1 (2001), pp. 29–41.

[58] ———, *Structural classification and stability of divergence-free vector fields*, Phys. D, 171 (2002), pp. 107–126.

[59] ———, *Topology of 2-d incompressible flows and applications to geophysical fluid dynamics*, Rev. R. Acad. Cien. Serie A. Mat. (RACSAM), 96 (2002), pp. 447–459.

[60] ———, *Attractor bifurcation theory and its applications to Rayleigh-Bénard convection*, Communications on Pure and Applied Analysis, 2 (2003), pp. 591–599.

[61] ———, *Rigorous characterization of boundary layer separations*, in Computational Fluid and Solid Mechanics 2003, ELSEVIER, 2003.

[62] ———, *Asymptotic structure for solutions of the Navier–Stokes equations*, Discrete and Continuous Dynamical Systems, Ser. A, 11:1 (2004), pp. 189–204.

[63] ———, *Boundary layer separation and structural bifurcation for 2-D incompressible fluid flows*, Discrete and Continuous Dynamical Systems, Ser. A, 10:1-2 (2004), pp. 459–472.

[64] ———, *Dynamic bifurcation and stability in the Rayleigh-Bénard convection*, Communication of Mathematical Sciences, 2:2 (2004), pp. 159–183.

[65] ———, *Dynamic bifurcation of nonlinear evolution equations and applications*, Chinese Annals of Mathematics, 26:2 (2004), pp. 185–206.

[66] ———, *Bifurcation Theory and Applications*, World Scientific, 2005.

[67] ———, *Block structure and block stability of incompressible flows*, Discrete Continuous Dynamical Systems, (2005).

[68] ———, *Periodic structure of 2-D Navier–Stokes equations*, J. Nonlinear Sciences, 15:3 (2005).

[69] A. J. MAJDA AND A. L. BERTOZZI, *Vorticity and incompressible flow*, vol. 27 of Cambridge Texts in Applied Mathematics, Cambridge University Press, Cambridge, 2002.

[70] C. MARCHIORO, *An example of absence of turbulence for any Reynolds number*, Comm. Math. Phys., 105 (1986), pp. 99–106.

[71] L. MARKUS AND R. MEYER, *Generic Hamiltonian systems are neither integrable nor ergodic*, Memoirs of the American Mathematical Society, 144 (1974).

[72] J. MILNOR, *Topology from the differentiable viewpoint*, University Press of Virginia, Charlottesville, 1965. based on notes by D. W. Weaver.

[73] A. MIRANVILLE AND M. ZIANE, *On the dimension of the attractor for the Bénard problem with free surfaces*, Russian J. Math. Phys., 5 (1997), pp. 489–502 (1998).

[74] J. MOSER, *Stable and Random Motions in Dynamical Systems*, Ann. Math. Stud. No. 77, Princeton, 1973.

[75] P. K. NEWTON, *The N-vortex problem*, vol. 145 of Applied Mathematical Sciences, Springer-Verlag, New York, 2001. Analytical techniques.

[76] O. OLEINIK AND V. SAMOKHIN, *Mathematical models in boundary layer theory*, Chapman and Hall, 1999.

[77] J. PALIS AND W. DE MELO, *Geometric theory of dynamical systems*, Springer-Verlag, New York, Heidelberg, Berlin, 1982.

[78] J. Palis and S. Smale, *Structural stability theorem*, in Global Analysis. Proc. Symp. in Pure Math., vol. XIV, 1970.

[79] A. Pazy, *Semigroups of linear operators and applications to partial differential equations*, vol. 44 of Applied Mathematical Sciences, Springer-Verlag, New York, 1983.

[80] J. Pedlosky, *Geophysical Fluid Dynamics*, Springer-Verlag, New-York, second ed., 1987.

[81] J. P. Peixoto and A. H. Oort, *Physics of Climate*, American Institute of Physics, New-York, 1992.

[82] M. Peixoto, *Structural stability on two dimensional manifolds*, Topology, 1 (1962), pp. 101–120.

[83] ———, *On the classification of flows on 2-manifolds*, in Dynamical systems, ed. by M. Peixoto, Academic Press, 1973.

[84] L. Prandtl, in Verhandlungen des dritten internationalen Mathematiker-Kongresses, Heidelberg, 1904, Leipeizig, 1905, pp. 484-491.

[85] C. C. Pugh, *The closing lemma*, Amer. J. Math., 89 (1967), pp. 956–1009.

[86] P. H. Rabinowitz, *Existence and nonuniqueness of rectangular solutions of the Bénard problem*, Arch. Rational Mech. Anal., 29 (1968), pp. 32–57.

[87] L. Rayleigh, *On convection currents in a horizontal layer of fluid, when the higher temperature is on the under side*, Phil. Mag., 32 (1916), pp. 529–46.

[88] C. Robinson, *Generic properties of conservative systems, I, II*, Amer. J. Math., 92 (1970), pp. 562–603 and 897–906.

[89] ———, *Structure stability of vector fields*, Ann. of Math., 99 (1974), pp. 154–175.

[90] A. M. Rogerson, P. D. Miller, L. J. Pratt, and C. K. R. T. Jones, *Lagrangian motion and fluid exchange in a barotropic meandering jet*, J. Phys. Oceanogr., 29 (1999), pp. 2635–2655.

[91] R. Salmon, *Lectures on geophysical fluid dynamics*, Oxford University Press, New York, 1998.

[92] H. Schlichting, *Boundary layer theory*, Springer, Berlin-Heildelberg, 8th edition ed., 2000.

[93] J. Shen, T. T. Medjo, and S. Wang, *On a wind-driven, double gyre, quasi-geostrophic ocean model: Numerical simulations and structural analysis*, Journal of Computational Physics.

[94] M. Shub, *Stabilité globale des systèmes dynamiques*, vol. 56 of Astérisque, Société Mathématique de France, Paris, 1978. With an English preface and summary.

[95] S. Smale, *An infinite dimensional version of Sard's theorem*, Amer. J. Math., 87 (1965), pp. 861–866.

[96] S. Smale, *Differential dynamical systems*, Bull. AMS, 73 (1967), pp. 747–817.

[97] V. A. Solonnikov and V. E. Scadilov, *A certain boundary value problem for the stationary system of Navier-Stokes equations*, Trudy Mat. Inst. Steklov., 125 (1973), pp. 196–210, 235. Boundary value problems of mathematical physics, 8.

[98] J. Sotomayor, *Generic bifurcation of dynamical systems*, in Dynamical systems, edited by M. Peixoto, Academic Press, (1973).

[99] ———, *Generic one parameter famlies of vector fields on two-dimensional manifolds*, Publ. Math. Inst. Hautes Études Sci., 43 (1973).

[100] S. Speich, H. Dijkstra, and M. Ghil, *Successive bifurcations in a shallow-water model, applied to the wind-driven ocean circulation*, Nonlin. Proc. Geophys., 2 (1995), pp. 241–268.

[101] S. Speich and M. Ghil, *Interannual variability of the mid-latitude oceans: a new source of climate variability?*, Sistema Terra, 3(3) (1994), pp. 33–35.

[102] R. Temam, *Navier-Stokes Equations, Theory and Numerical Analysis, 3rd, rev. ed.*, North Holland, Amsterdam, 1984.

[103] C. Truesdell, *The Kinematics of Vorticity*, Indiana University Press, Bloomington, 1954.

[104] X. Wang and R. Temam, *Asymptotic analysis of Oseen type equations in a channel at high Reynolds number*, Indiana Univ. Math. J., 45 (1996), pp. 863–916.

[105] S. Wiggins, *Introduction to Applied Nonlinear Dynamical Systems and Chaos*, Springer-Verlag, New York, Heidelberg, Berlin, 1990.

[106] V. I. Yudovich, *Free convection and bifurcation*, J. Appl. Math. Mech., 31 (1967), pp. 103–114.

[107] ———, *Stability of convection flows*, J. Appl. Math. Mech., 31 (1967), pp. 272–281.

[108] M. Ziane, *Optimal bounds on the dimension of the attractor of the Navier-Stokes equations*, Phys. D, 105 (1997), pp. 1–19.

Index

$B^r(TM)$, 51
$B_0^r(TM)$, 51
$B_\phi^r(TM)$, 67
$B_\varphi^r(TM)$, 51
$C^r(TM)$, 4, 17
$C_n^r(TM)$, 4, 17
D-block, 75
$D^r(TM)$, 19
$D_0^r(TM)$, 55
$D_1^r(TM)$, 55
$H_1(M, \partial M)$, homology group, 32
S-block, 75
T-block, 75
$\Phi(\cdot,\cdot)$,flow generated by velocity field, 19
$\alpha(x)$, α-limit set, 18
$\bigvee_n S^1$, 34
$\mathcal{H}^r(TM)$, 71
$\omega(x)$, ω-limit set, 18
∂-regular point, 61
∂-saddle, 61
∂-singular point, 61
  non-degenerate, 61

adverse pressure gradient, 185
  boundary layer separation, 11, 185
analytic semi-group, 146
asymptotic block stability, 123
asymptotic stability theorem, 148
attractor bifurcation, 14, 146
  definition of, 147
attractor bifurcation theorem, 147

basic vector field, 7, 84
  definition of, 87
Betti number, 32
bifurcated solution
  structure, 13
bifurcation
  in global structure, 157
  in local structure, 157
bifurcation time and location, 177
block decomposition, 76
block stability, 7, 81, 87, 123
block structure, 7, 75, 87
boundary condition
  Dirichlet, 1
  free-free, 144
  free-rigid, 144
  free-slip, 2
  no-normal flow, 1
  periodic, 2
  rigid-free, 144
  rigid-rigid, 144
boundary layer separation, 8, 177, 181
  adverse pressure gradient, 11, 185
  determination of, 182
  driven cavity flow, 210
  reattachment of, 11
  time and location, 11, 177
  vorticity crisis, 11, 186
Boussinesq equations, 143

canonical coordinate system, 68
center, 18
circle band, 31
circle cell, 31
connection lemma, 71, 72

divergence-free vector field
  Poincaré-Bendixson theorem for, 20
  structural classification of a, 31
  structural stability of, 51, 60
double-gyre ocean circulation, 203
  model, 205
  wind-driven, 205
driven cavity flow
  boundary layer separation, 210
dynamic bifurcation, 13

effective turbulent viscosity coefficient, 206
Ekman boundary layer, 206
Ekman pumping, 206
ergodic set, 31
  structure of, 31
Euler characteristic, 28, 204
Euler equations, 1
extended manifold, 32
extended orbit, 60

flow generated by v, $\Phi(\cdot,\cdot)$, 19
flow maps, 203
fluid flow
  representation of, 1
fluid flow maps, 203
focus, 18

genericity of stable steady states, 109
Ginzburg-Landau equations, 13
global oceanic flow, 204

Hamiltonian structural stability, 71
Hamiltonian vector field, 53
  block structure of, 75
  definition of, 53, 71
Hopf lemma, 185

index, 24
index formula, 203
index theorem, 27

kinematic condition, 189

Lagrange representation, 2
limit set theorem, 17, 21, 31, 34
limiting cycle, 18
Liouville-Arnold theorem, 74
local analyticity, 10

Navier-Stokes equations, 1, 117, 127
  periodic structure, 12
  separation equation of, 177
node, 18

orbit, 18
  closed, 18
  ending point of an, 18
  extended, 60
  saddle connection, 21
  starting point of an, 18

Peixoto, 51, 52
periodic structure, 127
Poincaré-Bendixson theorem, 20
  classical, 49
Poincaré-Hopf index theorem, 24, 27
  classical, 49
Prandtl condition, 8, 168
Prandtl number, 144
pseudo-manifold, 32

quasi-geostrophic model, 205

Rayleigh number, 144
  critical, 149
Rayleigh-Bénard convection, 13, 142, 155
  attractor bifurcation of, 149
regular point, 17
Reynolds number, 206
Rossby number, 203, 206

saddle, 18
  Ω-boundary, 33
  Ω-exterior, 33
  Ω-interior, 33
  self-connected, 52
saddle connection, 21, 31
  irretractable, 81
  retractable, 81
saddle connection diagram, 40
saddle connection set, 40
Sard-Smale theorem, 112
second bifurcation time, 217
self-connected saddle point, 52
self-connection vector field, 55
separation
  reattachment of, 11
separation equation, 177
singular point, 17, 24
  degenerate, 17
  non-degenerate, 17, 24
singularity classification theorem, 158
spectral manifold, 13
spectral-projection method, 206
standard manifold, 31
stratification, 203
structural bifurcation, 10, 13, 211
  Dirichlet boundary condition, 167
  driven by forcing, 179
  genericity, 196, 198
  integer index, 181
  interior, 187, 191
  necessary conditions for, 157
  no-normal flow condition, 160
structural bifurcation theorem, 161, 169, 173
structural classification, 4, 31
structural stability, 5
  asymptotic Hamiltonian, 117
  definition of, 52
  Hamiltonian, 71, 73
  local, 77
structural stability theorem, 52
structurally stable fields, 12
  genericity of, 12
superconductivity, 13
Sverdrup equation, 206

Taylor field, 98
Taylor vortex structure, 131
Taylor vortices, 13, 98
topologically equivalent, 42
tubular incompressible flow, 53

vector field, 17
  D-regular, 60
  basic, 84
  index of a, 24
  regular, 18
  self-connection, 55
vorticity crisis, 11, 186

Titles in This Series

121 **Anton Zettl,** Sturm-Liouville theory, 2005

120 **Barry Simon,** Trace ideals and their applications, 2005

119 **Tian Ma and Shouhong Wang,** Geometric theory of incompressible flows with applications to fluid dynamics, 2005

118 **Alexandru Buium,** Arithmetic differential equations, 2005

117 **Volodymyr Nekrashevych,** Self-similar groups, 2005

116 **Alexander Koldobsky,** Fourier analysis in convex geometry, 2005

115 **Carlos Julio Moreno,** Advanced analytic number theory: L-functions, 2005

114 **Gregory F. Lawler,** Conformally invariant processes in the plane, 2005

113 **William G. Dwyer, Philip S. Hirschhorn, Daniel M. Kan, and Jeffrey H. Smith,** Homotopy limit functors on model categories and homotopical categories, 2004

112 **Michael Aschbacher and Stephen D. Smith,** The classification of quasithin groups II. Main theorems: The classification of simple QTKE-groups, 2004

111 **Michael Aschbacher and Stephen D. Smith,** The classification of quasithin groups I. Structure of strongly quasithin K-groups, 2004

110 **Bennett Chow and Dan Knopf,** The Ricci flow: An introduction, 2004

109 **Goro Shimura,** Arithmetic and analytic theories of quadratic forms and Clifford groups, 2004

108 **Michael Farber,** Topology of closed one-forms, 2004

107 **Jens Carsten Jantzen,** Representations of algebraic groups, 2003

106 **Hiroyuki Yoshida,** Absolute CM-periods, 2003

105 **Charalambos D. Aliprantis and Owen Burkinshaw,** Locally solid Riesz spaces with applications to economics, second edition, 2003

104 **Graham Everest, Alf van der Poorten, Igor Shparlinski, and Thomas Ward,** Recurrence sequences, 2003

103 **Octav Cornea, Gregory Lupton, John Oprea, and Daniel Tanré,** Lusternik-Schnirelmann category, 2003

102 **Linda Rass and John Radcliffe,** Spatial deterministic epidemics, 2003

101 **Eli Glasner,** Ergodic theory via joinings, 2003

100 **Peter Duren and Alexander Schuster,** Bergman spaces, 2004

99 **Philip S. Hirschhorn,** Model categories and their localizations, 2003

98 **Victor Guillemin, Viktor Ginzburg, and Yael Karshon,** Moment maps, cobordisms, and Hamiltonian group actions, 2002

97 **V. A. Vassiliev,** Applied Picard-Lefschetz theory, 2002

96 **Martin Markl, Steve Shnider, and Jim Stasheff,** Operads in algebra, topology and physics, 2002

95 **Seiichi Kamada,** Braid and knot theory in dimension four, 2002

94 **Mara D. Neusel and Larry Smith,** Invariant theory of finite groups, 2002

93 **Nikolai K. Nikolski,** Operators, functions, and systems: An easy reading. Volume 2: Model operators and systems, 2002

92 **Nikolai K. Nikolski,** Operators, functions, and systems: An easy reading. Volume 1: Hardy, Hankel, and Toeplitz, 2002

91 **Richard Montgomery,** A tour of subriemannian geometries, their geodesics and applications, 2002

90 **Christian Gérard and Izabella Łaba,** Multiparticle quantum scattering in constant magnetic fields, 2002

89 **Michel Ledoux,** The concentration of measure phenomenon, 2001

88 **Edward Frenkel and David Ben-Zvi,** Vertex algebras and algebraic curves, second edition, 2004
87 **Bruno Poizat,** Stable groups, 2001
86 **Stanley N. Burris,** Number theoretic density and logical limit laws, 2001
85 **V. A. Kozlov, V. G. Maz′ya, and J. Rossmann,** Spectral problems associated with corner singularities of solutions to elliptic equations, 2001
84 **László Fuchs and Luigi Salce,** Modules over non-Noetherian domains, 2001
83 **Sigurdur Helgason,** Groups and geometric analysis: Integral geometry, invariant differential operators, and spherical functions, 2000
82 **Goro Shimura,** Arithmeticity in the theory of automorphic forms, 2000
81 **Michael E. Taylor,** Tools for PDE: Pseudodifferential operators, paradifferential operators, and layer potentials, 2000
80 **Lindsay N. Childs,** Taming wild extensions: Hopf algebras and local Galois module theory, 2000
79 **Joseph A. Cima and William T. Ross,** The backward shift on the Hardy space, 2000
78 **Boris A. Kupershmidt,** KP or mKP: Noncommutative mathematics of Lagrangian, Hamiltonian, and integrable systems, 2000
77 **Fumio Hiai and Dénes Petz,** The semicircle law, free random variables and entropy, 2000
76 **Frederick P. Gardiner and Nikola Lakic,** Quasiconformal Teichmüller theory, 2000
75 **Greg Hjorth,** Classification and orbit equivalence relations, 2000
74 **Daniel W. Stroock,** An introduction to the analysis of paths on a Riemannian manifold, 2000
73 **John Locker,** Spectral theory of non-self-adjoint two-point differential operators, 2000
72 **Gerald Teschl,** Jacobi operators and completely integrable nonlinear lattices, 1999
71 **Lajos Pukánszky,** Characters of connected Lie groups, 1999
70 **Carmen Chicone and Yuri Latushkin,** Evolution semigroups in dynamical systems and differential equations, 1999
69 **C. T. C. Wall (A. A. Ranicki, Editor),** Surgery on compact manifolds, second edition, 1999
68 **David A. Cox and Sheldon Katz,** Mirror symmetry and algebraic geometry, 1999
67 **A. Borel and N. Wallach,** Continuous cohomology, discrete subgroups, and representations of reductive groups, second edition, 2000
66 **Yu. Ilyashenko and Weigu Li,** Nonlocal bifurcations, 1999
65 **Carl Faith,** Rings and things and a fine array of twentieth century associative algebra, 1999
64 **Rene A. Carmona and Boris Rozovskii, Editors,** Stochastic partial differential equations: Six perspectives, 1999
63 **Mark Hovey,** Model categories, 1999
62 **Vladimir I. Bogachev,** Gaussian measures, 1998
61 **W. Norrie Everitt and Lawrence Markus,** Boundary value problems and symplectic algebra for ordinary differential and quasi-differential operators, 1999
60 **Iain Raeburn and Dana P. Williams,** Morita equivalence and continuous-trace C^*-algebras, 1998
59 **Paul Howard and Jean E. Rubin,** Consequences of the axiom of choice, 1998

For a complete list of titles in this series, visit the
AMS Bookstore at **www.ams.org/bookstore/**.